Edexcel A level

BIOLOGY B

2

Ann Fullick

LWAYS LEARNING

PEARSON

Published by Pearson Education Limited, 80 Strand, London WC2R 0RL.

www.pearsonschoolsandfecolleges.co.uk

Copies of official specifications for all Edexcel qualifications may be found on the website:
www.edexcel.com

Text © Ann Fullick 2015
Exam-style questions © Pearson Education Limited
Edited by Katie Wilkinson and Jo Egré
Designed by Elizabeth Arnoux for Pearson Education Limited
Typeset by Tech-Set Ltd, Gateshead
Original illustrations © Pearson Education Limited 2015
Illustrated by Tech-Set Ltd, Gateshead and Peter Bull Art Studio
Cover design by Elizabeth Arnoux for Pearson Education Limited
Picture research by Susie Prescott
Cover photo/illustration © Science Photo Library/British Library

The rights of Ann Fullick and Graham Hartland to be identified as authors of this work have been asserted by them in accordance with the Copyright, Designs and Patents Act 1988.

First published 2008
Second edition published 2015

19 18 17 16 15
10 9 8 7 6 5 4 3 2 1

British Library Cataloguing in Publication Data
A catalogue record for this book is available from the British Library

ISBN 9781447991151

Printed in Italy by Lego SpA

Acknowledgements

A note from the publisher
In order to ensure that this resource offers high-quality support for the associated Pearson qualification, it has been through a review process by the awarding body. This process confirms that this resource fully covers the teaching and learning content of the specification or part of a specification at which it is aimed. It also confirms that it demonstrates an appropriate balance between the development of subject skills, knowledge and understanding, in addition to preparation for assessment.

Endorsement does not cover any guidance on assessment activities or processes (e.g. practice questions or advice on how to answer assessment questions), included in the resource nor does it prescribe any particular approach to the teaching or delivery of a related course.

While the publishers have made every attempt to ensure that advice on the qualification and its assessment is accurate, the official specification and associated assessment guidance materials are the only authoritative source of information and should always be referred to for definitive guidance.

Pearson examiners have not contributed to any sections in this resource relevant to examination papers for which they have responsibility.

Examiners will not use endorsed resources as a source of material for any assessment set by Pearson.

Endorsement of a book does not mean that the resource is required to achieve this Pearson qualification, nor does it mean that it is the only suitable material available to support the qualification, and any resource lists produced by the awarding body shall include this and other appropriate resources.

Picture credits
The publisher would like to thank the following for their kind permission to reproduce their photographs:

(Key: b-bottom; c-centre; l-left; r-right; t-top)

123RF.com: Chris Elwell 130tl/B, Ewelina Kowalska 136–137, 154t, kishano 68tl/A, Peter Baxter 52bc, Richard Starkweather 85bl/A, Tim Hester 135cr; **Alamy Images:** age fotostock 239tr/C, Al Argueta 294bc/A, blickwinkel 74tr, Charles Stirling 263br/C, Chronicle 89c, Eric Lafforgue 153bl/A, Florida Images 256bl/D, Frans Lanting / Mint Images Limited 166tr/A, Karen Doody / Stocktrek Images 197br/C, Keystone Pictures USA 19c/B, Martin Shields 266r, Nigel Caitlin 184tc, 190cr/A, Pat Bennett Ghana 130br/C, Ragnar Th Sigurdsson 262tl, Rick & Nora Bowers 229cl/G, Tim Gainey 183bl/A, Wolfgang Weinh / Westend61 GMbH 296cl/F; **Anthony Short:** 11tc/D, 26, 36t/G, 156br/A, 160–162, 164tr/C, 174–175, 226tl/C (1), 226tl/C (2), 226tl/C (3), 233tr/B, 240l/A (1), 240r/A (2), 244–245, 246bl/B, 251cr/E, 258l, 259tl/C, 268–269, 272cl/E, 276bl/A, 292, 294bl/A, 298bl/C, 299cr/F, 300br/B; **BBC Panorama:** 216br/A; **CDC:** Janice Haney Carr / Sickle Cell Foundation of Georgia: Jackie George, Beverly Sinclair 173tl; **Corbis:** 66br/B, Alberto Lowe / Reuters 165tl/E, Aviral Mediratta 97cr/C, Carolina Biological / Visuals Unlimited 112cr/E, Darryl Estrine 94bl, Dr. Michael Gabridge / Visuals Unlimited 46bl/D; **Courtesy of The University of Alberta:** Paul Stothard 96br/B; **DK Images:** Tracey Morgan 138cl/A; **FLPA Images of Nature:** David T Grewcock 162tr/A, Tui De Roy / Minden Pictures 253cl/B, 256tr/F; **Fotolia.com:** Alexander Erdbeer 233bl/A, Keza 255br/C, schankz 278bl/D, Sima 206tl/C, todoryankov 238cl, Yoshitooo 282–283; **Getty Images:** Colin Monteath / Hedgehog House 21tl/B, Jan Bruder / iStock 47br/A, Kolett / Moment Open 162cr/B, Rudi Gobbo 44b/A, SAM YEH 122–123, skynesher 170br/A; **Harvard Apparatus Regenerative Technology:** 116tr/A; **Institut Pasteur:** Coll. Musee Pasteur 88br/A; **Martyn F. Chillmaid:** 48bc/B; **Minnesota Public Radio:** Photo by Jeffrey Thompson, Minnesota Public Radio News. © 2012 Minnesota Public Radio. Used with permission. All rights reserved. 60br/A; **NHS Blood and Transplant:** 144b/C; **PhotoDisc:** Photolink 246tl/A; **Phototake, Inc:** Dennis Kunkel Microscopy, Inc 203tr/B; **Press Association Images:** HO / AP 253cr/C, Khalil Senosi / AP 68cl/A, Mike Howarth 250br/B; **Public Health Image Library/CDC:** National Institute of Allergy and Infectious Diseases (NIAID) 53br/E; **Science and Plants for schools (SAPS):** This activity was developed through the Science and Plants for Schools (SAPS) programme. This and other resources supporting biology education can be downloaded for free from the SAPS website: www.saps.org.uk 38; **Science Photo Library Ltd:** 53cr/D, 86bl/C, Alan Carey 252cr, ASTRID & HANNS-FRIEDER MICHLER 220–222, Biophoto Associates 159t, Cavallini James / BSIP 67cl/C, CDC 70cl/A, Clouds Hill Imageing Ltd 64–65, CNRI 11tr/D, 11br/E, 25, DAVID SCHARF 52br/C, Dr Brad Morgan / Visuals Unlimited 126tl, DR GOPAL MURTI 74br, Dr Jeremy Burgess 29cr/B, Dr P Marazzi 78bl/A, Gastrolab 111cl/C, JIM VARNEY 56tr/B, Lowell Georgia 127br, NANCY KEDERSHA 194–195, NIBSC 106–107, Pascal Goetgheluck 98br/A, PROF MATTHIAS GUNZER 76–77, PROF. S.H.E. KAUFMANN & DR J.R GOLECKI 82tl/A, Steve Gschmeissner 17tr/E, 139cr/C, 278cr/E, TEK IMAGE 42–43, 92–93, Wally Eberhart / Visuals Unlimited 141bl/B, Wolfgang Hoffmann / AgstockUSA 45c/B, Zephyr 113br/B; **Shutterstock.com:** Damian Herde 68bl/B, Dr Alan Lipkin 164br/D, Floris Slooff 129tr/A, Hung Chung Chih 10tl, Image Point Fr 205br/B, JPC-PROD 79bl/B, Marco Uliana 183bc/A, Matej Hudovernik 303tl, Natursports 16c, 284tr, Naypong 110tr/A, Robynrg 173cr, Sari O'Neal 112cl/E, Smileus 28c/A, Stefan Schurr 8–9; **TopFoto:** PA 21tr; **U.S. Department of Agriculture:** Food Safety & Inspection Service (FSIS) 54bl; **U.S. National Library of Medicine:** 167c/C; **University of York:** Biology Graphics 132r/A

Cover images: *Front:* **Science Photo Library Ltd:** British Library

All other images © Pearson Education

Picture Research by: Susie Prescott

We are grateful to the following for permission to reproduce copyright material:

Figures
Figure on page 38 from SAPS poster PlantPower, http://www.saps.org.uk/attachments/article/1266/C4%20Rice%20-%20poster.pdf. This activity was developed through the Science and Plants for Schools (SAPS) programme. This and other resources supporting biology education can be downloaded for free from the SAPS website: www.saps.org.uk. Courtesy of Science and Plants for Schools; Figure on page 52 from Salmonella by month February 2014, http://www.wales.nhs.uk/sites3/documents/457/Salmonella%20by%20month%20Feb%202014.png, Public Health Wales NHS Trust; Figure on page 55 from Infectious diseases – Age standardised mortality rate (per 100,000), 1911–2005, http://www.nrscotland.gov.uk/files/statistics/scotlands-population-2005-the-register-generals-annual-review-151stedition/j9085e05.htm, contains public sector information licensed under the Open Government Licence v3.0; Figure on page 59 from http://www.ons.gov.uk/ons/resources/chart2_tcm77–323729.png, source: Office for National Statistics licensed under the Open Government Licence v.3.0; Figure on page 100 from GFI Laboratory, with permission from Zach Gaskin, Chief Scientific Officer; Figure on page 128 from Application A1018 - Food derived from High Oleic Acid Soybean Line DP-305423–1, http://www.foodstandards.gov.au/code/applications/pages/applicationa1018food4091.aspx © Food Standards Australia New Zealand, Creative Commons Attribution 3.0 Australia (CC by 3.0 Au); Figure on page 168 adapted from Te Ara, the Encyclopedia of New Zealand, http://www.teara.govt.nz/ © Crown Copyright 2005–2013 Manatu Taonga, The Ministry for Culture and Heritage, New Zealand. (Data originally sourced from Environment Canterbury); Figures on page 187, page 188, page 191 from http://www.saps.org.uk/attachments/article/543/SAPS%20Seeing%20Without%20Eyes%20Article.pdf, This activity was developed through the Science and Plants for Schools (SAPS) programme. This and other resources supporting biology education can be downloaded for free from the SAPS website: www.saps.org.uk. Courtesy of Science and Plants for Schools; Figure on page 201 republished with permission of John Wiley & Sons, Inc. from Active transport of cations in giant axons from Sepia and Loligo, *The Journal of Physiology*, Vol.128, Issue 1, 28 April, Fig 3, p.34 (Hodgkin A. L. and Keynes R. D. 1955), permission conveyed through Copyright Clearance Center, Inc; Figure on page 206 adapted from *Hong Kong Medical Journal* (C.P. Chan, F.L. Lau. Should lidocaine spray be used to ease nasogastric tube insertion? A double-blind, randomised controlled trial. Hong Kong Med J 2010;16(4):282–6 2010) Copyright 2010, Hong Kong Academy of Medicine; Figure on page 249 from Developments in plant colonization and succession on Surtsey during 1999–2008, http://www.surtsey.is/SRS_publ/2009-XII/low_res/2009-XII_057–076_Developments--lw.pdf, with permission from Icelandic Institute of Natural History; Figure on page 254 from Korpimäki, E. and Krebs, C.J., Predation and Population Cycles of Small Mammals: A reassessment of the predation hypothesis, *BioScience*, 1996, 46, 10, 754–64, by permission of Oxford University Press; Figures on page 255, page 274 from *Life: The Science of Biology*, 8th ed., Sinauer Associates, Inc. (Sadava, Heller, Orians, Purves, Hills 2008); Figure on page 259 from The Deer Initiative National Deer Collisions Project - Deer on our Roads – Counting the Cost published on www.deercollisions.co.uk, reproduced with permission from Dr. J Langbein on behalf of The Deer Initiative; Figure on page 286 from IPCC, 2013: Summary for Policymakers, *Climate Change 2013: The Physical Science Basis. Contribution of Working Group I to the Fifth Assessment Report of the Intergovernmental Panel on Climate Change* SPM.1, p.6 (Stocker, T.F., D. Qin, G.-K. Plattner, M. Tignor, S.K. Allen, J. Boschung, A. Nauels, Y. Xia, V. Bex and P.M. Midgley eds. 2013), with permission from IPCC; Figure on page 287 adapted from Climate Change 2001: Working Group I to the Third Assessment Report of the Intergovernmental Panel on Climate Change, Summary for Policymakers, Figure 1(b), with permission from IPCC; Figure on page 291 from Climate Change 2007: Synthesis Report, Contribution of Working Groups I, II and III to the Fourth Assessment Report of the Intergovernmental Panel on Climate Change, Figure SPM. 3(a), IPCC, Geneva, Switzerland; Figure on page 293 from IPCC, 2013: Summary for Policymakers, *Climate Change 2013: The Physical Science Basis. Contribution of Working Group I to the Fifth Assessment Report of the Intergovernmental Panel on Climate Change* SPM.3, p.10 (Stocker, T.F., D. Qin, G.-K. Plattner, M. Tignor, S.K. Allen, J. Boschung, A. Nauels, Y. Xia, V. Bex and P.M. Midgley eds. 2013), with permission from IPCC; Figure on page 294 © FAO, UN FAO Fisheries, *Fisheries at the Limit?*, http://www.theglobaleducationproject.org/earth/images/final-images/f-fish-stocks-status.gif; Figure on page 294 © FAO, UN FAO Fishstat database, http://www.theglobaleducationproject.org/earth/images/final-images/f-canad-atlantic-cod-catch.gif; Figure on page 295 from *Environment Programme Publication: Keeping Track of Our Changing Environment: From Rio to Rio+20 (1992–2012)* (UNEP 2011) © 2011 United Nations Environment Programme Publication, published October 2011; Figure on page 296 from Sustainable Development Indicators, July 2015, http://www.ons.gov.uk/ons/rel/wellbeing/sustainable-development-indicators/july-2015/art---sustainable-development-indicators.html#tab-Environment, source: Office for National Statistics licensed under the Open Government Licence v.3.0; Figure on page 297 adapted from http://www.iucnredlist.org/about/summary-statistics#Tables_1_2, with permission from IUCN; Figure on page 298 adapted from http://jr.iucnredlist.org/documents/summarystatistics/Fig_1_IUCN_Red_List_Index.jpg, with permission from IUCN.

Text
Extract on page 60 adapted from The Amazing Human Biome by Ann Fullick, an article from Society of Biology Bionet, http://societyofbiology.org/images/SB/Ann_Fullick-_Amazing_human_microbiome_article.pdf, with permission from Society of Biology; Extract on page 102 adapted from Main events in development of DNA sequencing, Wellcome Genome Campus Public Engagement/www.yourgenome.org; Extract on page 132 from a blog by Caroline Wood, http://scienceasadestiny.blogspot.co.uk/2014/04/explosive-plant-science.html, with permission from Caroline Wood; Extract on page 156 adapted from Human genome includes 'foreign' genes not from our ancestors, press release by Joel Winston, BioMed Central, http://www.cam.ac.uk/research/news/human-genome-includes-foreign-genes-not-from-our-ancestors, Joel Winston, BioMed Central; Extract on page 170 adapted from Are humans still evolving?, Wellcome Genome Campus Public Engagement/www.yourgenome.org; Extract on page 216 adapted from World first as man whose spinal cord was severed Walks by Ben Spencer, 21 October 2014, Daily Mail; Extract on page 240 from *Heinemann Advanced Science: Biology* by Ann Fullick, Pearson Education Ltd., © First edition Ann Fullick, 1994. © This edition Ann Fullick, 2000; Extract on page 300 adapted from Tortuguero green turtle nesting trend from 1986 – 2014, www.conserveturtles.org, with permission from Sea Turtle Conservancy.

The Publisher would like to thank Chris Curtis, Wade Nottingham and Janette Gledhill for their contributions to the Maths skills and Preparing for your exams sections of this book.

The Publisher would also like to thank Phil Arthur and Ian Honeysett for their collaboration in reviewing this book.

The author would like to acknowledge and thank the teams at Science and Plants for Schools (SAPS), the Wellcome Genome Campus Public Engagement Team, Cambridge University and the ABPI. The author would also like to thank the following for their support and individual contributions: Dr Liz Rylott, Caroline Wood, Alice Kelly, Amy Ekins-Coward, Matt Deacon, Tony Short, William Fullick, Thomas Fullick, James Fullick, Edward Fullick and Chris Short.

Contents

How to use this book

Welcome to Book 2 of your A level Biology B course. This book, covering Topics 5–10 of the specification, follows on from Book 1, which covered Topics 1–4. The exams that you will sit at the end of your A level course will cover content from both books. The following features are included to support your learning:

Chapter openers

Each chapter starts by setting the context for that chapter's learning:

- Links to other areas of biology are shown, including previous knowledge that is built on in the chapter, and future learning that you will cover later in your course.

- The **All the maths you need** checklist helps you to know what maths skills will be required.

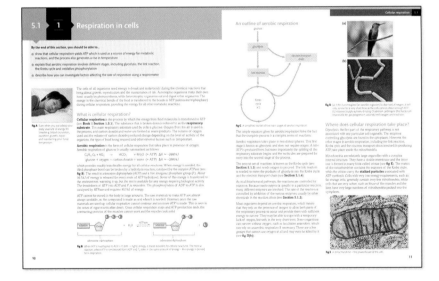

Main content

The main part of each chapter covers all the points from the specification that you need to learn. The text is supported by diagrams and photos that will help you understand the concepts.

Within each section, you will find the following features:

- **Learning objectives** at the beginning of each section, highlighting what you need to know and understand.

- **Key definitions** shown in bold and collated at the end of each section for easy reference.

- **Worked examples** showing you how to work through questions, and how your calculations should be set out.

- **Learning tips** to help you focus your learning and avoid common errors.

- **Did you know?** boxes featuring interesting facts to help you remember the key concepts.

- **Questions** to help you check whether you have understood what you have just read, and whether there is anything that you need to look at again.

Answers to the questions can be found on Pearson's website as a free resource.

Thinking Bigger

The book features a number of **Thinking Bigger** spreads that give you an opportunity to read and work with real-life research and writing about science. The timeline at the bottom of the spreads highlights which of the chapters the material relates to. These spreads will help you to:

- read real-life material that's relevant to your course
- analyse how scientists write
- think critically and consider the issues
- develop your own writing
- understand how different aspects of your learning piece together.

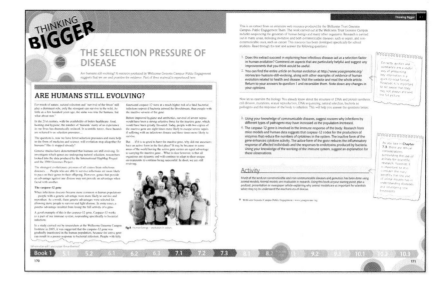

Exam-style questions

At the end of each chapter there are also **exam-style questions** to help you to:

- test how fully you have understood the learning
- practise for your exams.

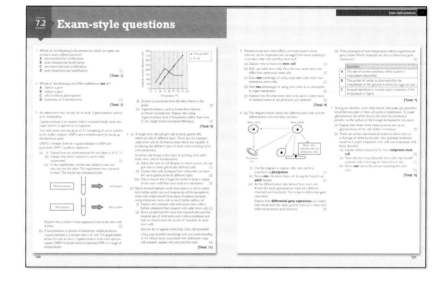

Getting the most from your online ActiveBook

This book comes with 3 years' access to ActiveBook* – an online, digital version of your textbook. Follow the instructions printed on the inside front cover to start using your ActiveBook.

Your ActiveBook is the perfect way to personalise your learning as you progress through your Edexcel A level Biology B course. You can:

- access your content online, anytime, anywhere
- use the inbuilt highlighting and annotation tools to personalise the content and make it really relevant to you
- search the content quickly using the index.

Highlight tool
Use this to pick out key terms or topics so you are ready and prepared for revision.

Annotations tool
Use this to add your own notes, for example links to your wider reading, such as websites or other files. Or make a note to remind yourself about work that you need to do.

*For new purchases only. If this access code has already been revealed, it may no longer be valid. If you have bought this textbook secondhand, the code may already have been used by the first owner of the book.

TOPIC 5
Energy for biological processes

5.1 > Cellular respiration

Introduction

Clostridium botulinum is a bacterium that produces one of the most powerful poisons known – 1 μg of botulinum toxin is enough to kill an adult human being. The bacteria are not common but they can grow in food that has not been preserved properly. It may seem surprising therefore that botulinum toxin is also widely used in the beauty industry. Botox injections, which prevent the wrinkling of the skin by paralysing facial muscles, contain around 5 ng of botulinum toxin!

C. botulinum is one of a group of organisms that respire only in the absence of oxygen. They are known as obligate anaerobes and they are killed by high levels of oxygen. They have a very specialised metabolism. Although most organisms, including human beings, can respire without oxygen for a short time if they need to, the majority of living things need oxygen to keep them alive long-term.

In this chapter you will be finding out about the main stages of the process of respiration, which take place in eukaryotic cells and most prokaryotic cells. You will discover how the anaerobic processes of glycolysis in the cytoplasm are linked to the aerobic processes of the Krebs cycle and oxidative phosphorylation in the mitochondria. You will learn how ATP is produced in oxidative phosphorylation and the importance of chemiosmosis, and you will develop an understanding of the different amounts of ATP produced in the different processes.

You will consider how scientists investigate the rate of respiration in organisms using respirometers, including versions that can be used in a school laboratory to measure the uptake of oxygen. You will also look at some of the techniques used to investigate the site where ATP is formed in the cell.

You will look in detail at how anaerobic respiration takes place in the cytoplasm of plant and animal cells, and consider the effect of lactate on muscle tissues and how they contract.

All the maths you need

- Recognise and make use of appropriate units in calculations (*e.g. kJ in exothermic reactions of cellular respiration*)
- Use fractions, ratios and percentages (*e.g. in considering the production of ATP during aerobic respiration*)
- Estimate results (*e.g. production of ATP in the processes of cellular respiration*)
- Construct and interpret frequency tables and diagrams, bar charts and histograms (*e.g. anaerobic exercise and the oxygen debt*)
- Translate information between graphical, numerical and algebraic forms (*e.g. oxygen debt*)

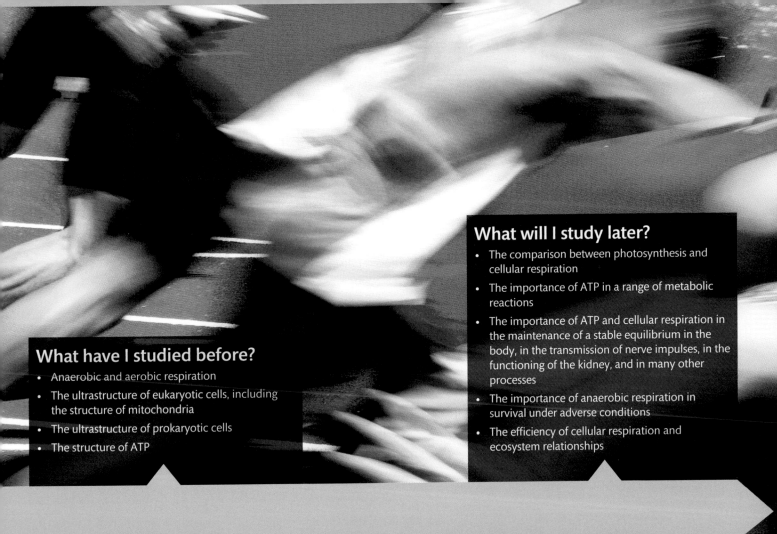

What will I study later?

- The comparison between photosynthesis and cellular respiration
- The importance of ATP in a range of metabolic reactions
- The importance of ATP and cellular respiration in the maintenance of a stable equilibrium in the body, in the transmission of nerve impulses, in the functioning of the kidney, and in many other processes
- The importance of anaerobic respiration in survival under adverse conditions
- The efficiency of cellular respiration and ecosystem relationships

What have I studied before?

- Anaerobic and aerobic respiration
- The ultrastructure of eukaryotic cells, including the structure of mitochondria
- The ultrastructure of prokaryotic cells
- The structure of ATP

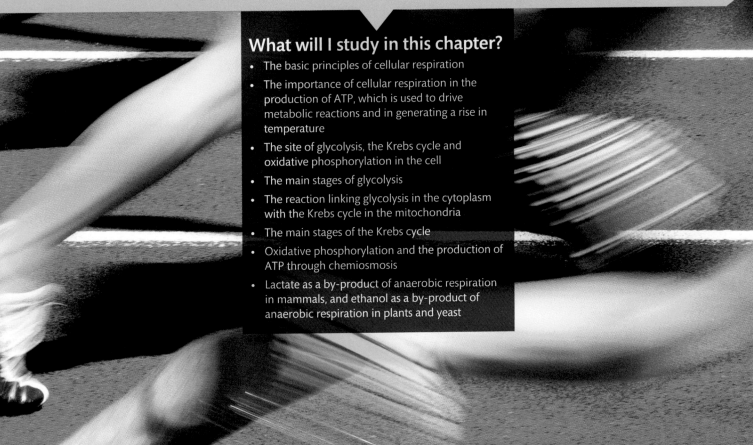

What will I study in this chapter?

- The basic principles of cellular respiration
- The importance of cellular respiration in the production of ATP, which is used to drive metabolic reactions and in generating a rise in temperature
- The site of glycolysis, the Krebs cycle and oxidative phosphorylation in the cell
- The main stages of glycolysis
- The reaction linking glycolysis in the cytoplasm with the Krebs cycle in the mitochondria
- The main stages of the Krebs cycle
- Oxidative phosphorylation and the production of ATP through chemiosmosis
- Lactate as a by-product of anaerobic respiration in mammals, and ethanol as a by-product of anaerobic respiration in plants and yeast

Respiration in cells

By the end of this section, you should be able to...

● show that cellular respiration yields ATP which is used as a source of energy for metabolic reactions, and the process also generates a rise in temperature

● explain that aerobic respiration involves different stages, including glycolysis, the link reaction, the Krebs cycle and oxidative phosphorylation

● describe how you can investigate factors affecting the rate of respiration using a respirometer

fig A Even when you are asleep your body uses lots of energy for breathing, blood circulation, excretion, growth, repair and maintaining your body temperature.

The cells of all organisms need energy to break and make bonds during the chemical reactions that bring about growth, reproduction and the maintenance of life. Autotrophic organisms make their own food, usually by photosynthesis, while heterotrophic organisms eat and digest other organisms. The energy in the chemical bonds of the food is transferred to the bonds in ATP (adenosine triphosphate) during cellular respiration, providing the energy for all other metabolic reactions.

What is cellular respiration?

Cellular respiration is the process by which the energy from food molecules is transferred to ATP (see **Book 1 Section 1.3.1**). The substance that is broken down is referred to as the **respiratory substrate**. The main respiratory substrate used by cells is glucose. Oxygen from the air is used in the process, and carbon dioxide and water are formed as waste products. The volume of oxygen used and the volume of carbon dioxide produced change depending on the level of activity of the organism, the type of food being respired and other external factors such as temperature.

Aerobic respiration is the form of cellular respiration that takes place in presence of oxygen. Aerobic respiration of glucose is usually summarised as follows:

$$C_6H_{12}O_6 + 6O_2 \rightarrow 6CO_2 + 6H_2O \ (+ \ ATP) \ \Delta H \approx -2880\,kJ$$

$$\text{glucose} + \text{oxygen} \rightarrow \text{carbon dioxide} + \text{water} \ (+ \ ATP) \ \Delta H \approx -2880\,kJ$$

which provides readily transferable energy for all cellular reactions. When energy is needed, the third phosphate bond can be broken by a hydrolysis reaction, catalysed by the enzyme ATPase (see **fig B**). The result is adenosine diphosphate (ADP) and a free inorganic phosphate group (P_i). About 30.5 kJ of energy is released for every mole of ATP hydrolysed. Some of this energy is transferred to the environment, warming it up, but the rest is available for any energy-requiring biological activity. The breakdown of ATP into ADP and P_i is reversible. The phosphorylation of ADP to ATP is also catalysed by ATPase and requires 30.5 kJ of energy.

ATP cannot be stored in the body in large amounts. The raw materials to make ATP are almost always available, so the compound is made as and when it is needed. However, once the raw materials are used up, cellular respiration cannot continue and no more ATP is made. This is seen in the onset of rigor mortis after death. Once cellular respiration stops and ATP production ends, the contracting proteins of the muscles cannot work and the muscles lock solid.

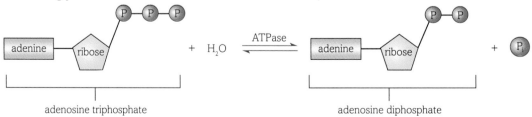

fig B When ATP is hydrolysed to ADP + P_i (left → right), energy is made available for cellular reactions. The reverse reaction, where ATP is synthesised from ADP and P_i, *takes in* the same amount of energy – this energy is derived from respiration.

An outline of aerobic respiration

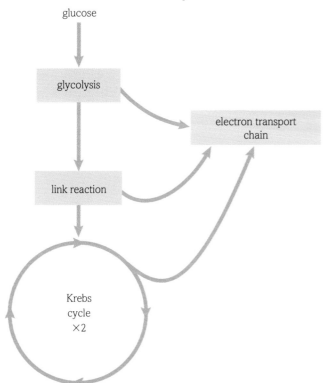

fig C A simplified model of the main stages of aerobic respiration.

The simple equation given for aerobic respiration hides the fact that the complete process is a complex series of reactions.

Aerobic respiration takes place in two distinct phases. This first stage is known as glycolysis and does not require oxygen. A little ATP is produced here, but more importantly the splitting of the respiratory substrate begins and the molecules are prepared for entry into the second stage of the process.

The second set of reactions is known as the Krebs cycle (see **Section 5.1.3**) and needs oxygen to proceed. The link reaction is needed to move the products of glycolysis into the Krebs cycle and the electron transport chain (see **Section 5.1.4**).

As in all biochemical pathways, the reactions are controlled by enzymes. Because each enzyme is specific to a particular reaction, many different enzymes are involved. The rate of the reaction is controlled by inhibition of the various enzymes, usually by other chemicals in the reaction chain (see **Section 5.1.2**).

Most organisms depend on aerobic respiration, which means that they rely on the presence of oxygen to allow both parts of the respiratory process to occur and provide them with sufficient energy to survive. They may be able to cope with a temporary lack of oxygen, but only in the very short term. Some organisms can survive without oxygen, such as facultative anaerobes, which can rely on anaerobic respiration if necessary. There are a few groups that cannot use oxygen at all and may even be killed by it (see **fig D(b)**).

(a)

(b)

fig D (a) If this hummingbird (an aerobic organism) is deprived of oxygen, it will only survive for a very short time as the cells cannot obtain enough ATP. (b) However, supply actively dividing *Clostridium perfringens* (the bacterium responsible for gas gangrene in wounds) with oxygen and it will die.

Where does cellular respiration take place?

Glycolysis, the first part of the respiratory pathway, is not associated with any particular cell organelle. The enzymes controlling glycolysis are found in the cytoplasm. However, the other stages in aerobic respiration, including the link reaction, Krebs cycle and the electron transport chain involved in producing ATP, take place inside the mitochondria.

Mitochondria are relatively large organelles with a complex internal structure. They have a double membrane and the inner one is formed in many folds called cristae (see **fig E**). The matrix of the mitochondrion contains the enzymes of the Krebs cycle, while the cristae carry the **stalked particles** associated with ATP synthesis. Cells with very low energy requirements, such as fat storage cells, generally contain very few mitochondria, while cells that are very active, such as those of the muscles and the liver, have very large numbers of mitochondria packed into the cytoplasm.

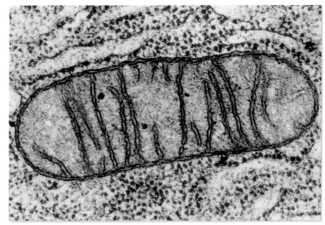

fig E A mitochondrion – the powerhouse of the cell.

The hydrogen acceptors

Simple representations of cellular respiration suggest that ATP is produced as a direct result of the breakdown of glucose, but this is not the case. Most of the ATP produced during cellular respiration is made through a series of oxidation and reduction reactions in the electron transport chain. **Reduction** is the addition of electrons to a substance and in the cell this is brought about by the addition of hydrogen or removal of oxygen. **Oxidation** is the removal of electrons from a substance and any compound that has oxygen added, or hydrogen or electrons removed, is oxidised.

During cellular respiration, hydrogen is removed from compounds and picked up by a **hydrogen acceptor**, which is therefore reduced. This happens several times during the reactions of respiration, as you will see later. The hydrogen is then passed to the next hydrogen acceptor and along the electron transport chain. A series of linked oxidation and reduction (redox) reactions takes place. Each redox reaction releases a small amount of energy which is used to drive the synthesis of a molecule of ATP (see **Section 5.1.4** for more details).

The most common hydrogen acceptor in cellular respiration is **NAD** (nicotinamide adenine dinucleotide). NAD is a coenzyme, one of the small molecules that assist in enzyme-catalysed reactions. When it accepts hydrogen atoms from a metabolic pathway it becomes reduced to form **reduced NAD** (NADH). The oxidised form of NAD is denoted as NAD^+. **FAD** (flavin adenine dinucleotide) is another hydrogen carrier (and coenzyme) that accepts hydrogen from reduced NAD and forms reduced FAD ($FADH_2$). Each time, a molecule of ATP is formed in the process.

Finding out about cellular respiration

Our understanding of the process of respiration has developed gradually over the years. In the early days the research was based on whole animals and plants. Now the work continues at the level of tiny cell fragments, evidence of which has become available to us using technology such as the electron microscope. We need to understand the factors that affect the rate of respiration and exactly how and where the reactions take place within the cell. Two examples are given below.

Investigating factors affecting the rate of respiration (using whole organisms)

It is not always easy to demonstrate the rate of cellular respiration without sophisticated biochemical techniques designed to measure the rate in isolated cell organelles. However, in a school lab, a **respirometer** can give some valuable information about the rate of cellular respiration by measuring the uptake of oxygen or the output of carbon dioxide by whole organisms. A basic respirometer consists of a sealed chamber containing one or more living organisms, such as germinating seeds (see **fig F**). A chemical such as soda lime or potassium hydroxide is used to absorb the carbon

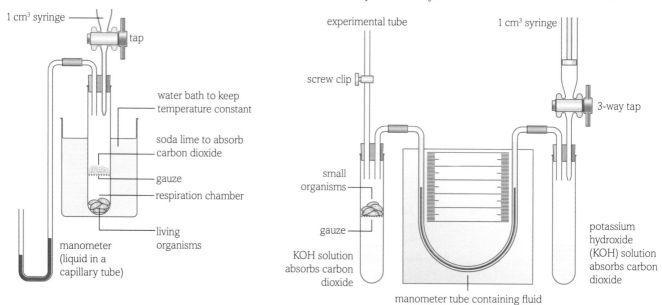

fig F Any movement of the liquid in the tube indicates an uptake of oxygen by the organisms in the respiration chamber. Some respirometers are more sophisticated than others.

dioxide produced by respiration, therefore any changes in volume will be due to the uptake of oxygen by the organisms. As the organisms use oxygen, the pressure reduces and so the fluid in the manometer moves towards the tube containing the organisms. The volume of gas needed to return the manometer to normal is measured using the syringe. This measurement can then be used to calculate the intake of oxygen per minute and give an approximate respiration rate for the organisms. By changing the external conditions (e.g. temperature) it is possible to measure their effect on the rate of respiration by recording changes in the uptake of oxygen. This simple apparatus has obvious limitations but it can be used to give an overall impression of the rate of respiration of organisms under differing conditions.

Investigating the site of ATP synthesis (using cell fragments)

There are a number of ways of investigating respiration at the cellular level. For example:

- You can break open cells and centrifuge the contents to obtain a fraction containing just mitochondria. If these are kept supplied with glucose and oxygen they will produce ATP.

- The high resolving power of electron microscopes has shown us that the surface of the inner membrane of the mitochondrion is covered in closely packed stalked particles. These provide a greatly increased surface area, which is an ideal site for enzymes to work.

- The stalked particles and the bits of membrane associated with them can be separated from the rest of the mitochondrial structure. Scientists have demonstrated that ATP synthesis only occurs here. As a result of this type of evidence, and more, the stalked particles have been accepted as vital for the formation of ATP.

Learning tip

As part of your study of this topic, you will carry out **Core Practical 9: Investigate factors affecting the rate of aerobic or anaerobic respiration using a respirometer, taking into account the safe and ethical use of organisms.** Make sure you have a good understanding of this practical as your understanding of the experimental method may be assessed in your examination.

Questions

1. Explain why cellular respiration is such an important reaction.

2. $C_6H_{12}O_6 + 6O_2 \rightarrow 6CO_2 + 6H_2O + ATP$ is sometimes presented as the equation of aerobic respiration. Explain why it is a summary, not an equation, and outline its strengths and limitations in representing aerobic respiration.

3. (a) Explain how respirometers are limited in what they can tell us about cellular respiration.
   (b) Evaluate the two pieces of apparatus in **fig F** to show which you think would deliver more reliable evidence and why.

4. Describe the kind of evidence that would be needed to identify the sites of the various stages of cellular respiration in a mitochondrion.

Key definitions

Cellular respiration is the process by which food is broken down to yield ATP, which is used as a source of energy for metabolic reactions.

The **respiratory substrate** is the substance oxidised during cellular respiration.

Aerobic respiration is the form of cellular respiration that takes place in the mitochondria in the presence of oxygen.

Stalked particles are structures where ATP production takes place on the inner mitochondrial membrane.

Reduction is the addition of electrons to a substance, e.g. by the addition of hydrogen or removal of oxygen.

Oxidation is the removal of electrons from a substance, e.g. by the addition of oxygen or removal of hydrogen.

A **hydrogen acceptor** is a molecule which receives hydrogen and becomes reduced in cell biochemistry.

NAD is a coenzyme that acts as a hydrogen acceptor.

Reduced NAD is NAD which has accepted a hydrogen atom in a metabolic pathway.

FAD is a hydrogen carrier and coenzyme. In cellular respiration it accepts hydrogen to form reduced FAD ($FADH_2$), driving the production of ATP.

A **respirometer** is a piece of apparatus used for measuring the rate of respiration in whole organisms or cultures of cells.

Glycolysis and anaerobic respiration

By the end of this section, you should be able to...

● describe the conversion of monosaccharides to pyruvate during glycolysis in the cytoplasm

● describe anaerobic respiration as the partial breakdown of hexoses to produce a limited yield of ATP in the absence of oxygen

● explain how lactate, as a by-product of anaerobic respiration, affects mammalian muscle contraction

● explain how anaerobic respiration in plants results in ethanol formation

To make understanding the biochemistry of respiration easier, you are going to look at the different stages separately, and then consider the overall situation. The first stage is **glycolysis**. In this initial part of the respiratory pathway, 6C glucose is split by a series of ten reactions into two molecules of the 3-carbon (3C) compound **pyruvate**. The pyruvate ions produced by glycolysis can either be used in aerobic respiration or anaerobic respiration.

Glycolysis

Glycolysis means 'sugar-splitting'. It takes place in the cytoplasm of the cell. The main stages of glycolysis are shown in **fig A**.

Glucose is a 6-carbon (6C), or hexose, sugar. The glucose for glycolysis may come directly from the blood or it may be produced by the breakdown of glycogen stores in muscle and liver cells (see **Book 1 Section 1.2.2**).

The first step in glycolysis actually uses ATP to provide the energy to phosphorylate the 6C sugar glucose, adding two phosphate groups. This phosphorylation makes the sugar more reactive and also makes it unable to pass through the cell membrane, so it becomes trapped within the cell.

The phosphorylated sugar is then split to give two molecules of a 3-carbon sugar, known as **glycerate 3-phosphate** (**GP**). GP is then converted by several steps into a molecule of pyruvic acid, which is found in solution as pyruvate ions. During these reactions a small amount of ATP is produced as follows:

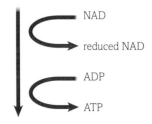

2 × 3C sugar (GP)

2 × pyruvate

fig A The main stages of glycolysis which result in the production of pyruvic acid in the form of pyruvate and a small quantity of ATP and reduced NAD. These reactions all happen within the cytoplasm of the cell.

* Two hydrogen atoms are removed from the 3C sugars and taken up by NAD, forming reduced NAD (NADH). This takes place in the cytoplasm of the cell. The reduced NAD then passes through the outer mitochondrial membrane into the electron transport chain, which is explained in detail in **Section 5.1.4**.

* A small amount of ATP is also made directly from the energy transfer when the 3C sugar is converted to pyruvate. The phosphorylation of the sugar at the beginning of glycolysis is reversed when the final intermediate compound is converted to pyruvate. The phosphate group released is used to convert ADP to ATP.

* If there is plenty of oxygen the pyruvate will enter the mitochondria and be used in the aerobic reactions of the Krebs cycle. If oxygen levels are low, the pyruvate remains in the cytoplasm and is converted into either **ethanol** (in plants) or **lactate** (in mammals) with only a little ATP produced. This is **anaerobic respiration** in the cytoplasm.

Anaerobic respiration

Anaerobic respiration takes place in the absence of oxygen. It takes place in most types of organism, but the final products differ between organism types.

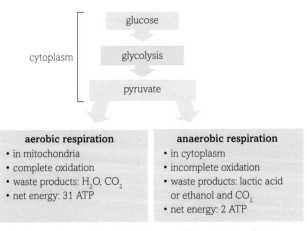

fig B The alternative routes for the products of glycolysis give very different yields of ATP.

Anaerobic respiration in mammals

If you exercise particularly hard, or the exercise lasts a long time, your muscles do not get enough oxygen to supply their needs. When this happens, the products of glycolysis cannot go on to the Krebs cycle (the aerobic stage of cellular respiration). The muscles will therefore respire anaerobically.

In anaerobic respiration in mammals the pyruvate from glycolysis is converted to lactic acid, another 3C compound that dissociates to form lactate and hydrogen ions. As a result of anaerobic respiration, only two molecules of ATP are produced per glucose molecule respired. In contrast, up to 8 ATP molecules are produced in glycolysis when pyruvate can feed on into the Krebs cycle. This very low yield of ATP in anaerobic respiration is because some of the reduced NAD is used to reduce pyruvate to lactate rather than entering the electron transport chain. The lactate moves out of the cells into the blood:

$$C_6H_{12}O_6 \rightarrow 2C_3H_6O_3 \ (+ \text{ATP}) \ \Delta H \approx -150\,\text{kJ}$$
$$\text{glucose} \rightarrow \text{lactic acid} \ (+ \text{ATP}) \ \Delta H \approx -150\,\text{kJ}$$

As levels of lactate and hydrogen ions increase during anaerobic respiration in the muscles, the pH falls and the muscle tissue becomes acidic. Traditionally this was thought to reduce the ability of the muscles to contract, so the contractions lost their force and eventually stopped completely. Modern research suggests this is not the case – neither lactate ions nor the fall in pH due to hydrogen ions seem to directly affect muscle contraction. It appears that the movement of lactate and hydrogen ions into the blood from the muscles lowers the pH of the blood, which in turn affects the central nervous system (see **Section 9.2.6**). It appears it may be reduced nervous stimulation from the central nervous system that reduces and eventually stops muscle contraction. Scientists think this is a protective adaptation to give the muscles time to recover and return to aerobic respiration, helping to raise the pH of the blood again.

When exercise stops, the levels of lactate in the blood remain raised. The lactate, which is toxic, must be oxidised back to pyruvate to enter the Krebs cycle to be respired aerobically, producing carbon dioxide, water and ATP. It takes oxygen to oxidise the accumulated lactate. This is why you continue to breathe deeply for some time after exercise has finished (see **fig C**).

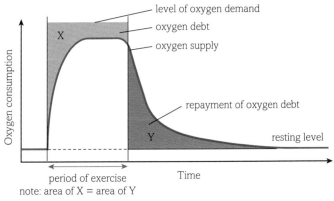

fig C This graph shows the difference between the oxygen demand during exercise and the available oxygen supply to the muscles.

Sprint athletes may run up to 95% of a race relying on the anaerobic respiration of their muscles. Long-distance runners have to maintain a much higher level of aerobic respiration because their muscles could not continue to work for the length of time needed to finish the race if the lactate levels were not kept to a minimum. Training allows athletes to get more oxygen to their muscles faster as a better blood supply develops, and to tolerate higher levels of lactate before the muscles fatigue. With repeated exposure to high lactate levels, more lactate transporter molecules develop in the mitochondrial membranes, resulting in faster processing of lactate to pyruvate when oxygen is available.

fig D Even elite athletes carry on breathing deeply for some time after a race is over, as their bodies recover from a period of anaerobic exercise.

Did you know?

The model of lactate metabolism after anaerobic respiration remained the same for many years, with a picture of lactate being oxidised back to pyruvate and fed into the Krebs cycle. However, recently a more complex picture has started to emerge. Careful measurements have shown that the magnitude of the oxygen debt and the amount of oxygen taken in as part of the recovery profile simply do not add up. We definitely take in more oxygen after exercising than we appear to need simply from our measured lactate levels. This is excess post-exercise oxygen consumption (epoc). There has been intense debate among scientists about the causes of this. The current consensus is that there are seven factors affecting our post-exercise oxygen needs:

* Lactate needs to be oxidised to pyruvate so it can move into the Krebs cycle.
* Lactate is removed from muscles and carried in the blood to the liver where it is converted back to pyruvate and then into glucose in a process called gluconeogenesis. The glucose is then carried around in the blood to replenish the glycogen stores in the muscles.
* ATP and phosphocreatine levels in the muscle fibres need to be restored, which takes oxygen from the Krebs cycle.
* Myoglobin in the muscles needs to be re-oxygenated.
* All chemical reactions, including those of respiration, go faster, so more oxygen is needed. This is the result of raised temperatures in the muscles, and hormones such as adrenalin that are released during exercise, both of which increase reaction rates.
* We need to breathe more deeply and rapidly during exercise and so the muscles of the ribs and diaphragm use more oxygen.
* The heart rate is elevated for a time after exercise so the heart muscle needs extra oxygen supplies for the increased respiration required to support this.

Anaerobic respiration in plants and fungi

Yeast is well known for anaerobic respiration, with ethanol and carbon dioxide as the main waste products. This is the basis of many processes in biotechnology, such as brewing and baking. However, plants also undergo anaerobic respiration, in particular the root cells in waterlogged soils. When plant cells respire anaerobically they also produce ethanol:

$$C_6H_{12}O_6 \rightarrow 2\ C_2H_5OH\ +\quad 2CO_2\qquad (+\ ATP)\quad \Delta H \approx -144\,kJ$$
$$\text{glucose} \rightarrow \quad \text{ethanol}\quad +\text{carbon dioxide}\ (+\ ATP)\quad \Delta H \approx -144\,kJ$$

Discovering the glycolysis pathway

It took many years for the pathways of glycolysis and the closely-associated processes of lactic and alcoholic fermentation to be understood. The main experimental organism used in much of the work on glycolysis from the early days onwards has been yeast, rather than plants. It is easy to grow, reproduces rapidly, there are no ethical issues and yeast contains all the enzymes of glycolysis.

By the 1940s, all the individual steps of the glycolysis pathway had been worked out. There were several landmarks along the way and many scientists were involved:

- In 1897, Eduard Buchner, a German chemist, discovered that an extract of yeast containing no cells could still convert glucose to ethanol. This showed that the enzymes of glycolysis and fermentation are not associated closely with the structure of the cell.

- In the early 1900s, British biochemists Arthur Harden and W.J. Young showed that phosphate was needed for the pathway to proceed and also that there were two elements of the yeast extract needed before fermentation could go ahead. One of these was inactivated by heat as it contained the enzymes. The other was not affected by heat in the same way and contained NAD, ADP and ATP.

- Work with inhibitors allowed some of the pathway intermediates to be studied, and then the German biochemists Gustav Embden and Otto Meyerhof worked out much of the rest of the sequence. Meyerhof received the Nobel Prize in 1922 for his work on glycolysis and lactate metabolism in muscles. The work of the Polish/Russian biochemist Jakub Parnas in discovering phosphorolysis was also key in developing our understanding of the process. Alternative names for glycolysis include the Embden-Meyerhof pathway and the Embden-Meyerhof-Parnas pathway.

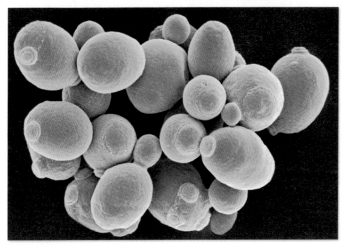

fig E Yeast – the main experimental organism for investigating glycolysis and anaerobic respiration.

Questions

1 Produce an annotated diagram of glycolysis in which the important biochemistry of each step is labelled.

2 Summarise how the anaerobic respiration of glucose releases useful energy.

3 Calculate the percentage efficiency of anaerobic respiration in mammallian muscles compared to aerobic respiration.

4 Explain why breathing rate and heart rate continue to be raised after exercise.

Key definitions

Glycolysis is the first stage of cellular respiration, which takes place in the cytoplasm and is common to both aerobic and anaerobic respiration.

Pyruvate is the end-product of glycolysis.

Glycerate 3-phosphate (GP) is a phosphorylated 3-carbon intermediate in the process of glycolysis.

Ethanol is an organic chemical with the formula C_2H_5OH produced as a result of anaerobic respiration (fermentation) in fungi and some plant cells.

Lactate is a 3-carbon compound which is the end-product of anaerobic respiration in mammals.

Anaerobic respiration is the form of cellular respiration that takes place only in the cytoplasm when there is no oxygen present.

By the end of this section, you should be able to...

● describe the link reaction and Krebs cycle taking place in the mitochondrial matrix

● explain how, during the complete oxidation of pyruvate, the events of the link reaction and Krebs cycle result in the removal of carbon atoms to produce carbon dioxide, NADH and ATP

When there is plenty of oxygen available, the pyruvate produced as the end-product of glycolysis is fed through into the mitochondria where it enters the **Krebs cycle** via the **link reaction**. The Krebs cycle is a series of biochemical steps that lead to the complete oxidation of glucose, resulting in carbon dioxide, water and relatively large amounts of ATP.

Like glycolysis, the Krebs cycle is a many-stepped process, with each individual step controlled and catalysed by a specific intracellular enzyme. The reactions of the cycle take place in the matrix of the mitochondrion, but ATP is produced in the stalked particles on the inner mitochondrial membranes in the presence of oxygen. You are going to be looking at the Krebs cycle but you do not need to learn the detailed biochemical steps that take place. Even without all the names and enzymes, **fig A** gives you a flavour of just how complex the Krebs cycle is.

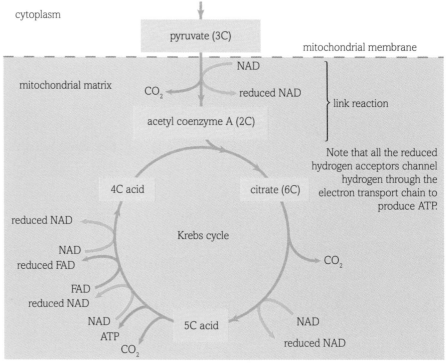

fig A The main stages of the Krebs cycle. It cycles continuously and the rate of its cycling is carefully controlled by regulatory enzymes to ensure that the right amount of ATP is produced to meet the demands of the body.

The link reaction

This is the name sometimes given to the reaction that links glycolysis to the Krebs cycle. The 3-carbon (3C) compound pyruvate crosses into the mitochondria through the mitochondrial membrane from the cytoplasm. An atom of carbon is removed from pyruvate (decarboxylation), resulting in the formation of a carbon dioxide molecule and a 2-carbon compound that joins with coenzyme A to form 2C **acetyl coenzyme A** (**acetyl CoA**). At the same time, pyruvate is oxidised, losing hydrogens to NAD (dehydrogenation), resulting in reduced NAD (NADH). The reduced NAD is used later in the electron transport chain to produce ATP. The energy contained in the acetyl CoA is released in the Krebs cycle. Enzymes known as **decarboxylases** remove carbon dioxide and enzymes known as **dehydrogenases** remove hydrogen. In summary:

pyruvate (3C) + CoA + NAD → acetyl CoA (2C) + CO_2 + reduced NAD (NADH)

The Krebs cycle

The key reactions of the Krebs cycle can be summarised as follows:

- 2C acetyl CoA combines with a 4C compound to form the 6C compound citric acid (citrate). At this point it has entered the Krebs cycle.

- The 6C citric acid now goes through a cyclical series of reactions during which the compound is broken down in a number of stages to give the original 4C compound. Two further molecules of carbon dioxide are removed in the process and are given off as a waste product.

- The 4C compound then combines with more 2C acetyl CoA and the cycle begins again.

For each molecule of pyruvate that enters the Krebs cycle, three molecules of reduced NAD (NADH), one of reduced FAD and one of ATP are produced. The reduced NAD (NADH) and reduced FAD then enter the electron transport chain. For each molecule of glucose that enters the glycolytic pathway, the Krebs cycle turns *twice* because the 6C glucose molecule produces two 3C pyruvate molecules, each of which pass through the Krebs cycle.

Did you know?

Working out the Krebs cycle

Hans Krebs first put forward his ideas for the now famous cycle in 1937. It was the result of brilliant reasoning and experimentation in the preceding years, by Krebs and by others.

In the period 1910–1920 several biochemists, including T. Thunberg, L.S. Stern and F. Batelli, used a blue dye that loses its colour when reduced to show that dehydrogenases are involved in transferring hydrogen atoms from certain organic acids known to occur in the cells of minced animal tissue.

In 1935 Albert Szent-Györgyi produced a sequence of enzymatic reactions showing the oxidation of several organic acids from succinic acid, which we now know is part of the Krebs cycle.

Krebs then carried out an elegant series of experiments to show that only certain organic acids are oxidised by cells, and that certain inhibitors could bring the oxidations to a halt. After much work he came up with the sequence we now know as the Krebs cycle. His master step was the discovery of the combination of the 2C molecule and the 4C molecule to form 6C citric acid. This was the 'missing link' that allowed him to show that the process was a cycle. Krebs also showed that all his suggested reactions could take place at a fast enough rate to account for the known pyruvate and oxygen use of the tissue. This suggested that his pathway was the main, if not the only, pathway for the oxidation of food molecules. He won the Nobel Prize in Physiology and Medicine in 1953 for his work – it changed perceptions of cell biology for ever.

fig B Sir Hans Krebs. Krebs was banned from practising medicine in Germany by the Nazis because he was Jewish. He moved to Cambridge University where, in a relatively short time, he discovered both the Krebs cycle and the ornithine cycle, which produces urea.

Questions

1 Summarise the differences between the Krebs cycle and glycolysis.

2 The Krebs cycle makes energy for the cell. Explain how this statement is incorrect.

3 Investigate the work of Krebs and write a short description of how he built up his model of the cycle of reactions that take place in the mitochondria.

Key definitions

The **Krebs cycle** is a series of biochemical steps that lead to the complete oxidation of glucose, resulting in the production of carbon dioxide, water and relatively large amounts of ATP.

The **link reaction** is the reaction needed to move the products of glycolysis into the Krebs cycle.

Acetyl coenzyme A (acetyl CoA) is the 2C compound produced in the link reaction which feeds directly into the Krebs cycle, combining with a 4C organic acid to form a 6-carbon compound.

Decarboxylases are enzymes that remove carbon dioxide.

Dehydrogenases are enzymes that remove hydrogen (carry out oxidation reactions).

The electron transport chain

By the end of this section, you should be able to...

- describe the electron transport chain taking place on the inner mitochondrial membrane and explain the importance of the membranes in this process
- explain the role of the electron transport chain in generating ATP (oxidative phosphorylation)
- explain the role of oxygen as a terminal electron acceptor forming water
- explain how ATP is synthesised by chemiosmosis
- explain the importance of mitochondrial membranes in this process
- explain the difference in the ATP yield from one molecule of hexose sugar in aerobic conditions compared with anaerobic conditions

In the final stage of respiration, reduced NAD or FAD from glycolysis and the Krebs cycle is used with oxygen to make ATP in a process called **oxidative phosphorylation**. It involves an **electron transport chain**, which is a series of electron carrier molecules, down which electrons from reduced NAD or FAD are passed. In a simultaneous process, the hydrogen ions (protons) left behind are used in **chemiosmosis** to supply the energy to synthesise ATP.

Although it is hydrogen atoms that are removed from the compounds in glycolysis and the Krebs cycle, and hydrogen atoms that eventually join up with oxygen atoms to form water, it is in fact mainly electrons that are passed along the carrier system. This is why the system is known as the electron transport chain. The hydrogen ions remain in solution. You can think of the various elements of the electron transport chain as being at different energy levels. The first member of the chain is the highest level, with subsequent steps down. Each electron is passed down from one energy level to another, driving the production of ATP (see **fig A**). The process of ATP production is known as oxidative phosphorylation because ADP is phosphorylated in a process that depends on the presence of oxygen. The electron transport chain is a model that can be used in different cellular processes. It describes the sequence of reactions by which living organisms make ATP.

There are four main electron carriers involved:

- The coenzymes NAD and FAD both act as hydrogen acceptors for hydrogen released in the Krebs cycle. One molecule of ATP is produced when the FAD is reduced and accepts hydrogen from the reduced NAD (NADH), which becomes oxidised in the process.
- **Cytochromes** are protein pigments with an iron group rather like haemoglobin. They are involved in electron transport and are reduced by electrons from reduced FAD and reduced NAD which is oxidised again. A molecule of ATP is produced at this stage.
- **Cytochrome oxidase** is an enzyme that receives the electrons from the cytochromes and is reduced as the cytochromes are oxidised. A molecule of ATP is also produced at this stage.
- Oxygen is the final hydrogen acceptor in the chain. When the oxygen is reduced, water is formed and the chain is at an end.

As a result of each molecule of hydrogen passing along the electron transport chain from reduced NAD, sufficient energy is released to make three molecules of ATP. When the hydrogen enters the chain from reduced FAD, only two molecules of ATP are produced.

fig A These are the main known components of the electron transport chain. As the carriers become reduced and then oxidised again, sufficient energy is released to drive the production of molecules of ATP.

Where is ATP actually made?

Glycolysis takes place in the cell cytoplasm and the other stages of respiration take place in the mitochondria. The link reaction and Krebs cycle take place in the matrix of the mitochondria. The electron transport chain and ATP production take place on the inner membrane of the mitochondria, which is folded up to form the cristae, which in turn are covered with closely packed stalked particles which seem to be the site of the ATPase enzymes.

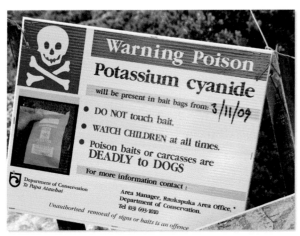

fig C Peter Mitchell.

fig B Cyanide is a poison that acts on cytochrome oxidase in the electron transport chain, preventing the production of ATP. The cells of the body cannot function without their energy supply, so the muscles spasm and the victim cannot breathe.

The chemiosmotic theory of ATP production

The link between the electrons that are passed down the electron transport chain and the production of ATP was first described by Peter Mitchell in 1961. He called it the **chemiosmotic theory** and it provides a very elegant explanation. The theory explains what happens to the hydrogen ions (protons) that are left behind when the electrons are passed along the electron transport chain, and how the movement of the hydrogen ions is coupled to the actual production of ATP.

Mitchell proposed that hydrogen ions are actively transported into the space between the inner and outer mitochondrial membranes using the energy provided as the electrons pass

along the transport chain (see **fig D**). The active transport of the hydrogen ions across the inner membrane results in a different hydrogen ion concentration on each side of the inner membrane. The membrane space has a higher concentration of hydrogen ions than the matrix, so there is a concentration gradient across the membrane. As a result of the different hydrogen ion concentrations there is also a pH gradient. And because positive hydrogen ions are concentrated in the membrane space there is an electrochemical gradient too.

All of these factors mean that there is a tendency for the hydrogen ions to move back into the matrix. The only way they can move back into the matrix is through special pores. These pores are found on the stalked particles and have an ATPase enzyme associated with them. As the hydrogen ions move along their electrical, concentration and pH gradients through these pores, their energy is used to drive the synthesis of ATP. So, ATP, the universal energy carrier, is produced in a universal process, found in all living things.

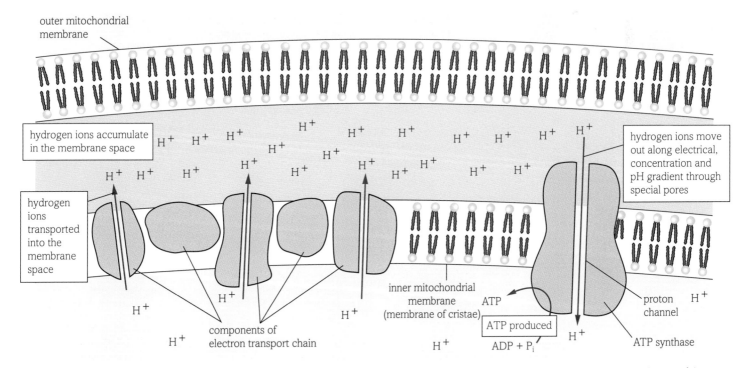

fig D Peter Mitchell's chemiosmotic theory explains the formation of ATP, not just in the mitochondria but also in chloroplasts and elsewhere. The use of the hydrogen ion gradient, produced using energy from the movement of electrons along the electron transport chain, explains all our current observations about the process.

The chemiosmotic theory of ATP production in the mitochondria was not accepted immediately. In the 1960s the widely held model for the formation of ATP in the cell was that a high-energy phosphate group was directly transferred to ADP from another intermediate, rather like the process at the end of glycolysis when pyruvate is formed. However, no one could find this intermediate compound. When Mitchell put forward his explanation for the production of ATP in the mitochondria using his chemiosmotic theory, other scientists were sceptical at first because the idea was so very different from the accepted model. Over the next 10 years evidence for Mitchell's ideas built up, and no high-energy intermediates were found. By 1978 Mitchell's chemiosmotic theory was widely accepted and he won the Nobel Prize for Chemistry.

How much ATP is gained?

Although we look at respiration in terms of the two stages, it is important to remember that they work together. Glycolysis continually feeds into the Krebs cycle, and the control of the whole process depends on various enzymes and the levels of some of the substrates and products of the reactions.

Cellular respiration has evolved to produce energy in the form of ATP for use in the cells. The fact that the process is the same in almost all living organisms suggests that it evolved at a very early stage in the development of organisms on Earth, and that it is a very effective method of producing available energy. If it was not, alternative successful life forms with a different system of respiration would doubtless have evolved long ago. But exactly how much ATP is gained during the oxidation of one molecule of glucose in its journey along the respiratory pathways?

The easiest way to look at this is to consider the whole process and where the ATP is produced (see **fig E**). For many years the average amount of ATP produced from one glucose molecule in aerobic respiration was said to be 36 molecules of ATP, assuming that glucose enters the cycle and that oxidation is complete. The actual total was taken as 38 molecules of ATP, but it takes two molecules of ATP to transport the reduced NAD molecules produced in glycolysis through the mitochondrial membrane, leaving 36 available for the body cells. If this is compared with the meagre two molecules of ATP that result when the breakdown of a glucose molecule is purely anaerobic, the importance of the oxygen-using process becomes abundantly clear.

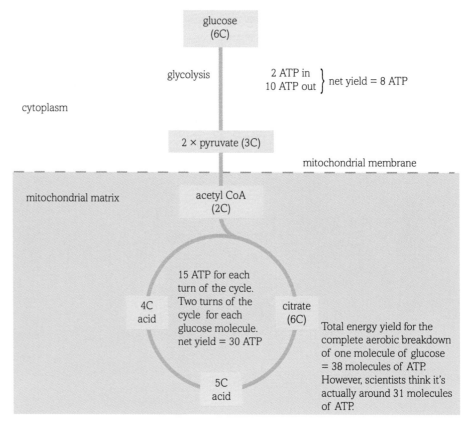

fig E This model of the 38 molecules of ATP gained by the complete oxidation of a molecule of glucose in cellular respiration has been accepted for a long time – but scientists are becoming aware that the figures are quite variable.

These figures were reached on the assumption that the yields of ATP are always in whole numbers. As our understanding of the processes of the electron transport chain has increased over the past 50 years, the figures have become less certain. Scientists now think that ATP yields may not always be in whole numbers. Currently, biochemists' best estimates are that the oxidation of two molecules of reduced NAD supplies enough energy to make five molecules of ATP. Similarly they now think that the oxidation of two molecules of reduced FAD produces about three molecules of ATP. This gives an overall yield for the process of aerobic cellular respiration of around 31 molecules of ATP.

In addition, the proton gradients in the mitochondria can be used to drive the active transport of several different molecules and ions through the inner membrane into the matrix, and NADH can be used as a reducing agent for many different reactions. So, although the functional yield of ATP is usually quoted as 36 (or 38, or 31), the amount of ATP resulting from one molecule of glucose going through complete oxidation and available to drive other reactions in aerobic respiration is probably rarely over 30.

Questions

1   Using **fig E** to help you, draw a large fully labelled diagram summarising the process of aerobic cellular respiration, starting with glucose.

2   Explain how the oxidation of glucose results in the formation of ATP.

3   (a) Explain why aerobic respiration produces so much more ATP than anaerobic respiration.
    (b) Explain why the quoted ATP yield of aerobic respiration is so variable.

4   What is Mitchell's chemiosmotic theory of ATP production and why is it so important to our understanding of cellular respiration?

Key definitions

Oxidative phosphorylation is the oxygen-dependent process in the electron transport chain where ADP is phosphorylated.

The **electron transport chain** is a series of electron-carrying compounds along which electrons are transferred in a series of oxidation/reduction reactions, driving the production of ATP.

Chemiosmosis is the process that links the electrons that are passed down the electron transport chain and the production of ATP, by the movement of hydrogen ions through the membrane along electrochemical, concentration and pH gradients.

Cytochromes are members of the electron transport chain. They are protein pigments with an iron group rather like haemoglobin which are reduced by electrons from reduced FAD, which is reoxidised, with the production of a molecule of ATP.

Cytochrome oxidase is an enzyme in the electron transport chain which receives the electrons from the cytochromes and is reduced as the cytochromes are oxidised, with the production of a molecule of ATP.

The **chemiosmotic theory** was developed by Peter Mitchell to explain the production of ATP in mitochondria, chloroplasts and elsewhere in living cells.

1 An electron from a glucose molecule would travel through the following processes:
   (1) Krebs cycle
   (2) glycolysis
   (3) electron transport chain
   (4) link reaction

   In which order would the electron travel through these processes?
   A 2–4–1–3
   B 2–1–3–4
   C 1–3–2–4
   D 1–4–3–2   [1]
                                                                   [Total: 1]

2 In cellular respiration, glucose is broken down to produce ATP. During this process, hydrogen atoms are removed from the glucose molecule and attach to which of the following substances?
   A water
   B oxygen
   C nitrogen
   D NAD   [1]
                                                                   [Total: 1]

3 What are the products of anaerobic respiration?
   A lactate in plants and ethanol in mammals
   B ethanoic acid in plants and lactate in mammals
   C ethanol in plants and lactate in mammals
   D lactate in plants and ethanoic acid in mammals   [1]
                                                                   [Total: 1]

4 (a) The diagram below shows some of the stages of anaerobic respiration in a muscle cell.

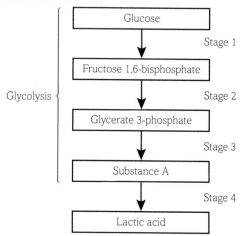

   (i) Name substance A.   [1]
   (ii) State which of the stages shown in the diagram:
     • Uses ATP
     • Produces ATP   [2]

   (b) The Krebs cycle occurs during aerobic respiration and is an example of a metabolic pathway.
     (i) Explain why the Krebs cycle is described as a metabolic pathway.   [1]
     (ii) State precisely where in the cell the Krebs cycle occurs.   [1]

   (c) The diagram below shows some of the stages that occur in the Krebs cycle.

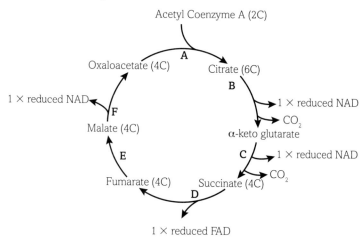

   Oxidoreductase enzymes are involved in some of the reactions in the Krebs cycle. Using the letters A to F and the information given in the diagram, give the letters of **all** the stages that involve an oxidoreductase enzyme.   [1]

                                                                   [Total: 6]

5 (a) The diagram below represents some of the stages of respiration in a yeast cell.

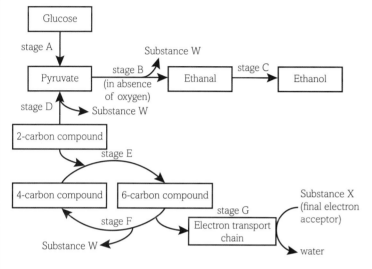

(i) Name the substances represented by the letters **W** and **X**. [2]

(ii) State which letter represents the stage during which most ATP is produced. [1]

(iii) State **two** letters that represent stages during which no ATP is made. [1]

(iv) State which letters represent stages that occur in the cytoplasm of a yeast cell. [1]

(b) During glycolysis, the coenzyme NAD$^+$ is reduced to NADH and H$^+$. State what happens to this reduced coenzyme when a yeast cell is respiring anaerobically. [1]

[Total: 6]

6 (a) The photograph below shows a mitochondrion as seen using an electron microscope.

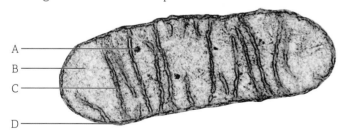

(i) Name the parts labelled **B** and **C**. [2]

(ii) Give the **letter** that represents the location of the electron transport chain. [1]

(b) Antimycin A is an inhibitor of the electron transport chain. It binds to one of the electron carriers in the chain. An experiment was carried out to investigate the effect of Antimycin A on the respiration of yeast cells.

Yeast cells were mixed with a buffer solution containing ADP, phosphate ions and glucose to form a suspension. This suspension was then placed in a waterbath at 30 °C and incubated for 30 minutes. During this time, the oxygen content of the suspension was measured.

The experiment was then repeated with Antimycin A added to the suspension 5 minutes after the start of the incubation.

The results are shown in the table below.

Time of incubation/ mins	Oxygen content of suspension/ arbitrary units	
	Without Antimycin A	With Antimycin A added 5 minutes after the start of incubation
0	6.4	6.4
5	3.7	3.7
10	2.4	3.7
15	1.6	3.7
20	0.9	3.7
25	0.5	3.7
30	0.5	3.7

(i) Comment on why the oxygen content of the suspension of cells without Antimycin A did not reach zero. [2]

(ii) Explain why the oxygen concentration of the suspension did not decrease after Antimycin A was added. [2]

(iii) Explain what effect the addition of Antimycin A will have on the production of ATP. [3]

[Total: 10]

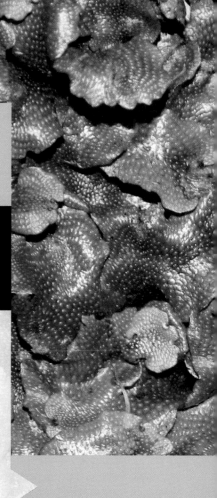

TOPIC 5
Energy for biological processes

CHAPTER

5.2 > Photosynthesis

Introduction

Plants make food by photosynthesis. The more food they make, the bigger they can grow. Because plants are the first stage in most food chains, photosynthesis is the reaction that underpins almost all ecosystems. The oxygen given off as a waste product is also vital for aerobic respiration for most organisms, so plants are key to the health of the planet.

In this chapter you will find out about photosynthesis. You will discover how chlorophyll pigments absorb light of different wavelengths to maximise their efficiency, and will see how the ultrastructure of the chloroplasts is adapted to their functions, in both the light-dependent and the light-independent reactions of photosynthesis.

You will discover how carbon dioxide is fixed from the air in the process of photosynthesis and learn about the enzymes and reactions that make up the Calvin cycle. You will discover the roles of cyclic and non-cyclic photophosphorylation in the production of the NADP and ATP needed in the process. You will also investigate different factors that affect the rate of photosynthesis and consider these effects on plants in their natural habitats.

All the maths you need

- Recognise and make use of appropriate units in calculations (*e.g. rate of photosynthesis in $mm^3\ CO_2\ cm^{-2}$ leaf h^{-1}*)
- Use fractions, ratios and percentages (*e.g. calculating R_f values for photosynthetic pigments*)
- Use the appropriate number of significant figures (*e.g. in R_f values for photosynthetic pigments*)
- Translate information between graphical, numerical and algebraic forms (*e.g. absorption spectra of chlorophyll*)
- Plot two variables from experimental or other data (*e.g. practical work on rates of photosynthesis*)
- Determine the intercept of a graph (*e.g. practical work on rates of photosynthesis*)
- Calculate rate of change from a graph showing a linear relationship (*e.g. the effect of temperature on the rate of photosynthesis*)
- Draw and use the slope of the tangent to a curve as a measure of rate of change (*e.g. the effect of carbon dioxide levels on the rate of photosynthesis*)

What have I studied before?

- Photosynthesis
- The structure and importance of ATP
- The principles of chemiosmosis
- The ultrastructure of plant cells and chloroplasts
- The structure of starch, chlorophyll and cellulose
- How enzymes catalyse and control metabolic pathways
- Factors affecting enzyme-controlled reactions
- The transport tissues in plants and how the products of photosynthesis are transported around the plant
- The production of ATP during cellular respiration

What will I study later?

- How pathogens can affect the metabolism of plants
- Gene technology and ways in which it can be used to make photosynthesis and other plant metabolic pathways more efficient
- Plant genetics
- Chemical control in plants and how plants have evolved responses that enable them to maximise their opportunities for photosynthesis
- The role of photosynthesis in ecosystems
- Net and gross primary productivity and the efficiency of biomass and energy transfer between trophic levels
- The importance of plants and photosynthesis in the colonisation of new land

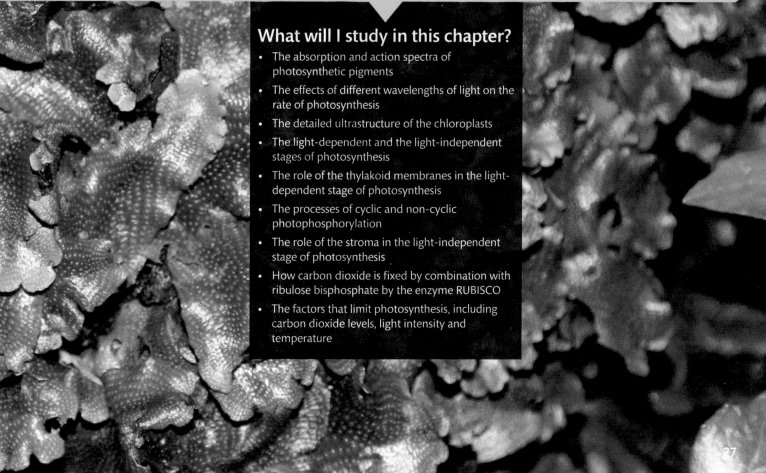

What will I study in this chapter?

- The absorption and action spectra of photosynthetic pigments
- The effects of different wavelengths of light on the rate of photosynthesis
- The detailed ultrastructure of the chloroplasts
- The light-dependent and the light-independent stages of photosynthesis
- The role of the thylakoid membranes in the light-dependent stage of photosynthesis
- The processes of cyclic and non-cyclic photophosphorylation
- The role of the stroma in the light-independent stage of photosynthesis
- How carbon dioxide is fixed by combination with ribulose bisphosphate by the enzyme RUBISCO
- The factors that limit photosynthesis, including carbon dioxide levels, light intensity and temperature

By the end of this section, you should be able to...

● describe the main photosynthetic pigments and explain what is meant by absorption and action spectra

● explain why plants have a variety of different photosynthetic pigments

● describe the structure of chloroplasts including the envelope, stroma, grana and lamellar structure

Energy is the currency of life. If the supply of energy to the cells of a living organism fails for any reason the organism will die. Massive amounts of energy continually flow through the biosphere and organisms can be classified according to where they get their energy from. **Autotrophic** organisms make energy-containing organic compounds out of inorganic sources, such as carbon dioxide and water. Most of them do this by **photosynthesis**.

Photosynthesis is the process by which living organisms, particularly plants, capture the energy of the Sun using chlorophyll and use it to convert carbon dioxide and water into simple sugars. This energy is transferred into the chemical bonds of organic molecules such as glucose and starch. These compounds are then used as an energy source by the cells of the plant, and as the building blocks of other important molecules such as proteins. There are some autotrophic bacteria that are not photosynthetic. They use energy from chemical reactions to synthesise their food. **Heterotrophic** organisms generally eat plants or other animals that have eaten plants. Therefore they use the products of photosynthesis indirectly, both for making molecules and as fuels to supply energy for a wide variety of activities. The Sun is therefore the ultimate source of energy for the chemical reactions in almost all organisms.

fig A The ability of plants to trap and use the light energy from the Sun in the process of photosynthesis underpins almost all life on Earth.

The process of photosynthesis

Photosynthesis can be summarised as follows:

$$\text{carbon dioxide} + \text{water} \xrightarrow[\text{chlorophyll}]{\text{light energy}} \text{glucose} + \text{oxygen} \qquad \Delta H \approx +2880\,\text{kJ}$$

$$6CO_2 + 6H_2O \xrightarrow[\text{chlorophyll}]{\text{light energy}} C_6H_{12}O_6 + 6O_2 \qquad \Delta H \approx +2880\,\text{kJ}$$

The energy from the light is used to split the strong H–O bonds in the water molecules. The hydrogen released is combined with carbon dioxide to form a fuel for the cells (glucose). Oxygen is released into the atmosphere as a waste product of this process.

Simple models of photosynthesis show a one-step process. They cover the most important points and make it relatively easy to understand. Detailed research into the biochemistry of photosynthesis has shown that this model is much too simplistic. Photosynthesis, like cellular respiration, is actually a sequence of many different reactions in two main stages, one of which does not need light to take place. You will be learning some of the details in the process, but even this is a simplification as photosynthesis is extremely complex.

The whole structure of plants has evolved around the process of photosynthesis. The different parts of the plant are adapted for efficiently obtaining the raw materials, carbon dioxide and water, and for trapping as much sunlight as possible.

<u>Learning tip</u>

The processes of photosynthesis and cellular respiration are closely linked.

- Photosynthesis takes in energy and uses carbon dioxide and water to synthesise glucose and oxygen. It is endothermic.
- Aerobic respiration uses glucose and oxygen to produce energy in the form of ATP for the cell, with carbon dioxide and water as the waste products. It is exothermic.

The importance of chloroplasts

Chloroplasts are relatively large organelles found in the cells of the green parts of plants (see **fig B**). An average green plant cell contains 10–50 chloroplasts which are uniquely adapted for the process of photosynthesis (see **Book 1 Section 2.1.6**). Each chloroplast is surrounded by an outer and an inner membrane with a space between the two, known as the chloroplast **envelope**. Inside the chloroplast there is a system of membranes that are arranged in stacks called **grana**. A single granum is made up of stacks of membrane discs known as **thylakoids**. This is where the green pigment chlorophyll is found. The pigment molecules are arranged on the membranes in the best possible position for trapping light energy. Evidence from electron micrographs shows that the granal membranes are covered in particles that seem to be involved in ATP synthesis. The grana are joined together by **lamellae**, extensions of the thylakoid membranes, which connect two or more grana. The lamellae act as a skeleton within the chloroplast, maintaining a working distance between the grana so that they get the maximum light and function as efficiently as possible.

These membrane stacks are surrounded by a matrix called the **stroma**. The stroma contains all the enzymes needed to complete the process of photosynthesis and produce glucose. Glucose can then be used in cellular respiration, converted to starch for storage, or used as an intermediate for the synthesis of other organic compounds such as amino acids and lipids.

Chlorophyll

The other major adaptation of the chloroplasts is their light-capturing, photosynthetic pigment 'chlorophyll', a mixture of closely-related pigments. These include **chlorophyll *a*** (blue-green), **chlorophyll *b*** (yellow-green), the chlorophyll **carotenoids** (orange carotene and yellow xanthophyll) along

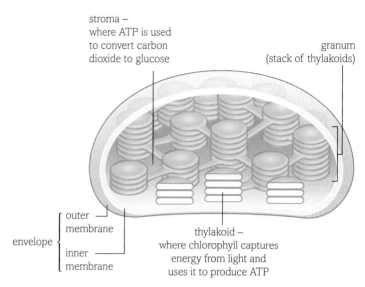

stroma – where ATP is used to convert carbon dioxide to glucose

granum (stack of thylakoids)

envelope { outer membrane / inner membrane

thylakoid – where chlorophyll captures energy from light and uses it to produce ATP

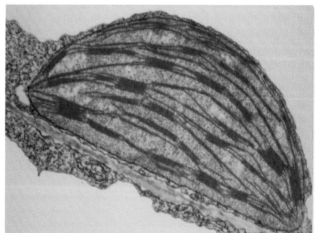

fig B Evidence from electron micrographs has helped scientists build up a realistic and complex picture of the way the structure of chloroplasts is adapted to their functions in photosynthesis.

with a grey pigment **phaeophytin**, which is a breakdown product of the others. Chlorophyll *a* is found in all photosynthesising plants and is the most abundant of the five. The other pigments are found in varying proportions, and it is these differences that give the leaves of plants their almost infinite variety of shades of green. Each of the pigments absorbs and captures light from particular areas of the spectrum. As a result, far more of the energy from the light falling on the plant can be used than if only one pigment was involved.

The different photosynthetic pigments can be demonstrated in a number of ways.

Absorption spectra and action spectra

The **absorption spectrum** describes the range of amount of light of different wavelengths that a photosynthetic pigment absorbs. It is usually represented as a graph. The absorption spectra of the different photosynthetic pigments are found by measuring their absorption of light of differing wavelengths (see **fig C**). It is also possible to produce an absorption spectrum for intact chloroplasts, with all the photosynthetic pigments combined.

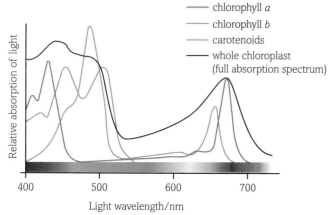

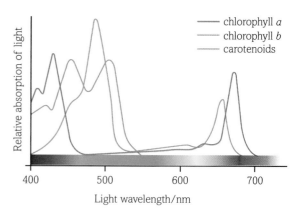

fig C The different photosynthetic pigments absorb light of different wavelengths, enabling the plant to make use of more of the available light.

You can compare the rate of photosynthesis with the wavelength of light. The first time this was done was by T.W. Engelmann in the late 1800s. He set up a strand of a filamentous alga in light of different wavelengths. He used bacteria that move towards oxygen to show where the most oxygen was given off, which is directly related to the amount of photosynthesis taking place. In this way he achieved an **action spectrum**, which is a way of demonstrating the rate of photosynthesis against the wavelength of light.

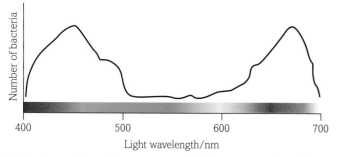

fig D Engelmann's first action spectrum was built up using numbers of bacteria to show where photosynthesis was taking place.

Modern action spectra use electronic data logging rather than bacterial movements to measure the rate of photosynthesis at different wavelengths of light, but the action spectra they produce still compare the rate of photosynthesis to the wavelength of light. Action spectra show us that the rate of photosynthesis is very closely related to the combined absorption spectrum of all the photosynthetic pigments in a plant. This demonstrates that the range of photosynthetic pigments makes a much bigger portion of the wavelength of light available to plants and so gives them an adaptive advantage.

Chromatography

Plants look green. If you extract the pigments from a plant by grinding up leaves with propanone and filtering, the filtrate looks green. So how can you show that there are several different pigments? The answer is paper chromatography. With a suitable solvent, the pigments travel up the paper at different speeds and are readily separated (see **fig F**).

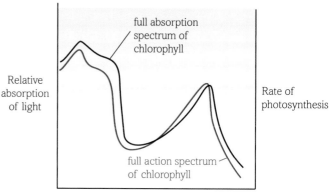

fig E The absorption spectrum and action spectrum of chlorophyll in a chloroplast compared with the absorption spectrum of the individual photosynthetic pigments shows clearly the importance of having more than one pigment available to absorb the light.

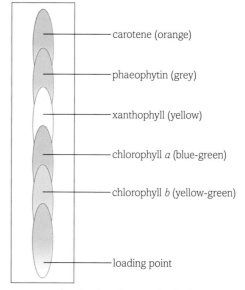

fig F Chromatogram showing five photosynthetic pigments.

Once you have carried out chromatography on the photosynthetic pigments you can work out their R_f values and compare them to the R_f values of known pigments in the same solvent, as the R_f value may be very different for the same pigments using different solvents. The R_f value is the ratio of the distance travelled by the pigment to the distance travelled by the solvent alone. The R_f value is always somewhere between 0 and 1 and it is calculated as follows:

$$R_f \text{ value} = \frac{\text{distance travelled by solute (photosynthetic pigment)}}{\text{distance travelled by solvent}}$$

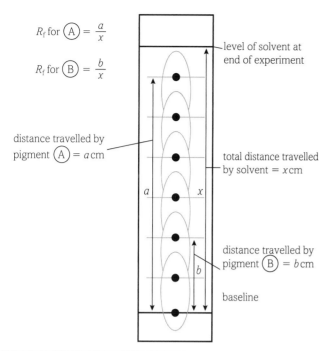

R_f for Ⓐ $= \dfrac{a}{x}$

R_f for Ⓑ $= \dfrac{b}{x}$

distance travelled by pigment Ⓐ $= a$ cm

level of solvent at end of experiment

total distance travelled by solvent $= x$ cm

distance travelled by pigment Ⓑ $= b$ cm

baseline

Photosynthetic pigment	R_f value for spinach leaves extracted with hexane and chromatographed with 3:1:1 petroleum ether-propanone-chloroform solvent on silica gel
carotene	0.98
chlorophyll a	0.59
chlorophyll b	0.42
phaeophytin	0.81
xanthophyll 1	0.28
xanthophyll 2	0.15

fig G In this example, a method for calculating the R_f values for two photosynthetic pigments has been shown.

Learning tip

As part of your study of this topic, you will carry out **Core Practical 11: Investigate the presence of different chloroplast pigments using chromatography**. Make sure you have a good understanding of this practical as your understanding of the experimental method may be assessed in your examination.

Photosystems

The photosynthetic pigments carry out the absorption of light in two distinct chlorophyll complexes known as **photosystem I (PSI)** and **photosystem II (PSII)**. Each system contains a different combination of chlorophyll pigments and so absorbs light in a slightly different area of the spectrum (wavelength 700 nm for PSI and 680 nm for PSII). The different photosystems have been identified in electron micrographs as different-sized particles attached to the membranes in the chloroplasts. PSI particles are mainly on the intergranal lamellae, while PSII particles are on the grana themselves. They have different functions in photosynthesis, as you will see later.

Questions

1 Some people describe photosynthesis as the most important reaction in living organisms. Why might it deserve this title – and why not?

2 (a) Chloroplasts are not present in all plant cells – why not?
 (b) Summarise the adaptations of chloroplasts for their role in photosynthesis.

3 Using the data in **fig C**, explain why plant leaves usually appear green.

4 Measure the distances travelled by the solute and solvent and calculate the R_f values for the pigments labelled A and B in **fig G**. Identify these two pigments, assuming they were extracted with hexane and chromatographed with 3:1:1 petroleum ether-propanone-chloroform solvent on silica gel.

Key definitions

Autotrophic organisms make complex organic compounds from simple compounds in their environment.

Photosynthesis is the process by which living organisms, particularly plants and algae, capture the energy of the Sun using chlorophyll and use it to convert carbon dioxide and water into simple sugars.

Heterotrophic organisms obtain complex organic molecules by feeding on other living organisms or their dead remains.

The **envelope** of a chloroplast is the outer and inner membranes along with the intermembrane space.

Grana are stacks of thylakoid membranes within a chloroplast.

A **thylakoid** is a membrane disc found in the grana of a chloroplast.

Lamellae are extensions of the thylakoid membranes which connect two or more grana, acting as a supporting skeleton in the chloroplast, maintaining a working distance between the grana so that they get the maximum light and function as efficiently as possible.

The **stroma** is the matrix which surrounds the grana and contains all the enzymes needed to complete the process of photosynthesis and produce glucose.

Chlorophyll a is a blue-green photosynthetic pigment, found in all green plants.

Chlorophyll b is a yellow-green photosynthetic pigment.

Carotenoids are photosynthetic pigments made up of orange carotene and yellow xanthophyll.

Phaeophytin is a grey pigment which is a breakdown product of the other photosynthetic pigments.

An **absorption spectrum** is a graph of the amount of light absorbed by a pigment against the wavelength of light.

An **action spectrum** is a graph demonstrating the rate of photosynthesis against the wavelength of light.

Photosystem I (PSI) is a combination of chlorophyll pigments which absorbs light of wavelength 700 nm and is involved in cyclic and non-cyclic photophosphorylation.

Photosystem II (PSII) is a combination of chlorophyll pigments which absorbs light of wavelength 680 nm and is involved only in non-cyclic photophosphorylation.

By the end of this section, you should be able to...

- describe the role of the thylakoid membranes in the light-dependent stage of photosynthesis
- explain the processes of cyclic and non-cyclic photophosphorylation
- describe the role of the stroma in the light-independent stage of photosynthesis
- explain how carbon dioxide is fixed, including the role of RUBISCO
- explain how reduced NADP and ATP from the light-dependent stage of photosynthesis are used in the light-independent stages
- explain how GALP is used as a raw material in the production of monosaccharides, amino acids and other molecules
- explain the factors that limit the rate of photosynthesis including carbon dioxide, light intensity and temperature

Photosynthesis is a two-stage process involving a complex series of reactions. The reactions in the first stage only occur in light, while those of the second stage occur independently of light. The **light-dependent reactions** produce materials that are then used in the **light-independent reactions**. The whole process takes place all the time during the hours of daylight. However, the light-independent reactions can also continue when it is dark.

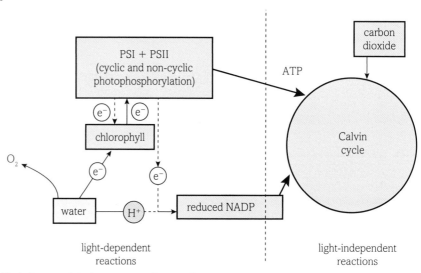

fig A Simplified diagram of the key stages in photosynthesis.

The light-dependent stage of photosynthesis

The light-dependent stage of photosynthesis takes place on the thylakoid membranes of the chloroplasts. It has two main functions. One is to split water molecules in a **photochemical reaction**, providing hydrogen ions to reduce carbon dioxide and produce carbohydrates in the light-independent stage. The other is to produce ATP, which is needed to supply the energy to produce those carbohydrates.

Light is a form of electromagnetic radiation and the smallest unit of light is a photon. When a photon of light hits a chlorophyll molecule, the energy is transferred to the electrons of that molecule. The electrons are excited and are raised to higher energy levels. If an electron is raised to a sufficiently high energy level, it will leave the chlorophyll molecule completely. The excited electron can be picked up by an electron acceptor (carrier molecule) and result in the synthesis of ATP by one of two processes: **cyclic** and **non-cyclic photophosphorylation**. Both take place at the same time and in both cases ATP is formed as the excited electron is passed along an electron transport chain. In non-cyclic photophosphorylation, reduced NADP is also produced.

Did you know?

A two-stage process

How do we know that photosynthesis is a two-stage process? There are several strands of evidence for this current model.

1 Photochemical reactions get the energy they need from light, so temperature should not affect the rate of the reaction. However, when the rate of photosynthesis is investigated experimentally, temperature can be shown to have a clear effect (see **fig B**). Initially, photochemical (light-dependent) reactions are limiting the rate of the overall process and so temperature has no effect. However, once there is plenty of available light, the process seems to be limited by different, temperature-sensitive reactions. This suggests there are two distinct phases to photosynthesis, one dependent on light and the other controlled by temperature-sensitive enzymes (see **Book 1 Section 1.4.2**).

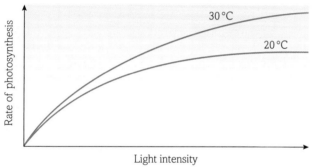

fig B The effect of temperature on the rate of photosynthesis suggests that two different processes are involved.

2 A plant that is given alternating periods of dark and light forms more carbohydrate than a plant in continuous light. The best explanation is that the light-dependent reactions produce a chemical that feeds into the light-independent stage. In continuous light this product builds up, as the light-independent stage cannot keep up. As the concentration rises, it inhibits the enzymes controlling the light-independent reactions from making carbohydrates. A period of darkness ensures that all of the light stage products are converted into carbohydrate without the concentration getting too high. This system is very efficient in a natural environment with periods of light and dark (day and night). More recent techniques have allowed regions of the chloroplast to be isolated. The reactions occurring on the grana have been shown to depend on the presence of light, while those in the stroma do not.

Cyclic photophosphorylation

Cyclic photophosphorylation involves only photosystem I (PSI) and drives the production of ATP. When light hits a chlorophyll molecule in PSI, a light-excited electron leaves the molecule. It is taken up by an electron acceptor and passed directly along an electron transport chain to produce ATP. When an electron returns to the chlorophyll molecule in PSI, it can then be excited in the same way again.

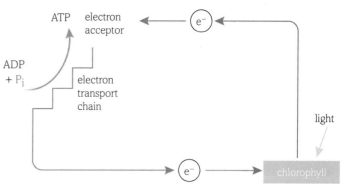

fig C Cyclic photophosphorylation.

Non-cyclic photophosphorylation

During non-cyclic photophosphorylation, water molecules are split, providing hydrogen ions to reduce NADP. ATP is also produced. It involves both PSI and PSII (photosystem II).

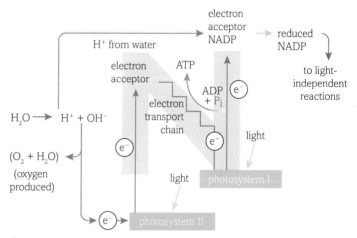

fig D Non-cyclic photophosphorylation: one electron leaves a chlorophyll molecule in PSI and moves into the light-independent stage of the process. A different electron is returned to PSI from PSII by an electron transport chain, driving the production of more ATP as it does so.

Under light conditions, photons are constantly hitting chlorophyll molecules in both PSI and PSII, exciting the electrons to a high enough level for them to be lost and picked up by electron acceptors. An excited electron from PSII is picked up by another electron acceptor and passed along an electron transport chain to PSI, driving the synthesis of one molecule of ATP. PSI receives an electron to replace one that was lost to the light-independent reactions (see **fig D**). Now the chlorophyll molecule in PSII is short of an electron and unstable. The original electron cannot be returned to the chlorophyll because it has continued on to PSI. So an electron has to be found from somewhere to restore the chlorophyll to its original state. This electron comes from the splitting of water, a process that is known as **photolysis** as it depends on light. Water dissociates spontaneously into hydrogen (H^+) ions and hydroxide (OH^-) ions, resulting in plenty of these ions being present in the cell, including in the chloroplasts, and also replacing the lost electrons. Once PSII's chlorophyll molecule has received an electron it is restored to its original state, ready to be excited again when hit by another photon of light.

Meanwhile, electrons in PSI are also being excited by light and lost to an electron acceptor. Electrons are carried down an electron transport chain and taken up by the electron acceptor, nicotinamide adenine dinucleotide phosphate (NADP). The NADP also takes up a hydrogen ion from the dissociated water at the same time to form reduced NADP.

The reduced NADP and ATP produced during non-cyclic photophosphorylation go on to be used as a source of reducing power and energy respectively in the light-independent reactions of photosynthesis to make glucose.

Photosynthesis is a reaction that occurs millions of times in every chloroplast. This means many hydrogen ions are removed by NADP, and many hydroxide ions are 'left behind'. The hydroxide ions react together to form oxygen and water. As a result of the reaction, electrons are freed and taken up by chlorophyll. Four chlorophyll molecules regain electrons in the production of one molecule of oxygen:

$$4OH^- - 4e^- \text{ (lost to chlorophyll)} \rightarrow O_2 + 2H_2O$$

This clever piece of biochemistry was worked out by Robert Hill and Fay Bendall at Cambridge University in 1960 and is often referred to as the Z scheme, though nowadays the Z is usually turned round so it looks like an N (see the central section of **fig D**).

Learning tip

Make sure you are clear about the differences between PSI and PSII. Remember that both photosystems rely on light. Do not get them confused with the light-dependent and light-independent stages of photosynthesis.

The light-independent stage of photosynthesis

The light-independent stage of photosynthesis uses the reducing power (reduced NADP) and ATP produced by the light-dependent stage to build carbohydrates. This stage consists of a series of reactions known as the **Calvin cycle** and takes place in the stroma of the chloroplast. A series of small steps results in the reduction of carbon dioxide from the air to bring about the synthesis of carbohydrates (see **fig E**). Each stage of the cycle is controlled by enzymes.

The Calvin cycle

In the first step, carbon dioxide from the air combines with the 5-carbon compound **ribulose bisphosphate (RuBP)** in the chloroplasts. The carbon dioxide is said to be fixed, so this is carbon fixation. The enzyme **ribulose bisphosphate carboxylase/oxygenase** (usually known as **RUBISCO**) is necessary for this vital step. Research has shown that RUBISCO is the rate-limiting enzyme in the process of photosynthesis.

The result of the reaction between RuBP and carbon dioxide is, in theory, a 6-carbon compound. Scientists are convinced that this theoretical compound exists but it is highly unstable and no one has been able to isolate it. It immediately splits to give two molecules of glycerate 3-phosphate (GP), a 3-carbon compound. GP is then reduced (hydrogen is added) to form **glyceraldehyde**

3-phosphate (GALP), a 3-carbon sugar. The hydrogen for this reduction comes from reduced NADP, and the energy required comes from ATP, both produced in the light-dependent stage.

Much of the 3-carbon GALP passes through a series of steps to replace the RuBP needed in the first step of the cycle. However, some of it is synthesised into the 6-carbon sugar glucose or passed directly into the glycolysis pathway where it may be used for the synthesis of other molecules needed by the plant (see the next page). The reactions of the Calvin cycle take place both in the light and in the dark. These reactions only stop in the dark when the products of the light reaction run out, leaving no reduced NADP or ATP available in the chloroplasts.

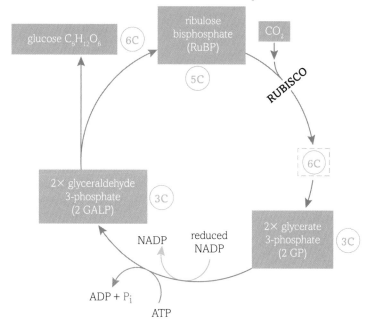

fig E The Calvin cycle: here the products of the light-dependent stage of photosynthesis are used in a continuous cycle to fix carbon dioxide. The end result is new carbohydrates.

Did you know?

RUBISCO and photorespiration

RUBISCO makes up about 30% of the total protein of a leaf, so it is probably the most common protein on Earth. It is also possibly the most important enzyme because of its role in fixing carbon dioxide during photosynthesis. But RUBISCO is very inefficient. The active site cannot distinguish between the carbon–oxygen double bonds of CO_2 molecules and the oxygen–oxygen double bonds of O_2 molecules. As a result there is competitive inhibition (see **Book 1 Section 1.4.3**) between the two:

- RUBISCO as a carboxylase: in high levels of CO_2/relatively low O_2, RUBISCO binds to the carbon dioxide and combines it with RuBP, giving two molecules of 3C GP which feed into the Calvin cycle.

- RUBISCO as an oxygenase: in low levels of CO_2/relatively high O_2, RUBISCO binds to the oxygen and combines it with RuBP to form one molecule of GP and one molecule of glycolate-2-phosphate. This is converted into GP in a reaction that uses products of the Calvin cycle, and ATP and releases carbon dioxide. Because it uses oxygen and releases carbon dioxide, it is known as **photorespiration**.

Fortunately, RUBISCO has an affinity for CO_2 that is 80 times higher than its affinity for oxygen, but in spite of this, photorespiration wastes both carbon and energy. About 25% of the products of the Calvin cycle are lost in photorespiration. So in many plants photosynthesis is 25% less efficient than it might be.

Why is this key enzyme so inefficient? All the evidence suggests that when RUBISCO evolved the atmosphere was high in carbon dioxide with very little oxygen, so photorespiration never occurred and there was no selection pressure against it. Even today, with a high oxygen/low carbon dioxide atmosphere, photorespiration is not a problem for plants. There is no selection pressure for the enzyme to evolve to become more specific to carbon dioxide. A more efficient version of RUBISCO would, however, be extremely useful to people, because our crop plants could become 25% more productive!

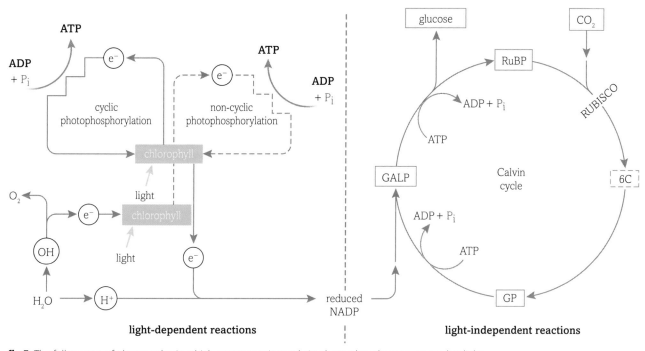

fig F The full process of photosynthesis, which occurs continuously in plants when they are exposed to light.

Using the products of photosynthesis

GALP is the primary end-product of the process of photosynthesis and it is the key molecule for the synthesis of everything else needed in the plant. Some of the GALP is used directly in glycolysis and so fed on into the Krebs cycle. Some of the GALP produced in the Calvin cycle is used to produce glucose in a process called **gluconeogenesis**. This glucose may then be converted into disaccharides such as sucrose for transport round the plant and into polysaccharides such as starch for energy storage and cellulose for structural support (see **Book 1 Section 1.2.2**).

The GALP that feeds into the glycolysis and Krebs cycle pathways is used as a fuel to provide energy in the form of ATP for the functions of the cell. Compounds from these pathways are also used as the building blocks of amino acids, combining the molecules with nitrates from the soil. If GALP continues round the Calvin cycle, it can also be used for the production of nucleic acids with the addition of phosphates from the soil.

Some of the GALP taken into the glycolysis pathway is converted to acetyl coenzyme A, which in turn can be used to synthesise fatty acids needed for the production of phospholipids for membranes, and lipids needed for storage and other functions within the plant.

GP is also part of this process, but GALP is regarded as the main molecule leading to the synthesis of all the other molecules needed by the plant (see **fig G**).

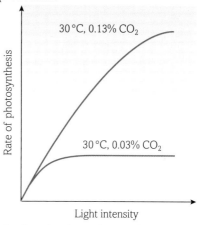

fig G The body of a plant is made up of an enormous range of different chemical compounds, most of which are built up from the products of photosynthesis.

Alternative routes

The majority of plants use this photosynthetic pathway, where the carbon dioxide is fixed directly into 3-carbon compounds, so they are known as C_3 plants. There are some plants that have alternative ways of fixing carbon dioxide that help to minimise the amount of photorespiration carried out (see Did you know? box on the previous page). They produce a compound containing four carbon atoms and use this to raise the carbon dioxide levels inside the cells at all times. This in turn minimises photorespiration, making photosynthesis more efficient. They are known as C_4 plants. Another group of plants have evolved a way of fixing carbon dioxide in the dark, releasing it during the day. These are called CAM plants and they have evolved in hot, dry ecosystems. This adaptation enables them to keep their stomata closed during the day, reducing water loss, and still have plenty of carbon dioxide for efficient photosynthesis.

Limiting factors in photosynthesis

Once you understand the process of photosynthesis, you can see how certain factors affect the ability of a plant to photosynthesise. Photosynthesis is limited by the factor that is nearest to its minimum value.

Light

The amount of light available affects the amount of chlorophyll which is excited and therefore the amount of reduced NADP and ATP produced in the light-dependent stage of the process. If there is a low level of light, insufficient NADP and ATP will be produced to allow the reactions of the light-independent stage to progress at their maximum rate. In this situation light is said to be the **limiting factor** for the process.

Carbon dioxide

Carbon dioxide levels are very important in photosynthesis – if there is not enough carbon dioxide available for fixing in the Calvin cycle, the reactions cannot proceed at the maximum rate. When this is the case, carbon dioxide is the limiting factor. In the natural situation of plants, carbon dioxide is the most common limiting factor. Changes in the level of carbon dioxide have a clear effect on the rate of photosynthesis (see **fig H**). In some areas – Europe and the US for example – commercial growers of some fruits and vegetables such as tomatoes grow their crops in greenhouses with a carbon dioxide enriched atmosphere to increase their production.

fig H The effect of carbon dioxide as a limiting factor in photosynthesis.

Temperature

The other main factor which limits the rate of photosynthesis is temperature. All of the Calvin cycle reactions and many of the light-dependent reactions of photosynthesis are controlled by enzymes and are therefore sensitive to temperature. This means that even when the light and carbon dioxide levels are suitable for a very high rate of photosynthesis, unless the temperature is also satisfactory, the plant will be unable to take advantage of the conditions. The rate of photosynthesis in a wild plant is often determined by a combination of these factors, some or all of them limiting the process to some extent.

Photosynthesis, limiting factors and real plants

Photosynthesis and its limiting factors are relatively easy to investigate in the laboratory – but what about in the real world?

Plants do not usually grow in pots in a controlled environment – they are found in woods, gardens, ponds, mountains, bogs, savannah and desert. The way in which plants grow and the ecosystems that develop are governed by competition by the plants for the factors that can limit photosynthesis and growth. For example, carbon dioxide levels do not generally vary much in the air, but plants compete for light and warmth. They also compete for the nutrients that are needed to convert the carbohydrate into proteins and fats. Growth in height, spreading of leaves into a mosaic pattern, climbing, and developing large leaves are all ways in which plants are adapted to get as much light as possible so that photosynthesis is not limited. Even methods of seed dispersal have evolved to reduce competition by ensuring that seedlings do not develop in the shade of their parents. The biochemistry investigated in the artificial situation of the laboratory is vitally important in the lives of plants in their natural habitats.

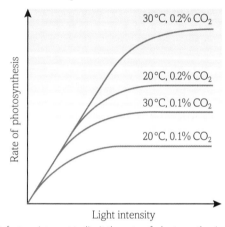

fig I Different factors interact to limit the rate of photosynthesis.

Learning tip

As part of your study of this topic, you will carry out **Core Practical 10: Investigate the effects of different wavelengths of light on the rate of photosynthesis.** Make sure you have a good understanding of this practical as your understanding of the experimental method may be assessed in your examination.

Questions

1   Make a table to compare what happens in cyclic and non-cyclic photophosphorylation.

2   Calvin cycle reactions are also known as the light-independent reactions of photosynthesis. Explain why this name is appropriate, yet in some ways inaccurate.

3   Why is GALP sometimes referred to as the primary product of photosynthesis?

4   In many greenhouses and polytunnels, which are used to grow food crops and flowers as economically and quickly as possible, there is careful monitoring and control of light levels, temperature and carbon dioxide. Explain this in terms of limiting factors.

Key definitions

Light-dependent reactions are reactions that take place in the light on the thylakoid membranes of the chloroplasts. They produce ATP and split water molecules in a photochemical reaction, providing hydrogen ions to reduce carbon dioxide and produce carbohydrates.

Light-independent reactions are reactions that use the reduced NADP and ATP produced by the light-dependent stage of photosynthesis in a pathway known as the Calvin cycle, which take place in the stroma of the chloroplast and result in the reduction of carbon dioxide from the air to bring about the synthesis of carbohydrates.

A **photochemical reaction** is a reaction initiated by light.

Cyclic photophosphorylation is a process that drives the production of ATP. Light-excited electrons from PSI are taken up by an electron acceptor and passed directly along an electron transport chain to produce ATP, with the electron returning to PSI.

Non-cyclic photophosphorylation is a process involving both PSI and PSII in which water molecules are split using light energy to provide reducing power to make carbohydrates and at the same time produce more ATP.

Photolysis is the splitting of a molecule using light.

The **Calvin cycle** is a series of enzyme-controlled reactions that take place in the stroma of chloroplasts and result in the reduction of carbon dioxide from the air to bring about the synthesis of carbohydrate.

Ribulose bisphosphate (RuBP) is a 5-carbon compound that joins with carbon dioxide from the air in the Calvin cycle to fix the carbon dioxide and form a 6-carbon compound.

Ribulose bisphosphate carboxylase/oxygenase (RUBISCO) is a key, rate-controlling enzyme that catalyses the reaction between carbon dioxide/oxygen and ribulose biphosphate.

Glyceraldehyde 3-phosphate (GALP) is a 3-carbon sugar produced in the Calvin cycle using reduced NADP and ATP from in the light-dependent stage. GALP is the key product of photosynthesis. It is used to replace the RuBP needed in the first step of the cycle, in glycolysis and the Krebs cycle, and in the synthesis of amino acids, lipids, etc. for the plant cells.

Photorespiration is the alternative reaction catalysed by RUBISCO in a low carbon dioxide environment which uses oxygen and releases carbon dioxide, making photosynthesis less efficient.

Gluconeogenesis is the synthesis of glucose from non-carbohydrates.

A **limiting factor** is the factor needed for a reaction such as photosynthesis to progress that is closest to its minimum value.

C3 OR C4?

There are different types of photosynthesis, each with differing biochemistry. Some types of photosynthesis are more efficient than others. Scientists are looking for ways to use this to make crop plants more productive.

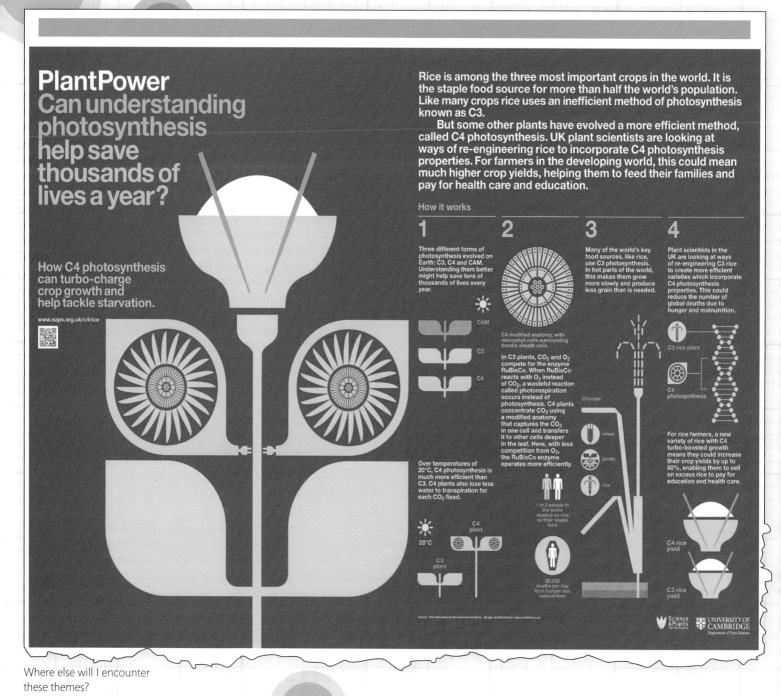

PlantPower
Can understanding photosynthesis help save thousands of lives a year?

How C4 photosynthesis can turbo-charge crop growth and help tackle starvation.

www.saps.org.uk/c4rice

Rice is among the three most important crops in the world. It is the staple food source for more than half the world's population. Like many crops rice uses an inefficient method of photosynthesis known as C3.

But some other plants have evolved a more efficient method, called C4 photosynthesis. UK plant scientists are looking at ways of re-engineering rice to incorporate C4 photosynthesis properties. For farmers in the developing world, this could mean much higher crop yields, helping them to feed their families and pay for health care and education.

How it works

1 Three different forms of photosynthesis evolved on Earth: C3, C4 and CAM. Understanding them better might help save tens of thousands of lives every year.

2 C4 modified anatomy, with mesophyll cells surrounding bundle sheath cells

In C3 plants, CO_2 and O_2 compete for the enzyme RuBisCo. When RuBisCo reacts with O_2 instead of CO_2, a wasteful reaction called photorespiration occurs instead of photosynthesis. C4 plants concentrate CO_2 using a modified anatomy that captures the CO_2 in one cell and transfers it to other cells deeper in the leaf. Here, with less competition from O_2, the RuBisCo enzyme operates more efficiently.

Over temperatures of 20°C, C4 photosynthesis is much more efficient than C3. C4 plants also lose less water to transpiration for each CO_2 fixed.

3 Many of the world's key food sources, like rice, use C3 photosynthesis. In hot parts of the world, this makes them grow more slowly and produce less grain than is needed.

1 in 2 people in the world depend on rice as their staple food

25,000 deaths per day from hunger and malnutrition

4 Plant scientists in the UK are looking at ways of re-engineering C3 rice to create more efficient varieties which incorporate C4 photosynthesis properties. This could reduce the number of global deaths due to hunger and malnutrition.

For rice farmers, a new variety of rice with C4 turbo-boosted growth means they could increase their crop yields by up to 50%, enabling them to sell on excess rice to pay for education and health care.

CAM
C3
C4

20°C
C4 plant
C3 plant

C3 crops
wheat
potato
rice

C3 rice plant
C4 photosynthesis

C4 rice yield
C3 rice yield

Source: The International Rice Research Institute Design and illustration: www.smithltd.co.uk

Science & Plants for Schools

UNIVERSITY OF CAMBRIDGE
Department of Plant Sciences

Where else will I encounter these themes?

Book 1   5.1   5.2   YOU ARE HERE 6.1   6.2   6.3   7.1   7.2

This is part of a poster produced by Science and Plants for Schools (SAPS), to introduce students to different types of photosynthesis. It is an infographic, which is a very popular way of communicating information.

Visit http://www.saps. org.uk/attachments/ article/1266/C4%20 Rice%20-%20poster.pdf to see the whole poster.

1. What is an infographic?
2. Does this poster help you understand why C4 photosynthesis is an exciting alternative to the C3 photosynthesis that you have studied in **Chapter 5.2**? Explain your answer.

Now let us examine the biology. You have looked at C3 photosynthesis in some detail. This poster gives you some information about another form of photosynthesis that uses a rather different biochemical pathway.

3. Make a flow diagram to summarise the process of C3 photosynthesis.
4. Using **only** the information on the poster:
   a. Summarise the process of C4 photosynthesis.
   b. Explain why scientists are attempting to genetically modify rice plants so they use C4 photosynthesis instead of C3.

Activity

Look into the science behind the poster to find out more about C4 photosynthesis and the way it might be used to help feed the growing population of the world. The best place to start is with the resources produced by the team at SAPS, who work from the Botanic Gardens at the University of Cambridge. Visit http://www.saps.org.uk/students/further-reading/1266 to get you started. You can look up other sources as well. Now choose one of the following activities:

- Produce your own poster to inform other A level students who are not studying biology about C4 photosynthesis and its potential to improve the yield of crops including rice.
- Find out more about C4 and CAM photosynthesis and make a table to compare the three types of photosynthesis.

The resources produced by SAPS are a great place to start when looking into plants and photosynthesis. Remember to always reference your sources.

● Science and Plants for Schools (SAPS)

1 The following substances are all involved in photosynthesis:
   (1) glyceraldehyde phosphate
   (2) glycerate 3-phosphate
   (3) carbon dioxide
   (4) glucose

In which order of substances will a carbon atom be transferred?
  A 2–1–4–3
  B 3–1–2–4
  C 3–2–1–4
  D 1–2–4–3   [1]
               [Total: 1]

2 What are the products of the light-dependent stage of photosynthesis?
  A ATP and reduced NAD
  B ATP and oxidised NADP
  C ATP and oxidised NAD
  D ATP and reduced NADP   [1]
               [Total: 1]

3 How many molecules of ribulose bisphosphate need to be broken down to produce two molecules of glucose?
  A 10
  B 12
  C 14
  D 18   [1]
               [Total: 1]

4 One of the reactions of photosynthesis can be summarised as shown below.

    water → hydrogen ions + oxygen gas + electrons

(a) Name the reaction shown.   [1]
(b) Give **one** other factor, not shown above, that would be required for this reaction to occur in a chloroplast.   [1]
(c) Describe the role of the electrons in the light dependent reaction of photosynthesis.   [4]
(d) Explain how the products of the light dependent reaction are involved in the production of glyceraldehyde 3-phosphate (GALP).   [4]
(e) GALP does not accumulate in a chloroplast during photosynthesis. Explain how GALP is used following its production.   [2]
               [Total: 12]

5 The diagram below summarises the light-dependent reactions of photosynthesis.

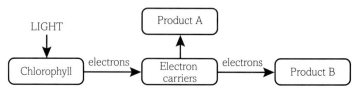

(a) Give the precise location within a chloroplast where this sequence of reactions occurs.   [2]
(b) Give the names of product A and product B.   [2]
(c) Give the name of the process that provides electrons to replace those lost by chlorophyll.   [1]
(d) A chemical called atrazine prevents the flow of electrons to the electron carriers. Describe the likely effect of atrazine on the production of carbohydrate in a chloroplast. Explain your answer.   [4]
(e) Atrazine can be used as a weedkiller.
  (i) Explain how the presence of weeds can reduce the yield of crop plants.   [2]
  (ii) A change in a single gene can alter the electron carriers so that atrazine is ineffective. Discuss how crop plants unaffected by atrazine could be used to increase crop yields.   [2]
               [Total: 13]

6 The diagram below shows what happens to electrons during part of the light-dependent reactions of photosynthesis. Any excited electrons that are not taken up by electron carriers follow pathway A and release energy as light in a process called fluorescence. The excited electrons that are taken up by electron carriers follow pathway B.

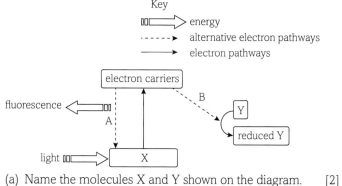

(a) Name the molecules X and Y shown on the diagram.   [2]
(b) Explain the importance of reduced Y in the process of photosynthesis.   [3]

(c) A light was shone on a leaf and left switched on.

The graph below shows changes in the amount of light given off as fluorescence by the leaf.

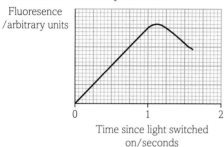

Time since light switched on/seconds

(i) Suggest an explanation for the increase in fluorescence. [2]

(ii) Suggest a reason for the fall in fluorescence. [1]

(d) Explain why an inhibitor of carbon dioxide fixation would lead to an increase in fluorescence. [4]

[Total: 12]

7 The diagram below shows structures found in a chloroplast.

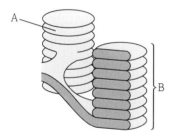

(a) Name the structures **A** and **B**. [2]

(b) Describe the role of each of the following molecules in the process of photosynthesis.
  (i) RuBP (ribulose bisphosphate)
  (ii) GP (glycerate 3-phosphate)
  (iii) GALP (glyceraldehyde 3-phosphate) [6]

(c) Name **one** region of the chloroplast in which all three of these molecules are found. [1]

[Total: 9]

8 The carbohydrates in green plants are formed during the light-independent stage of photosynthesis. They are synthesised from glycerate 3-phosphate (GP).

(a) State precisely where the synthesis of carbohydrates takes place during the light-independent stage of photosynthesis. [1]

(b) Name the products of the light-dependent stage of photosynthesis used during the synthesis of carbohydrates. [2]

(c) Describe the role of ribulose bisphosphate (RuBP) in the light-independent stage of photosynthesis. [2]

(d) An investigation of photosynthesis in cells taken from a green alga was carried out. Samples of the algal cells were taken at 1 minute intervals over a period of 6 minutes. The quantities of GP and RuBP in these cell samples were measured.

At the start of the investigation, the algal cells were kept in an atmosphere with 1% carbon dioxide. After 3 minutes, the concentration of carbon dioxide was decreased to 0.003%.

The graph below shows the results of this investigation.

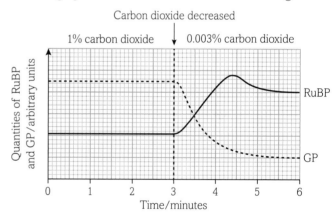

(i) Describe the effects of the decrease in the concentration of carbon dioxide on the quantities of GP and RuBP. [2]

(ii) Deduce explanations for the effects you have described in part (i). [2]

[Total: 9]

CHAPTER

6.1 > Bacteria and disease

Introduction

The 2014 Longitude Prize offered a reward of £10 million to the scientist or scientists who developed an easy-to-use test for bacterial infections. The last time the prize was offered it was for the development of the technology that allowed us to measure latitude and safely navigate the oceans of the world.

Why is a test for bacterial infections so important? Globally, bacteria cause many serious and even fatal diseases. The development of antibiotics has enabled people to survive many of these once-dreaded infections. But in recent years, more and more bacteria have become resistant to commonly-used antibiotics. At the same time, few new antibiotics have been developed.

Fewer people are researching antimicrobials, partly because they generate less income than drugs for chronic conditions that are used for a lifetime. Yet new drugs with new modes of actions are badly needed. Once we have these new drugs it is important only to use them when they are really needed in order to avoid antibiotic resistance in the future. This is why a test to show if an infection is caused by bacteria is so important.

In this chapter, you will learn more about how we can grow microorganisms in the laboratory and how we can calculate the number of bacteria in a sample. You will discover how bacteria cause disease by producing different types of toxins or invading body tissues. You will learn how bacteria can be spread from one person to another, and how awareness of this can help us avoid spreading diseases. The bactericidal and bacteriostatic actions of different antibiotics are explained, and you will discover more about how antibiotic resistance develops. Finally you will look at ways in which the problems of antibiotic-resistant diseases can be tackled and perhaps eventually overcome.

All the maths you need

- Estimate results (*e.g. size of bacterial populations*)
- Use ratios, fractions and percentages (*e.g. calculating serial dilutions*)
- Use exponential and logarithmic functions (*e.g. growth of bacterial colonies*)
- Use logarithms in relation to quantities that range over several orders of magnitude (*e.g. growth of bacterial colonies*)
- Translate information between graphical, numerical and algebraic forms (*e.g. calculating the exponential growth rate constant*)
- Plot two variables from experimental data (*e.g. the growth of bacterial colonies*)
- Determine the intercept of a graph (*e.g. the growth of bacterial colonies*)

What have I studied before?

- The ultrastructure of prokaryotic cells
- The difference between Gram-positive and Gram-negative bacterial cell walls and why each type reacts differently to antibiotics
- The importance of staining specimens in microscopy
- How mitosis contributes to growth and asexual reproduction
- Gene mutations and adaptations
- The evolutionary race between pathogens and the development of medicines to treat the diseases they cause

What will I study later?

- How other organisms can act as pathogens, including fungi, viruses and protozoa
- The social, economic and ethical implications of different control methods for endemic diseases and the role of the scientific community in validating those methods
- The response of the body to infection, including the action of the macrophages, neutrophils and lymphocytes
- The development of the humoral and the cell-mediated immune responses
- The role of T and B memory cells in the secondary immune response
- How immunity can be natural or artificial, and active or passive
- How vaccination is used in the control of communicable diseases and the development of herd immunity, and the potential problems in a population where a proportion of people choose not to vaccinate their children

What will I study in this chapter?

- The aseptic techniques used in culturing organisms
- The principles and techniques involved in culturing microorganisms including the use of different media
- Different methods for measuring the growth of a bacterial culture, including cell counts, dilution plating, and mass and optical methods
- The different phases of a bacterial growth curve and the calculation of the exponential growth rate constant using the equation $k = \dfrac{\log_{10}N_t - \log_{10}N_0}{0.301 \times t}$
- Bacteria as pathogens, and the way that pathogenic effects can be produced by exotoxins, endotoxins and the invasion of the host tissue
- The action of bactericidal and bacteriostatic antibiotics
- The development and spread of antibiotic resistance in bacteria
- The methods used to control the spread of antibiotic resistance, and the difficulties involved

By the end of this section, you should be able to...

- explain the principles and describe the basic aseptic techniques used in culturing microorganisms

- explain the use of different media including broth cultures, agar and selective media

In **Book 1** you met the following four groups of microorganisms – viruses, bacteria, fungi and protozoa. These organisms are of almost universal importance in biology and have many positive roles in the ecology both of the environment and of individual organisms. However, microorganisms are also responsible for the enormous range of communicable diseases that kill or damage millions of plants and animals, including people, every year. Microorganisms that cause disease are known as **pathogens**. In this chapter you will be concentrating on bacteria.

Culturing microorganisms

Most microorganisms are so small that they are impossible to see with the naked eye. To investigate microorganisms, for example to diagnose a disease or for scientific experiments, we need to **culture** them. This involves growing large numbers of the microorganisms so they can be measured in some way. They need to be provided with the right level of nutrients and oxygen along with the ideal pH and temperature for them to grow. Bacteria and fungi are the most commonly cultured organisms.

It is important to take great care when culturing microorganisms because:

- even if the microorganism you are planning to culture is completely harmless there is always the risk of a mutant strain arising that may be pathogenic

- there is a risk of contamination of the culture by pathogenic microorganisms from the environment

- when you grow a pure strain of a microorganism, the entry of any other microorganisms from the air or your skin into the culture will contaminate it.

Health and safety precautions must always be followed scrupulously when handling, culturing or disposing of microorganisms. All of the equipment must be **sterile** before the culture is started. It is particularly important that once a culture has grown it does not leave the lab. All cultures should be disposed of safely by sealing them in plastic bags and sterilising them at 121 °C for 15 minutes under high pressure, before throwing them away. There are no ethical issues associated with the culturing of microorganisms from the perspective of the microorganisms themselves. However, the danger of infecting other people inadvertently with pathogens should always be considered.

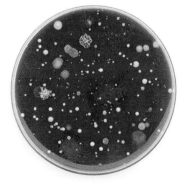

fig A Different types of bacteria cultured on an agar plate.

Aseptic culture techniques

Culturing microorganisms involves a number of steps. First, you need to decide which microorganisms you want to culture, and obtain a culture of them. Then the microorganisms need to be provided with the right nutrients in order to grow. Most microorganisms require a good source of carbon and nitrogen as well as specific minerals. The **nutrient medium** can be in the form of **nutrient broth** (nutrients in liquid form) or in a solid form, usually **nutrient agar**. Agar is a jelly extracted from seaweed. It is very useful because although it sets as a jelly at 50 °C, it does not melt again until it is heated to 90 °C. Both solid and liquid media must be kept sterile until ready for use.

Some microorganisms will grow on pure agar but most need added nutrients. The majority of microorganisms grow on or in a medium enriched with good protein sources such as blood, yeast extract or meat extract. Some need a very precise balance of nutrients. By producing a nutrient medium with very specific ingredients you provide a **selective medium**, a medium in or on which only a select group of microorganisms with those particular requirements will grow. Selective media are important in identifying particular mutant strains of microorganisms and antibiotic resistance. For example, YM media have a low pH, which encourages the growth of fungi and moulds but discourages bacteria, while MacConkey agar is designed to grow Gram-negative bacteria (see **Book 1 Section 2.2.1**). Selective media are also useful for identifying microorganisms that have been genetically modified. This is because antibiotic resistance or the requirement for a particular nutrient is often engineered along with a desired gene as a marker.

fig B Cultures grown in broth are usually prepared in flasks or test tubes, while agar-based cultures are prepared in Petri dishes. The hot sterile liquid agar is poured onto the plates which are immediately resealed and cooled so they set ready for use.

Once you have prepared a suitable medium for your culture, the next step is to introduce your microorganisms, for example, bacteria. Getting the bacteria onto your agar or into your broth is called **inoculation**. Many inoculations are done with an inoculating loop, scraping off bacteria from one solid media surface either into a liquid medium, or streaking across another solid medium plate. Inoculating broth can be used, which involves making a suspension of the bacteria to be grown and mixing a known volume with the sterile nutrient broth in the flask. The flask is then stoppered again as quickly as possible with cotton wool to prevent contamination from the air and clearly labelled. The flask is incubated at a suitable temperature and is often shaken or agitated regularly to make sure that the broth is aerated, allowing oxygen to the growing bacteria. Inoculating solid media also involves a suspension of bacteria but the process is slightly more complicated and is shown in **fig C**.

(a) Sterilise the inoculating loop by holding it in the Bunsen burner until it glows red hot and then leave it to cool.

(b) Dip the sterilised loop in the suspension of the bacteria. Streak the loop across the surface of the agar, avoiding digging into the agar. Replace the petri dish lid, tape closed and label. Turn dish upside down.

fig C Inoculating an agar plate.

Learning tip

Safety warning
Always conduct a risk assessment prior to handling and culturing microorganisms.

Growing a pure culture

To get a pure culture of just a single type of bacterium or fungus, the desired microorganism needs to be isolated. Isolating an organism is most commonly achieved by using information either about its specific needs or about the requirements of possible contaminating organisms:

- Growing a culture under anaerobic conditions will ensure that only anaerobic bacteria will survive. Similarly, growing organisms with oxygen means that only aerobic organisms can survive. However, some bacteria will grow under both conditions. This may not allow you to complete the separation of microorganisms necessary for a pure culture, but it will reduce the variety considerably.

- The nutritional requirements of different microorganisms vary greatly. You can produce a medium that will favour the growth of the organism you wish to culture and inhibit the growth of others. This allows you to identify the colony you want and then re-inoculate it to produce a single pure culture. You may need to control the range of nutrients available, or introduce selective growth inhibitors, antibiotics or antifungal chemicals that will reduce or prevent the growth of all but the desired microorganisms.

- There are indicator media that cause certain types of bacteria to change colour. Colonies that change colour – or do not – can be isolated and cultured (see **fig D**).

- We can only culture around 1% of the known species of bacteria – we do not yet know how to 'persuade' the remaining 99% of species to grow in the lab.

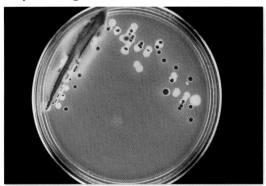

fig D Colonies of red-stained lactose fermenting *E. coli* bacteria show up clearly on this indicator agar, while the other, non-lactose fermenting, gut bacteria in the mixture form colourless colonies. It would be a simple step from here to isolate and further culture either type of bacteria.

It is particularly important to be able to isolate disease-causing organisms from those of the normal body flora, so that diagnosis can be made and appropriate treatment planned. The techniques described above make this possible.

Questions

1 Make a flowchart to explain how to produce a culture of microorganisms:
   (a) in a flask of broth
   (b) on a plate of nutrient agar.

2 Describe three safety precautions that should always be taken when culturing microorganisms and explain their importance.

3 Explain how different media may be used to produce a pure sample of microorganisms for culturing.

Key definitions

Pathogens are microorganisms that cause disease.

In a **culture** microorganisms are provided with the nutrients, level of oxygen, pH and temperature that they require to grow large numbers so they can be observed and measured.

Sterile is a term used to describe something that is free from living microorganisms and their spores.

Nutrient medium is a substance used for the culture of microorganisms, which can be in liquid form (nutrient broth) or in solid form (usually nutrient agar).

Nutrient broth is a liquid nutrient for culturing microorganisms, commonly used in flasks, test tubes or bottles.

Nutrient agar is a jelly extracted from seaweed and used as a solid nutrient for culturing microorganisms, commonly used in Petri dishes.

Selective medium is a growth medium for microorganisms containing a very specific mixture of nutrients, so only a particular type of microorganism will grow on it.

Inoculation is the process by which microorganisms are transferred into a culture medium under sterile conditions.

By the end of this section, you should be able to...

● explain the different methods of measuring the growth of bacterial cultures as illustrated by cell counts, dilution plating, mass and optical methods (turbidity)

To be able to measure the changes in a bacterial culture and the growth in the population, we need to be able to measure the number of cells present in the culture at various time intervals. Bacteria are so small that it is impossible to do this with the naked eye, and so we need to use one of a variety of different methods to count bacteria either directly or indirectly.

Cell counts

Bacteria and single-celled fungi cultured in nutrient broth can be counted directly using a microscope and a **haemocytometer**, so named as it was originally designed for counting blood cells. A haemocytometer consists of a specialised thick microscope slide with a rectangular chamber that holds a standard volume of liquid – 0.1 mm³. The chamber is engraved with a grid of lines.

The sample of nutrient broth is diluted by half with an equal volume of trypan blue, a dye that stains dead cells blue so you can identify and count only the living cells. The cells are viewed using a microscope and counted.

Each corner of the haemocytometer grid has a square divided into 16 smaller squares. The number of cells in each of these four sets of 16 squares is usually counted (see **fig A**) and the mean calculated. The haemocytometer is calibrated so that the number of bacterial or fungal cells in one set of 16 squares equates to the number of cells $\times 10^4$ per cm³ of broth. This enables you to calculate the number of microorganisms in a standard volume of broth. By taking measurements at regular time intervals throughout the life of a bacterial colony, a picture of the changing cell numbers can be built up.

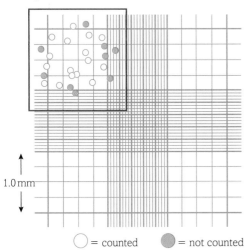

1.0 mm

◯ = counted ● = not counted

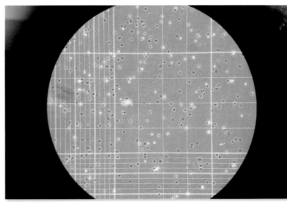

fig A Using a haemocytometer. You need to decide for each line whether you will count cells which are touching the line or not. In this case, the cells that are on or touching the top and left lines are counted, but the ones on or touching the right or bottom lines are ignored.

Optical methods (turbidity)

An alternative way of measuring the number of cells in a culture is by **turbidimetry**, a specialised form of colorimetry. As the numbers of bacterial cells in a culture increase it becomes increasingly cloudy looking or **turbid**. As a solution gets more turbid, it absorbs more light, so less light passes through it. A colorimeter measures how much light passes through a sample, showing how much light is absorbed and therefore, indirectly, how many microorganisms are present. A calibration curve is produced by growing a control culture and taking samples at regular time intervals. The turbidity of each sample is measured and a cell count using a haemocytometer (see previous page) is made for each sample. This gives us a relationship between the turbidity of the culture and the number of bacterial cells present. Using this calibration curve we can then measure the number of microorganisms simply using turbidimetry; for example, if we then wanted to investigate the effect of different conditions on the growth rate of the microorganism.

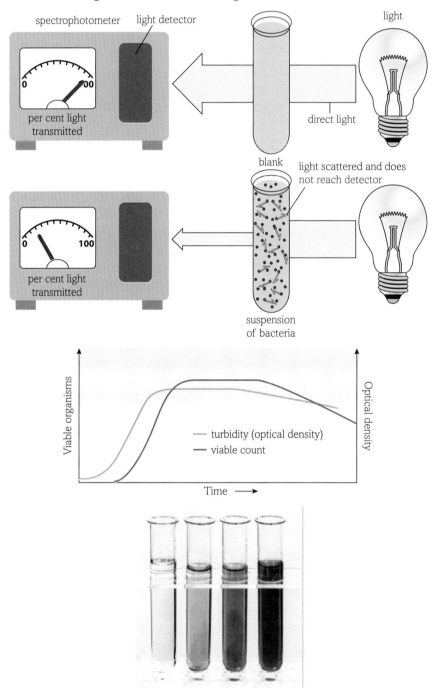

fig B Turbidimetry is a quick and easy way to measure the cell population of a culture once the initial calibration curve for an organism has been set up.

Dilution plating

Another way of counting the microorganisms in a culture involves the technique of **dilution plating**, which is used to find the **total viable cell count**. This technique is based on the idea that each of the colonies on an agar plate has grown from a single, viable microorganism on the plate. So if you have two bacterial colonies after culturing, it is presumed that there were two initial living bacteria on the plate. For example, if you count 30 patches of fungal mycelia, it can be said that 30 fungal cells were on the plate when it was inoculated. However, a solid mass of microbial growth is often present after culturing and it is not possible to work out the individual colonies. This problem is solved by diluting the original culture in stages until a point is reached when the colonies can be counted. If the number of colonies is multiplied by the dilution factor, then a total viable cell count for the original sample can be determined. Because there are often two or more plates where counting individual colonies is possible, it is possible to reach a mean, giving a reasonably accurate number of the cells in a particular sample. The accuracy can be checked using a haemocytometer to count the cells in the original culture.

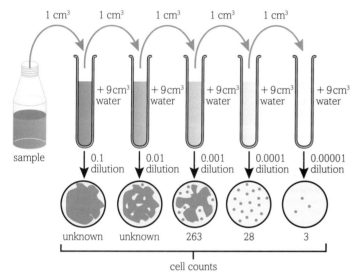

fig C The technique of dilution plating.

Learning tip

As part of your study of this topic, you will carry out **Core Practical 12: Investigate the rate of growth of bacteria in liquid culture taking into account the safe and ethical use of organisms**. Make sure you have a good understanding of this practical as your understanding of the experimental method may be assessed in your examination.

Area and mass of fungi

When fungi rather than bacteria are being cultured, one simple way to assess growth is to measure the diameter of the patches of mycelium. This can be used to compare growth rates in different

conditions. For example, to find the optimum temperature for growth, identical Petri dishes containing identical growth medium are inoculated with the same number of spores of a fungus. The Petri dishes are cultured at different temperatures, with several identical dishes grown at each temperature. After a set period of time, the diameter of each fungal colony is measured and the mean value for the diameter at each temperature is calculated. The temperature that has resulted in the largest mean diameter is the optimum temperature for growth. The same technique can be used for bacteria but, because the microorganisms themselves are so small, the growth of the colony tends to be slower and therefore less easy to measure.

One very effective way to discover the best array and concentration of nutrients or the optimum pH at which to grow fungi is to test the dry mass of the microorganism. This is best done using a liquid growth medium. Samples of broth can be removed at regular intervals and the fungi separated from the liquid by centrifugation or filtering. The material is then dried thoroughly to the point that no more loss of mass is recorded, for example, in an oven overnight at around 100 °C. This gives a measure of the dry mass of biological material in a certain volume of the culture medium, and an increase or decrease in the dry mass gives an indication of the increase or decrease in the mycelial mass. The conditions which produce the greatest dry mass of fungus are the optimum ones for growth.

Questions

1. In **fig B**, how do you explain the difference between the curve for turbidity and the curve for viable cells?

2. Express each dilution in **fig C** as a ratio, e.g. 0.1 = 1 : 10.

3. Using **fig C**, work out the number of cells in the original sample using the fourth and fifth tubes.

4. Use your answers to question 3 and the third tube in **fig C** to work out the mean cell count in the original sample.

Key definitions

A **haemocytometer** is a thick microscope slide with a rectangular indentation and etched grid of lines that is used to count cells.

Turbidimetry is a method of measuring the concentration of a substance by measuring the amount of light passing through it.

Turbid is a term used to describe something that is opaque, or thick with suspended matter.

Dilution plating is a method used to obtain a culture plate with a countable number of bacterial colonies.

A **total viable cell count** is a measure of the number of cells that are alive in a given volume of a culture.

By the end of this section, you should be able to...

● explain the different phases of a bacterial growth curve and calculate exponential growth rate constants

Bacteria can reproduce at astonishing speed. The small size and relative simplicity of these prokaryotic cells means that they can undergo asexual reproduction – binary fission or splitting in two – every 20 minutes when conditions are favourable, although it often takes longer.

The growth of bacterial colonies

The time between bacterial divisions is known as the **generation time**. The implications of the rapid rate of reproduction in bacterial cells are enormous. If a single bacterium had unlimited space and nutrients, and if all its offspring continued to divide at the same rate, then after 48 hours there would be 2.2×10^{43} bacteria, which would weigh 4000 times the weight of the Earth! Clearly this sort of growth never takes place, because although the number of bacteria in a colony can increase exponentially at first, limited nutrients and a build-up of waste products always act as a brake on reproduction and growth.

Using a logarithmic scale

When you are considering the growth of bacteria, the numbers involved are enormous. They rapidly become completely unwieldy and it is impossible to show what is happening. If we use a logarithmic scale, the data become much easier to manage. Log numbers are used to represent the bacterial population because the difference in numbers from the initial organism to the billions of descendants is too great to represent using standard numbers. Growth on the y-axis is logarithmic, time is kept on a normal scale (on the x-axis), so this is a semi-logarithmic graph.

In a logarithmic scale, the numbers on the scale are actually logarithms – also known as powers – of a base number. So in our example the time scale stays the same (non-logarithmic) but the numbers of bacteria are represented on a $\log_{10}$ scale. This means that every number on the y-axis represents a power of 10. So 1 is actually 10^1 (10), 2 is 10^2 (100), 3 is 10^3 (1000) and so on. Look at the simple bacterial growth curve over 12 hours compared with the logarithmic graph in **fig A** to see the difference this makes.

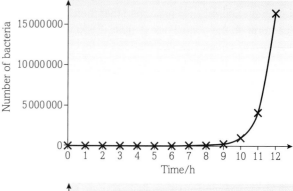

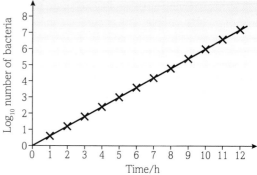

fig A Non-log and log scales of bacterial growth, assuming bacteria divide every 30 minutes.

Exponential growth rate constants

You can calculate the number of bacteria in a population using the following formula:

$N_t = N_0 \times 2^{kt}$ where:

N_t = the number of organisms at time t

N_0 = the number of organisms at time 0 (the beginning of the experiment)

k = the exponential growth rate constant

t = the time the colony has been growing

This allows you to work out the number of bacteria in a colony at any point if you know the initial number of bacteria and the time the colony has been growing under constant conditions. However, to use the equation, you need to know the exponential growth rate constant. This constant equals the number of times that the population will double in one unit of time. You can calculate it using the following formula:

$$k = \frac{\log_{10} N_t - \log_{10} N_0}{\log_{10} 2 \times t}$$

Learning tip

This formula can be derived from the equation for population growth. You do not need to be able to derive the formula yourself, but if you are confident using logarithms you might like to give it a try!

WORKED EXAMPLE

Using the data from **fig A**, if the number of bacteria at the beginning is 1 and the colony has 1024 bacteria after growing for 5 hours, calculate k:

$$k = \frac{\log_{10}(1024) - \log_{10}1}{\log_{10}2 \times 5}$$

$$k = \frac{3.010}{0.301 \times 5}$$

$$k = 2$$

You can then substitute k into the equation to find out the population at any time.

For example, after 15 hours:
$N_{15} = 1 \times 2_{15k}$ so $N_{15} = 2^{30} = 1\,073\,741\,824$

This answer makes intuitive sense. If the bacteria divide every 30 minutes, the size of the population will double twice each hour. So in 15 hours, the initial population size of 1 will have doubled 30 times.

Analysing the data

When the numbers of cells in a bacterial culture are measured over time, by any one of the methods you considered in **Section 6.1.2**, a pattern emerges. There are four stages to this growth curve:

- the **lag phase** – when bacteria are adapting to their new environment and are not yet reproducing at their maximum rate
- the **log phase** or exponential phase – when the rate of bacterial reproduction is close to or at its theoretical maximum, repeatedly doubling in a given time period
- the **stationary phase** – when the total growth rate is zero as the number of new cells formed by binary fission is equal to the number of cells dying
- the **death phase** or decline phase – when reproduction has almost ceased and the death rate of cells is increasing.

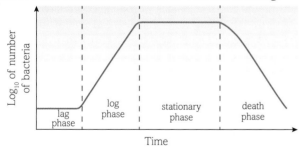

fig B Graph showing pattern of bacterial growth.

Exponential growth in a bacterial culture does not continue. Reasons for the slowing down of the rate of growth include:

- A reduction in the amount of nutrients available. At the start of the culture there are more than enough nutrients for all the microorganisms but as the numbers multiply exponentially in the log phase of growth the excess is used up. Unless fresh nutrients are added, the level of nutrients available will become insufficient to support further growth and reproduction and so will limit the growth of the organism.

- A build-up of waste products. At the beginning of the growth cycle of a bacterial culture the waste products are minimal, but as cell numbers rise the build-up of toxic material becomes sufficient to inhibit further growth and even to poison and kill the culture. In particular, as carbon dioxide produced by the respiration of the bacterial cells builds up, the pH of the colony falls to a point where the bacteria can no longer grow.

Did you know?

From psychrophiles to hyperthermophiles...

Temperature has long been recognised as one of the limiting factors on the growth of a colony of bacteria. Very little growth occurs at low temperatures. High temperatures are regularly used to destroy bacteria and make sure that culture media and medical instruments are sterile. However, this view has been somewhat shaken by the discovery of bacteria living in the most extreme conditions in the depths of the ocean and deep in cores of ice.

In the mineral-rich waters of the deep-sea hydrothermal vents known as 'black smokers', almost 3000 m below the surface of the ocean, the pressure is high and the temperature is around 350 °C. Not only do 'black smoker' bacteria thrive in these conditions, they stop reproducing if the temperature falls much below 100 °C!

Bacteria have also been found under almost 4000 m of ice, at temperatures and pressures that scientists would not have thought suitable for life and where most enzymes would no longer be active.

These bacteria challenge many of our accepted theories about the biochemistry of enzymes and cells, and illustrate clearly the ability of bacteria to take advantage of environments in which nothing else can survive.

Questions

1. If a single bacterium starts dividing every 30 minutes, how long will it take for the colony to contain 16 777 216 bacteria? Show your workings to get a record of the bacterial numbers at each stage.

2. Use your answer to question 1 to:
   (a) explain the use of log scales in graphs to show the rate of bacterial growth
   (b) plot a graph of the data that you have generated
   (c) sketch another graph to show what would have happened if the culture had been allowed to continue in a closed culture.

Key definitions

Generation time is the time span between bacterial divisions.

The **lag phase** is when bacteria are adapting to their new environment and are not yet reproducing at their maximum rate.

The **log phase** is when the rate of bacterial reproduction is close to or at its theoretical maximum, repeatedly doubling in a given time period.

The **stationary phase** is when the total growth rate is zero as the number of new cells formed by binary fission is equalled by the numbers of cells dying due to factors including competition for nutrients, lack of essential nutrients, an accumulation of toxic waste products and possibly lack of, or competition for, oxygen.

The **death phase** is when reproduction has almost ceased and the death rate of cells is increasing so that the population number falls.

By the end of this section, you should be able to...

- describe how bacteria can act as agents of infection, invading and destroying host tissues and producing toxins

- explain that pathogenic effects can be produced by exotoxins, endotoxins and invasion of the host tissue

Bacteria cause many human diseases, from gastroenteritis and tonsillitis to tuberculosis and pneumonia. They are important human pathogens. They also cause disease in other animals and in plants; for example, crown gall, leaf blight, and soft rots are bacterial diseases in many different species of plants.

If we can understand how bacteria act as agents of infection it can help us to avoid infections and treat bacterial diseases effectively. Once bacteria get into the body they will often grow in a localised area unless they get into the blood, in which case they can be carried all around the system and cause a systemic infection.

How do bacteria cause disease?

Once pathogenic bacteria get into the body, they cause the signs and symptoms of disease in a number of ways. Some bacteria cause the symptoms of disease simply as a result of the way they invade and destroy the host tissues. However, most bacteria make people unwell through the toxins they make as a by-product of their metabolism. The toxins may incapacitate the whole host organism or just its immune system. Toxins are classified as **endotoxins** and **exotoxins**.

Endotoxins

Endotoxins are **lipopolysaccharides** that are an integral part of the outer layer of the cell wall of Gram-negative bacteria. Scientists have discovered it is the lipid part of the lipopolysaccharides that acts as the toxin, while the polysaccharide stimulates an immune response.

Endotoxins have effect around the site of infection by the bacteria. The pathogenic effects produced by bacterial endotoxins tend to be symptoms such as fever, vomiting and diarrhoea, seen with strains of *Salmonella* spp. and *E. coli*. Although the diseases caused by bacterial endotoxins are often not fatal themselves, the symptoms they cause can lead indirectly to death, for example by dehydration as a result of severe diarrhoea. Antibiotic treatments that destroy the bacterial cells by lysis of the cell wall can also lead to further endotoxin release, due to the lipopolysaccharide component of the cell wall.

Case study: *Salmonella* spp.

The World Health Organization (WHO) estimates that every year tens of millions of people around the world suffer from salmonellosis –

gastroenteritis caused by one of the many different strains of *Salmonella* bacteria. For most people the disease is unpleasant but not serious, but it is estimated that globally around 100 000 people die as a result of these infections every year. The majority of these bacteria act in a very similar way. The bacteria invade the lining of the intestine and the endotoxins cause inflammation. The cells no longer absorb water, so the faeces become liquid. The gut then goes into spasms of peristalsis that result in diarrhoea.

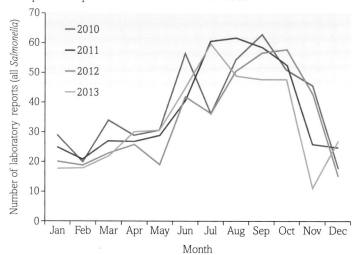

fig A The pattern of *Salmonella* spp. infections reported in Wales between 2010 and 2013. A clear seasonal pattern emerges.

Salmonella spp. is spread by ingestion of food and water contaminated with infected faeces. *Salmonella* bacteria live in the guts of many food animals, and can easily contaminate the meat. If food is not cooked properly the bacteria survive and pass into your gut when you eat the food. If someone handles raw meat and does not wash their hands before preparing other foods, they can transmit the bacteria. Salads washed in contaminated water, or contaminated drinking water can also cause infection. These bacteria survive the stomach acid that destroys many pathogens, passing into the small intestine where they invade the lining cells and their endotoxins produce the symptoms of disease. Salmonella infections are relatively rare now in the UK as chickens are vaccinated against the disease.

fig B Ingesting food contaminated by raw meat, or meat that is not cooked properly are two of the most common causes of infection with *Salmonella* spp.

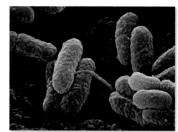

fig C *Salmonella enteritidis.*

Treating these bacterial infections with antibiotics is not very useful unless the patient is very young, very old or has a compromised immune system. In most people antibiotics just reduce symptoms, allowing the person to feel better but act as a carrier for longer.

The best way to deal with the diseases caused by *Salmonella* spp. is to avoid them, simply by cooking all meat thoroughly, washing hands after handling raw meat and after using the toilet, and avoiding drinking water that may be contaminated. Some strains of *Salmonella* cause more serious diseases, for example *S. typhi*, which causes typhoid.

Exotoxins

Exotoxins are usually soluble proteins that are produced and released into the body by bacteria as they metabolise and reproduce in the cells of their host. Exotoxins are produced by both Gram-positive and Gram-negative bacteria, and their effect can be more widespread than that of endotoxins as they often act at sites at a distance from the infecting bacteria. There are many different types with specific effects. Some damage cell membranes causing cell breakdown or internal bleeding, some act as competitive inhibitors to neurotransmitters, while others directly poison cells. Exotoxins rarely cause fevers but they include some of the most dangerous and fatal bacterial diseases. For example, *Clostridium botulinum* produces botulinum toxin, one of the most toxic natural substances known. It has been estimated that 1 mg of pure toxin could kill a million guinea pigs!

Exotoxin	Lethal dose /mg	Host animal	Lethal toxicity compared with:		
			Strychnine	Snake venom	Bacterial endotoxin
botulinum toxin	0.8×10^{-8}	mouse	3×10^{6}	3×10^{5}	3×10^{7}
tetanus toxin	4×10^{-8}	mouse	1×10^{6}	1×10^{5}	1×10^{7}
diphtheria toxin	6×10^{-5}	guinea pig	2×10^{3}	2×10^{2}	2×10^{4}

table A Comparison of bacterial exotoxins with other toxins and bacterial endotoxins.

Case study: *Staphylococcus* spp.

There are around 40 different types of *Staphylococcus* bacteria and about one-third of us carry at least some of them around all the time in our normal skin and gut flora. They only cause disease if they get inside the tissues of the body, if the normal skin flora is changed, or if the person has a compromised immune system due to another disease, or has been receiving treatments such as chemotherapy. *Staphylococcus* spp. are Gram-positive bacteria. They produce exotoxins that can cause anything from mild skin diseases to rapid death. The most common are *Staphylococcus aureus* and *Staphylococcus epidermis,* both commonly found on the skin. *S. aureus* causes skin infections such as styes, boils and impetigo. It can cause infections of the joints in septic arthritis, infections of the membranes around the heart, gastroenteritis, and diseases such as septicaemia, toxic shock syndrome and necrotising pneumonia, all of which can rapidly be fatal. Both *S. aureus* and *S. epidermis* can cause bacterial meningitis if they get access to the brain as a result of surgery or through the brain membranes in poorly controlled diabetes. *Staphylococcus* infections can be treated with antibiotics if they are diagnosed quickly, but the exotoxins they produce are often very powerful and can cause death very rapidly. Worryingly, strains of *S. aureus* are becoming increasingly resistant to common antibiotics (see methicillin-resistant *Staphylococcus aureus* in **Section 6.1.6**).

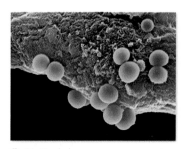

fig D *Staphylococcus aureus.*

Host tissue invasion

Invading host tissues and damaging the cells is the third way bacteria act as pathogens. It is the response of the host organism to the cell damage that causes the symptoms of disease. Often this cell damage is also linked to the production of exotoxins or the presence of endotoxins in the bacterial cell walls.

Case study: *Mycobacterium tuberculosis*

The bacterium *Mycobacterium tuberculosis* causes **tuberculosis** (TB), one of the most common human infections on a global scale. WHO figures show that in 2012 there were around 8.6 million new cases of TB worldwide, and approximately 1.3 million people died of the disease. Up to a third

fig E *Mycobacterium tuberculosis* causes tuberculosis in millions of people globally every year.

of the population of the world have been infected by TB, although many of them have no symptoms. At least one in ten of these symptomless people will go on to develop full-blown active TB.

TB is most commonly caused by *Mycobacterium tuberculosis*, spread by droplet infection. Crowded living or working conditions increase the likelihood of it spreading as people breathe, cough and sneeze near to each other. People who are malnourished, ill or have problems with their immune systems are more vulnerable to the disease and are much more likely to develop active TB than healthy, well-fed individuals. People living with HIV/AIDS are particularly vulnerable to infection with *M. tuberculosis* as a result of their reduced immune response. Globally, the other common source of infection is from the bacterium *Mycobacterium bovis*, which affects cattle. People become infected by drinking infected milk or living and working in close contact with cattle.

TB often affects the respiratory system, damaging and destroying lung tissue. It also suppresses the immune system, making the body less able to fight the disease. The well-known symptoms of TB, coughing up blood and weakness, come at the end of the disease process.

Only about 30% of the people exposed to TB will actually develop the disease. Once the bacteria are inhaled, they invade the cells of the lungs and multiply slowly. This **primary infection** often causes no obvious symptoms. If you have a healthy immune system there will be a localised inflammatory response forming a mass of tissue called a **tubercle** (hence tuberculosis) containing dead bacteria and macrophages. In about 8 weeks the immune system controls the bacteria, the inflammation dies down and the lung tissue heals. Primary TB infections often take place in childhood and the majority (about 90%) get better without the person ever realising they have had TB.

Mycobacterium tuberculosis, however, has an adaptation that enables it to avoid the immune system, allowing some bacteria to survive the primary infection stage. The bacteria produce a thick waxy outer layer that protects them from the enzymes of the macrophages. Bacteria with an effective coating will remain deep in the tubercles in the lungs, dormant or growing slowly for years until the person is malnourished, weakened or their immune system does not work well. These bacteria then cause active tuberculosis. In this way the most effective bacteria are selected and will be passed on. Once they become active again they can grow and reproduce very rapidly, causing serious damage and disease.

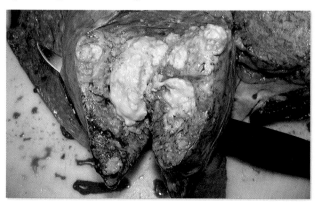

fig F The damage caused by TB infection can clearly be seen in the lungs of this cow.

Questions

1 Bacterial endotoxins usually have a relatively local effect but exotoxins can affect many areas of the body. Explain this observation.

2 Make a table to compare the way *Staphylococcus* spp., *Salmonella* spp. and *Mycobacterium tuberculosis* cause disease.

3 The figures for the numbers of people infected with a particular bacterium or deaths from specific diseases are often preceded with terms such as 'approximately' or 'estimated'. Suggest a reason for this.

4 Use the graph in **fig A** to answer the following questions:
   (a) Which are the peak months for *Salmonella* spp. infections in Wales?
   (b) Why do you think the infections are so much higher in these months?
   (c) What additional information would be helpful to explain the way the pattern varies slightly from year to year?

Key definitions

Endotoxins are lipopolysaccharides that are an integral part of the outer layer of the cell wall of Gram-negative bacteria.

Exotoxins are soluble proteins that are produced and released into the body by bacteria as they metabolise and reproduce in the cells of their host.

Lipopolysaccharides are large molecules containing a lipid element and a polysaccharide element.

Tuberculosis is a lung disease caused by *Mycobacterium tuberculosis* and *M. bovis*.

Primary infection is the initial stage of tuberculosis when *M. tuberculosis* has been inhaled into the lungs, invaded the cells of the lungs and multiplied slowly, often causing no obvious symptoms.

A **tubercle** is the result of a healthy immune response to an infection by *M. tuberculosis*. A localised inflammatory response forms a mass of tissue containing dead bacteria and macrophages.

By the end of this section, you should be able to...

● describe the action of bactericidal and bacteriostatic antibiotics

The idea that diseases can be caused by microorganisms is central to the way we look at health and medical treatments. However, people have only known about bacteria, viruses and other microorganisms for about 150 years. For thousands of years people used simple remedies to try to overcome the ravages of disease. In the developed world, from the middle of the nineteenth century, research into known and recognised pathogens became more focused. Not surprisingly, the battle against disease became more successful.

Drugs against microorganisms

The big breakthrough in the treatment of bacterial infections was the discovery and manufacture of the first **antibiotic** in the first half of the twentieth century. Antibiotics are medicines that either destroy microorganisms or prevent them from reproducing. They are used to treat patients suffering from bacterial infections and they have had a dramatic impact. It is hard for us in the twenty-first century to imagine how many people died of communicable diseases just over 100 years ago. In 1901, 36.2% of all deaths (and 51.5% of childhood deaths) in the UK were from communicable diseases. By the year 2000, 11.6% of all UK deaths and only 7.4% of childhood deaths were from communicable diseases.

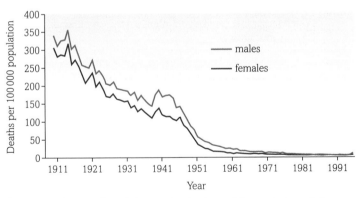

fig A Many factors affected the way the death rate fell over the twentieth century, but these data demonstrates the impact of antibiotics. They first became available to the public after the end of the Second World War in 1945. It was the introduction of the National Health Service in 1948 that made the big difference, however, when medical help was available to everyone.

All modern antimicrobial drugs work against microorganisms by the principle of **selective toxicity**. They interfere with the metabolism or function of the pathogen, with minimal damage to the cells of the human host. The most commonly used and best known antimicrobials are the antibiotics. They can also be used, but to a limited degree, in the treatment of some fungal infections.

The first antibiotic to be widely and successfully used in the treatment of bacterial diseases was **penicillin**. Penicillin does not work against all pathogens, so scientists searched for more antibiotics – a search which continues today. And although many other antibiotics have been discovered, globally penicillin is still commonly used to combat bacterial infections.

Antibiotic action

Antibiotics are effective because they disrupt the biochemistry of the bacterial cells. Different classes of antibiotics interrupt different processes. The main classes of antibiotic drugs and the way they affect microorganisms are summarised in **table A**.

Antimicrobial action	Examples of antibiotics
antimetabolites interrupt metabolic pathways, such as blocking nucleic acid synthesis, causing death	sulfonamides
protein synthesis inhibitors interrupt or prevent transcription and/or translation of microbial genes, so protein production is affected	tetracyclines, chloramphenicol
cell wall agents prevent formation of cross-linking in cell walls, so bacteria are killed by lysis (bursting)	beta-lactams, e.g. penicillins
cell membrane agents damage the cell membrane, so metabolites leak out or water moves in, killing the bacteria	some penicillins, cephalosporins
DNA gyrase inhibitors stop bacterial DNA coiling up, so it no longer fits within the bacterium	quinolone

Key: [Bacteriostatic]   [Bactericidal]

table A Some methods of antibiotic action.

When an antibiotic is taken, it may have one of two different effects. It may be **bacteriostatic**, which means that the antibiotic used completely inhibits the growth of the microorganism. This level of treatment is usually sufficient for the majority of everyday infections because, combined with the actions of our immune systems, it will ensure that the pathogen will be completely destroyed. **Tetracycline**, which is used to treat acne, urinary tract infections, respiratory tract infections and *Chlamydia*, is a bacteriostatic antibiotic.

However, sometimes a particular drug, or the dose of a drug which is given, will be **bactericidal**. This means it will destroy almost all of the pathogens present. This type of treatment is often, but not always, used in severe and dangerous infections. It is also used for treating infections where the immune system of the patient is suppressed such as in transplant patients, where immunosuppressant drugs protect the transplanted organ, or in certain diseases such as TB and HIV/AIDS. Penicillins (drugs related to the original penicillin) are bactericidal antibiotics used to treat skin infections, chest infections and urinary tract infections.

In fact, the terms bacteriostatic and bactericidal are deceptive. Often it depends on the dose given. Bacteriostatic antibiotics often kill a lot of the bacteria, and will usually kill them all at high enough doses. Bactericidal antibiotics do not kill all of the bacteria – just 99% within a given time frame. Lower doses of bactericidal antibiotics are often bacteriostatic – so the definitions are flexible.

A *broad-spectrum antibiotic* destroys a wide range of harmful bacteria, pathogens, and neutral and good bacteria alike. A *narrow-spectrum antibiotic* targets one or two specific pathogens.

The effectiveness of any antimicrobial drug depends on many factors. These include:

* the concentration of the drug in the area of the body infected, which will be affected by how easily the drug can reach the tissue and how quickly it is excreted
* the local pH
* whether either the pathogen or the host tissue destroy the antibiotic
* the susceptibility of the pathogen to the particular antibiotic used.

If the standard dose of a drug (in other words, what a doctor would normally prescribe) successfully destroys the pathogen and cures the disease then the pathogen is sensitive to that antibiotic. If the disease is cured only by using a dose of antibiotic that is much higher than the standard dose, then the pathogen is regarded as moderately sensitive to that antibiotic. However, there are increasing numbers of cases where a particular microorganism is not affected at all by an antibiotic, sometimes even one that may have been effective in the past. In these cases the microorganism has become resistant to the antibiotic – it is **antibiotic resistant** (see **Section 6.1.6**).

You can investigate the effect of different antibiotics on bacteria using standard microbiological techniques. An agar plate is inoculated with a known bacterial culture. Filter paper discs containing different antibiotics, or different concentrations of the same antibiotic, are placed on the agar and the plates covered, ensuring the oxygen is not excluded. A control culture of microorganisms with known sensitivity to the antibiotic is grown at the same time under the same conditions. The level of inhibition of bacterial growth gives a measure of the effectiveness of the drugs.

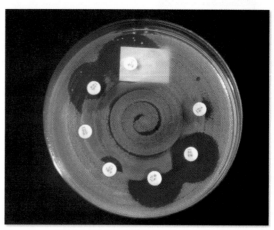

fig B By using discs of known concentrations of different antibiotics, or different concentrations of the same antibiotic, the sensitivity of a pathogen to a particular antibiotic can be determined.

Learning tip

Safety warning
Always conduct a risk assessment prior to handling microorganisms.

Questions

1 Using the information in the text on page 55, calculate the percentage reduction in childhood deaths due to communicable diseases between 1901 and 2000.

2 (a) Give two examples of how antibiotics affect bacteria.
  (b) Some antibiotics are bactericidal and others are bacteriostatic. Explain the difference.
  (c) A single antibiotic may be both bacteriostatic and bactericidal, depending on the conditions.
      Which factor is most likely to influence the way it works? Explain your answer.

3 Investigate the discovery and development of penicillin. Write a paragraph on each of the following to explain their role in this: Ronald Hare, Cecil Paine, Norman Heatley, Mary Hunt.

Key definitions

An **antibiotic** is a drug that either destroys microorganisms or prevents them from growing and reproducing.

Selective toxicity means that a substance is toxic against some types of cells or organisms but not others.

Penicillin was the first antibiotic discovered. It affects the formation of bacterial cell walls, and it is bactericidal.

Bacteriostatic antibiotics inhibit the growth of bacteria.

Tetracycline is a bacteriostatic antibiotic that inhibits protein synthesis.

Bactericidal antibiotics kill bacteria.

An **antibiotic-resistant** microorganism is not affected by an antibiotic, even one that may have been effective in the past.

By the end of this section, you should be able to...

● explain the development and spread of antibiotic resistance in bacteria

● describe the methods and explain the difficulties of controlling the spread of antibiotic resistance in bacteria

Antibiotic drugs provide an essential approach to controlling bacterial diseases, but they are not the whole answer to the problem. New problems are now arising that result directly from the over-use of these valuable drugs.

Creating drug-resistant bacteria

There is a constant 'evolutionary race' between pathogens and us, their hosts. We keep developing new medicines, and bacteria keep evolving resistance to these drugs. An antibiotic is only effective if the microorganism has a binding site for the drug and a metabolic process or biochemical pathway with which the antibiotic interferes. However, during bacterial reproduction there is always the chance of a mutation occurring. Some mutations may help the microorganism resist the effects of the antibiotic – for example, by making the cell wall impermeable to the drug. As a result of natural selection, these mutations become more common and the bacterial population becomes increasingly resistant to the drug (see **Book 1 Section 3.2.3**). Mutations can also result in new biochemical pathways, or switch on or create a gene for the production of an antibiotic-destroying enzyme. These mutations will also become more common as a result of natural selection and the pathogens will edge ahead in the evolutionary race.

Widespread use of antibiotics accelerates this process. As different antibiotics are used to tackle increasing resistance, this increases the selection pressure for the evolution of bacteria that are resistant to all of them. This 'evolutionary race' is creating what are known as 'superbugs', for example, **methicillin-resistant _Staphylococcus aureus_ (MRSA)**, and in the absence of an effective antibiotic these resistant bacteria are quite capable of causing death. The only way to prevent this trend continuing is to reduce the selection pressure for resistance by using antibiotics sparingly, only when they are strictly necessary, to ensure that people understand that they must complete each course of antibiotics, and to use as few different antibiotics as possible, holding some in reserve for use only when all else has failed.

To reduce the development of antibiotic resistance we need to vary the antibiotics we use and introduce new antibiotics when possible. The problem here is that the numbers of new antibiotics being developed and approved for use is dropping fast. This means bacteria get repeated exposure to existing antibiotics which increases the likelihood that they will develop resistance (see **fig A**).

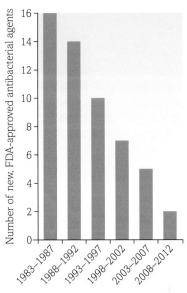

fig A The number of new antibacterial drugs available for treatment is falling steadily.

Healthcare-associated infections

'Superbugs' are commonly found in hospitals and care homes, where people are ill or have had surgery and where antibiotic use is at its highest. Methicillin-resistant _Staphylococcus aureus_ (MRSA) and **_Clostridium difficile_** are causing particular problems. Properly known as **healthcare-associated infections (HCAIs)**, 'superbugs' have a high profile in public awareness and are often referred to in the media as hospital-acquired infections. Patients who become infected have to stay in hospital much longer. This, combined with the treatment needed to overcome the infection, costs on average an extra £4000–10 000 per patient, as well as causing suffering to both the people involved and their families. Some people even die as a result of these infections. A 2012 study of NHS trusts found that around 6% of patients in English hospitals have healthcare-associated infections, including pneumonia and norovirus (diarrhoea and sickness) as well as MRSA and antibiotic-resistant _C. difficile_. This is down from 8.2% in 2006, but still represents many thousands of patients.

MRSA

About one-third of people have the bacterium _S. aureus_ on their skin or in their nasal passages, without it causing problems. If the bacterium gets into the body it can cause boils or even infections throughout the body such as septicaemia. Many _Staphylococcus_ infections have been treated very effectively with methicillin (a penicillin-related antibiotic). However, a mutation has occurred in some of the bacteria that enables them to produce a penicillinase enzyme that breaks down methicillin. In hospitals and care homes, patients weak with other infections often develop opportunistic _S. aureus_ infections and have been treated with methicillin, and with other antibiotics when this has not worked. The result is MRSA. In hospitals and care homes *S. aureus* is now winning the evolutionary race as almost all the bacteria produce penicillinases. The infections can be treated, but only with high doses of a very small number of antibiotics which are used sparingly to prevent the same thing happening again. Control measures reduced the incidence of MRSA from 1.8% of patients being infected in 2006 to less than 0.1% by 2012.

Clostridium difficile

Clostridium difficile is an anaerobic bacterium that is found in small numbers in the large intestine of about 5% of the population. It is not affected by many of the commonly used antibiotics and produces extremely tough spores that can survive for months outside of the human body. In a healthy person *C. difficile* causes no problems, as its numbers are limited by competition with the normal gut flora. Some common broad-spectrum antibiotics used to treat bacterial infections destroy the normal gut flora as well as the pathogens they are treating. *C. difficile* can sometimes increase rapidly in number as a result. These bacteria produce two different toxins that damage the lining of the intestines, causing severe diarrhoea that can lead to bleeding from the gut and even death.

Infection prevention and control

Codes of practice have been drawn up for doctors, nurses and all healthcare workers to try to prevent the spread of healthcare-associated infections, and to control them as effectively as possible when they do occur:

- Controlling the use of antibiotics: The careful use of antibiotics reduces the likelihood of resistant bacteria evolving. Antibiotics should only be used when absolutely necessary and every course of antibiotics should be completed. If a patient takes only part of a course, the immune system is unable to cope with the numbers of bacteria remaining and does not destroy them all. Those bacteria that have some resistance will escape and can go on to infect other people. Use of different antibiotics just encourages a faster evolution of multiple resistance. It is important to educate the general public, who still tend to demand antibiotics for minor infections and often stop taking the medicine as soon as they feel better.

- Hygiene measures: Good hygiene within hospitals and care homes can have a major impact on healthcare-associated infections. Examples of good practice include doctors, nurses and other healthcare professionals washing their hands or using alcohol-based gels between patients to destroy as many pathogens as possible. For example, MRSA can easily be spread by personal contact, but it is destroyed by alcohol and washing. However, the spores of *C. difficile* are not destroyed by the alcohol gels and need a chlorine-based disinfectant to destroy them. Another aspect of good hygiene includes new guidelines about the clothing doctors and other staff should wear. For example, long ties which might dangle and carry bacteria from one patient to another are now banned, as are wristwatches and even long-sleeved shirts as the cuffs can carry bacteria although there is little evidence that this makes any difference. In the US, all healthcare workers wear scrubs (sterile surgical outfits). In the UK, some hospitals use scrubs and some do not. There is no discernible difference in the levels of HCAIs between them. Thorough cleaning of hospital wards, toilets and equipment such as bedpans can also control the spread of disease. This is particularly important with diseases spread through faecal traces, such as *C. difficile*. Regular thorough cleaning is needed as well as the normal daily cleaning that occurs.

- Isolation of patients: Patients affected by a healthcare-associated infection need to be isolated from other patients as quickly as possible. By nursing infected individuals in separate rooms with high levels of hygiene and infection control, the spread of disease to other patients can be minimised. This can be hard for patients though, as they can feel cut off from other people.

- Prevention of infection coming into the hospital: Screening patients as they come into hospital allows people who are carrying MRSA (around 3% of the general public) or other infections to be treated immediately and isolated until the bacteria have been destroyed. This reduces the chance of the person becoming clinically ill during treatment or of spreading the bacteria to others. Another problem is that people visiting patients may bring infections into the wards. This is particularly the case with MRSA and many viral infections, such as norovirus (which can cause sickness and diarrhoea). Hospitals advise people not to visit if they are unwell but people often want to see seriously unwell relatives and ignore the advice. People visiting patients should follow good hygiene procedures and keep their hands clean and washed. Most hospitals provide alcohol gels for people to use as they come in and out of the hospital, but the biggest problem is getting people to use them!

- Monitoring levels of healthcare-associated infections: It has been made mandatory that hospitals measure and report levels of MRSA and *C. difficile* infections. These results are published and available to the general public as well as to the government. This has focused attention on the problem and led to increased efforts to reduce and overcome the infections. For example, government figures showing numbers of deaths involving (but not caused by) MRSA in England and Wales indicated that they had dropped from 40.2 per million population in 2005 to 6.3 per million in 2013.

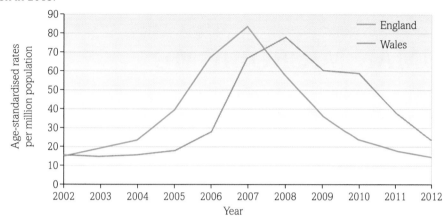

fig B Rates of *S. aureus* and MRSA per million population from 2002 to 2012 in England and Wales.

As a result of hospital trusts following these codes of practice, there has been a marked decrease in the incidence of MRSA and *C. difficile* in hospitals in recent years. Although some other HCAIs seem to be increasing, particularly different resistant gut bacteria, the overall total is reducing. Hopefully, as good practice of infection control continues to be given a high priority, levels of healthcare-associated infections and, perhaps most importantly, deaths resulting from them, will continue to fall.

Questions

1   Explain how bacteria develop resistance to common antibiotics.

2   (a) Find out the current code of practice on the prescription of antibiotics.
    (b) Explain how this can help to reduce the development of further antibiotic-resistant strains of bacteria.

3   If 15.1 million people were admitted to hospital in England in 2012–13, how many would you expect to be affected by HCAIs? How many would you expect to be carrying MRSA when they came into hospital?

4   (a) Describe how the use of broad-spectrum antibiotics is implicated in the spread of *C. difficile*.
    (b) Explain how minimal use of narrow-spectrum antibiotics can help to avoid the problems of HCAIs.
    (c) Discuss how the data in **fig A** has serious implications for the development of HCAIs.

5   Explain how an understanding of the causes and methods of spread of HCAIs relates to the codes of practice on hospital hygiene.

Key definitions

Methicillin-resistant *Staphylococcus aureus* (MRSA) is a strain of *S. aureus* that is resistant to several antibiotics, including methicillin.

Clostridium difficile is a type of bacteria that often exists in the intestines and causes no problems unless it becomes dominant as a result of the normal gut flora being removed/damaged by antibiotic treatment.

Healthcare-associated infections (HCAIs) are infections that are acquired by patients while they are in hospitals or care facilities. They may be the result of poor hygiene between patients or the result of antibiotic treatment, and may be antibiotic resistant.

GOOD BACTERIA, BAD BACTERIA

Read the following article to find out about a surprising medical treatment.

THE AMAZING HUMAN MICROBIOME

As a human being, you may well think that you are made up of human cells. But scientists have discovered that in healthy people, microbial cells outnumber human cells by about 10 : 1 – although they only make up between 1–3% of the body mass. It appears that the human body is a rich and varied ecosystem for microorganisms, and that those microorganisms carry out some important, if not vital, functions in the body.

For many years our understanding of the microorganisms which live in and on our bodies (our microflora) has been limited to the ones we can culture in the laboratory – and about 80% of our microflora hasn't yet been persuaded to grow outside the human body. But in the last few years techniques in DNA sequencing and genome analysis have improved enormously, becoming not only fast and accurate but also relatively cheap. Projects […] have been set up to identify all of the microorganisms that intimately share our lives, and to learn more about their role in both health and disease.

Microbial medicine

Antibiotics have saved millions of lives since they were developed for general use in the 1940s. However most antibiotics knock out part of your healthy microbiome when they are destroying the pathogenic bacteria which have caused disease. The effects of this can be seen in the gut for months afterwards and sometimes they can cause serious illness and may even be fatal – for example *Clostridium difficile* (*C. difficile*)… [*This bacterium can have a devastating effect on the guts of ill or immune-compromised people after antibiotic treatment for other diseases*]… It is resistant to most antibiotics – the only treatment that sometimes works is vancomycin, a powerful antibiotic normally used when all others have failed.

In a mind-blowing experiment in the Netherlands, a team of researchers took 3 groups of patients all with recurrent *C. difficile* infections. In the initial experiment, 13 people were given vancomycin for 14 days. A second group of 13 patients were given vancomycin and also bowel lavage (washing out the colon). The third group of 16 was given just four days of vancomycin and bowel lavage followed by an infusion of faeces from a donor down a nasogastric tube. In other words, a solution of poo from a healthy person was fed into the infected guts! After 13 of this group recovered completely, the other three were given a second treatment from another donor and 2 more recovered. The overall results were astonishing.

The study size was very small – but only because the trial was halted. The success of the faecal transfusion was so marked it became unethical not to give it to all the patients affected by repeated *C. difficile* infections.

Treatment	% of patients fully recovered
Vancomycin only	31
Vancomycin and bowel lavage	23
Vancomycin, bowel lavage and faecal transplant	94

In the US this treatment is rapidly gaining ground … An Australian doctor, Professor Tom Borody not only treats *C. difficile* patients with faecal transplants on a regular basis, he believes the same treatment will be effective for patients with inflammatory bowel diseases such as Crohn's disease and ulcerative colitis – both chronic and painful conditions of the colon. Not everyone is convinced – this is an area of science and medicine which is very new – but in Australia a nationwide trial of faecal transplants as a treatment for colitis is about to start.

fig A Scientists and doctors are trying to identify the key microorganisms which out-compete *C. difficile* and culture something that looks and sounds more scientific than faeces to do the job.

Where else will I encounter these themes?

| Book 1 | 5.1 | 5.2 | 6.1 | YOU ARE HERE 6.2 | 6.3 | 7.1 | 7.2 |

This is an extract from an article published on BioNet, a website produced by the Royal Society of Biology to support school and college students who are interested in biology. BioNet articles aim to introduce students to new and exciting aspects of biological research in a way that is engaging and adds depth to topics covered as part of the examination courses.

1. Find elements of this extract that illustrate or extend the content of your Biology A level studies and explain the link.

2. **a.** How does the content of this article differ from the material in a textbook? Give examples of differences in writing styles.

   b. Discuss whether this excerpt succeeds in engaging the reader and suggest how it succeeds or fails.

> If a question asks you to 'discuss', you need to explore all aspects of an issue or argument using well thought-out reasoning.

Now let us examine the biology. You already know about bacteria, their role as pathogens, the use of antibiotics and some of the problems linked with antibiotic-resistant bacteria. This will help you answer the questions below.

3. What does the general information given about the human biome tell you about the role of bacteria in healthy people?

4. Look at the data collected from the investigation into the experimental treatment for *C. difficile*.

   a. Vancomycin is not widely used. Doctors mainly use it only for the most severely ill patients. Why do you think this antibiotic is prescribed so rarely?

   b. Suggest why the treatment involving vancomycin with bowel lavage might have had a lower success rate than simply giving the patients vancomycin.

   c. The trial was halted because it was deemed unethical for it to continue. What does this mean?

   d. From the information given here, what would you suggest are the main limitations of this trial?

5. Find out what you can about ulcerative colitis and Crohn's disease. What features of these diseases suggest they might be treatable by faecal transfusion?

6. Research teams in a number of countries are trying to identify the bacteria in faecal samples that are most effective at recolonising diseased colons with 'good' bacteria and restoring health. They hope to culture them and produce capsules or suppositories that can be used to recolonise guts without needing faecal transplants. Discuss why people might feel this research is necessary.

Activity

Faecal transfusion, also known as a faecal transplant, is at one level a very new idea and its effectiveness and safety are still being investigated. However, there are descriptions of this as an effective therapy going back to fourth century China. Records from the sixteenth and seventeenth century also describe effective faecal transfer therapies.

There is now a growing body of scientific evidence that shows this treatment can be an effective tool against *C. difficile* infections, and may also be effective against other chronic gut disorders.

Choose one of the following activities:

· Put together a timeline on the use of faecal transplants through history to treat gut disorders.

· Produce a poster presentation summarising the evidence for the success of faecal transfusions in treating *C. difficile* infections.

> You can look at online resources including encyclopaedias, scientific magazines and journals, scientific papers, books, and so on. In each case, judge the reliability of your source before you use it. In an academic presentation like this, it is vital to reference all your sources.

Adapted from *The Amazing Human Biome* by Ann Fullick, an article from Society of Biology, http://societyofbiology.org/images/SB/Ann_Fullick-_Amazing_human_microbiome_article.pdf

1 The table below refers to some of the stages involved in Gram staining and the appearance of Gram-negative and Gram-positive bacteria after each stage. Copy and complete the table by writing the most appropriate word or words in the empty boxes.

Stage of Gram staining	Appearance of Gram-negative bacteria	Appearance of Gram-positive bacteria
Cells heat fixed onto slide	Colourless	Colourless
Slide flooded with crystal violet		
Slide flooded with Gram's iodine		
Slide rinsed with alcohol or acetone		
Slide counterstained with safranin/carbol fuchsin		

[5]
[Total: 5]

2 An investigation was carried out to study the effect of pH on the growth of *Escherichia coli* (*E. coli*) and *Lactobacillus bulgaricus* (*L. bulgaricus*).

Two sets of agar plates were prepared using agar at five different pHs (5, 6, 7, 8 and 9).

A stock suspension of *E. coli* was prepared and a 0.1 cm³ sample was spread out over each of the agar plates. A stock suspension of *L. bulgaricus* containing the same number of cells as the suspension of *E. coli* was also prepared and 0.1 cm³ samples were spread over a second set of agar plates.

Both sets of agar plates were incubated for two days and the number of colonies on each plate were counted.

The graph below shows the results of this investigation.

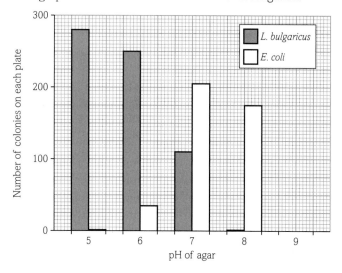

(a) (i) Describe the effect of pH on the growth of *E. coli* and *L. bulgaricus*. [3]
   (ii) Explain the effect of pH on the growth of both types of bacteria. [3]

(b) Give **two** precautions that should have been taken during this investigation to ensure that the results were valid. For each precaution given, explain how the results would have been affected if the precaution had not been taken. [4]

[Total: 10]

3 (a) During the production of yoghurt, the number of cells of the bacterium *Lactobacillus bulgaricus* change. The graph below shows the changes in the $\log_{10}$ numbers of this bacterium and the changes in the pH during incubation.

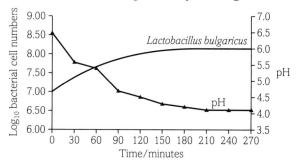

(i) Describe the relationship between the number of cells of *Lactobacillus bulgaricus* and the changes in the pH in the yoghurt. [3]

(ii) Calculate the number of generations of *Lactobacillus bulgaricus* produced during the first 120 minutes, using the formula below.

$$n = \frac{\log_{10}N_1 - \log_{10}N_0}{\log_{10}2}$$

Where n is the number of generations
N_0 = number of bacteria at 0 hours
N_1 = number of bacteria at 120 minutes
$\log_{10}2 = 0.301$
Show your working. [3]

(b) Describe the role of bacteria in the production of yoghurt. [3]

[Total: 9]

4 The graph below shows the number of cases of food poisoning caused by *Salmonella* and *Staphylococcus* in Japan between 1985 and 1998.

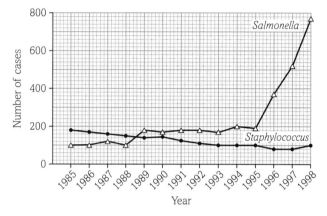

(a) For the years between 1985 and 1998, compare the number of cases of food poisoning caused by *Salmonella* with the number of cases of food poisoning caused by *Staphylococcus*. [3]

(b) Food poisoning can be caused by the endotoxins of *Salmonella* and the exotoxins of *Staphylococcus*. State **two** differences between endotoxins and exotoxins. [2]

[Total: 5]

5 The bacterium *Mycobacterium tuberculosis* causes a disease called tuberculosis (TB). TB can be a disease of the lung (pulmonary TB) or other parts of the body (non-pulmonary TB). In some cases, this can cause death.

(a) Explain how this bacterium causes TB. [2]

(b) The graph below shows the number of new cases of TB and the number of deaths from TB reported each year in Scotland, between 1980 and 2004.

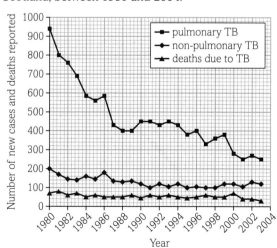

(i) Describe how the number of new cases of pulmonary TB changed between 1980 and 2004. [2]

(ii) Explain why the number of deaths from TB remained fairly constant between 1980 and 2004. [2]

[Total: 6]

6 The graph below shows the changes in population size of bacterial cultures grown in the presence of three antibiotics, A, B and C. In each case the antibiotic was added at 7 hours.

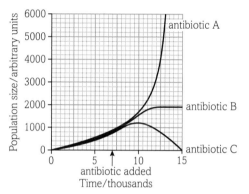

(a) Use examples from the graph to explain the differences between bactericidal and bacteriostatic antibiotics. [3]

(b) A previous investigation on the same bacterium using antibiotic A had produced a curve similar to that for antibiotic B. Explain the change in the response to antibiotic A. [4]

(c) Outline a technique that could demonstrate the effectiveness of antibiotics on bacteria. [4]

[Total: 11]

7 MRSA is a strain of the bacterium *Staphylococcus aureus*. MRSA can survive treatment with several antibiotics. An infection with MRSA is difficult to treat.

(a) Explain how some strains of bacteria have become able to survive treatment with antibiotics. [4]

(b) It is important to use an antibiotic that is effective against specific bacteria.

Outline how you could test the effectiveness of an antibiotic on a specific bacterium in the laboratory. Include aspects of the method that ensure safe working. [5]

[Total: 9]

TOPIC 6
Microbiology and pathogens

CHAPTER **6.2** > # Non-bacterial pathogens

Introduction

In sub-Saharan Africa, sleeping sickness is a disease that affects both humans and cattle. It is caused by protozoan parasites of the genus *Trypanosoma*. It is carried from one host to another by the tsetse fly that acts as a universal vector between animals and humans. Sleeping sickness might sound gentle, but it is anything but. The trypanosomes pass into the blood from the bite of an infected insect and then reproduce rapidly in the tissues under the skin, in the blood and in the lymph. At first, the patient may have few symptoms or may have fevers, headaches, joint pains and itching. In the second stage of the disease, the parasites pass over the blood-brain barrier and infect the central nervous system. This affects behaviour, makes the patient confused, interferes with basic coordination and disturbs the sleep cycle. Without treatment it is fatal.

However, the treatment of human sleeping sickness in Africa is a medical success story. In some areas of Africa it used to kill more people than HIV/AIDS each year. Now, as a result of major control efforts that involved screening populations for the disease and treating anyone affected, the number of cases has dropped since 2000 by over 73%. In 2009 there were fewer than 10 000 cases for the first time. By 2012 this had fallen to 7216 new cases. The World Health Organization, along with several pharmaceutical companies and the Bill and Melinda Gates Foundation, are working together to entirely eradicate human sleeping sickness in Africa in the very near future.

In this chapter you will discover that bacteria are not the only organisms that cause disease. You will learn about the strange structures and life cycles of different types of viruses, and the way they take over the cells of their host organisms to cause disease in both animals and plants. You will learn that many plant diseases are fungal, how these fungi are transmitted from one plant to another and how some of the problems of controlling fungal plant diseases are linked to the ways in which they are spread. Finally, you will look at how protozoa also cause some of the diseases that are a major threat to human and animal health around the world. You will begin to understand some of the social, ethical and economic implications of the different control methods used for endemic malaria, and will investigate the role of the scientific community in validating the different methods.

All the maths you need

*   Translate information between graphical, numerical and algebraic forms (*e.g. considering mortality rates during pandemic disease*)

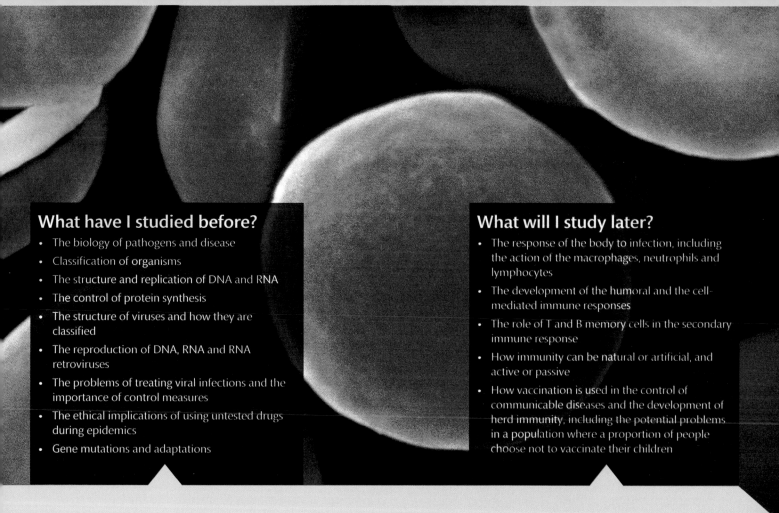

What have I studied before?

- The biology of pathogens and disease
- Classification of organisms
- The structure and replication of DNA and RNA
- The control of protein synthesis
- The structure of viruses and how they are classified
- The reproduction of DNA, RNA and RNA retroviruses
- The problems of treating viral infections and the importance of control measures
- The ethical implications of using untested drugs during epidemics
- Gene mutations and adaptations

What will I study later?

- The response of the body to infection, including the action of the macrophages, neutrophils and lymphocytes
- The development of the humoral and the cell-mediated immune responses
- The role of T and B memory cells in the secondary immune response
- How immunity can be natural or artificial, and active or passive
- How vaccination is used in the control of communicable diseases and the development of herd immunity, including the potential problems in a population where a proportion of people choose not to vaccinate their children

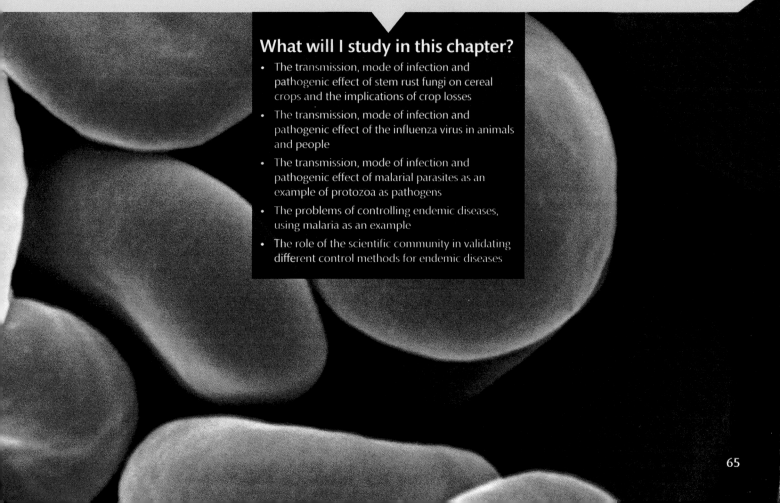

What will I study in this chapter?

- The transmission, mode of infection and pathogenic effect of stem rust fungi on cereal crops and the implications of crop losses
- The transmission, mode of infection and pathogenic effect of the influenza virus in animals and people
- The transmission, mode of infection and pathogenic effect of malarial parasites as an example of protozoa as pathogens
- The problems of controlling endemic diseases, using malaria as an example
- The role of the scientific community in validating different control methods for endemic diseases

By the end of this section, you should be able to...

● explain the transmission, mode of infection and pathogenic effect of the influenza virus

Viruses cause disease in animals and plants, and viruses known as bacteriophages even attack and destroy bacteria. You learned about the life cycles of viruses in **Book 1 Section 2.2.2**. The main way in which viruses cause the symptoms of disease is when they cause the lysis of the host cells. The protein coat of some viruses acts as a toxin, causing disease. When they take over the metabolism of a cell, some viruses cause the cell itself to make toxins.

Viral diseases

Viral infections are often specific to particular tissues. For example, adenoviruses which cause colds affect the tissues of the gas exchange system but do not affect the cells of the brain or the intestine. This specificity is the result of the presence or absence of antigenic markers on the surface of host cells. Each type of cell has its own markers, and each type of virus can only bind to particular antigens on the host cells. The presence or absence of these antigens can affect the vulnerability of whole groups of living organisms to attack by viruses. For example, angiosperms (the true flowering plants) are commonly affected by viruses but the gymnosperms (the conifers and their relatives) are virtually free of viral diseases.

Some viral pathogens are fast and deadly in their effect. Others remain dormant inside the body of their host for many years before causing the symptoms of disease.

Influenza – a viral killer

Influenza, more commonly referred to as flu, is a relatively common respiratory disease that can be a killer. It is highly infectious and has a very short incubation period. The virus mutates frequently so it can overcome the immune response of the body (see **Section 6.3.2**). As a result flu causes huge epidemics and even extensive pandemics. In an epidemic, a higher than average number of people are infected with a disease in a particular area or country. In a pandemic, a higher than average number of people are infected with a disease across a number of countries. Four flu pandemics have been recorded since 1900.

The 1918 pandemic, hitting populations exhausted by the rigours of the First World War, caused more deaths in a few months than had occurred throughout the whole of the war. At times people would develop symptoms in the morning and be dead by the evening. It had been calculated that between 20 and 40% of the world's population was infected and up to 50 million people died of the flu or secondary infections. This flu, unlike many other strains, killed young healthy adults in great numbers.

Years	Number of deaths
1918–19	~50 million
1957–58	~1.5 million
1968–69	~1 million
2009–10	~200 000

table A The deaths from flu in different pandemics.

Transmission

There are three major strains of the influenza virus – influenza A, B and C. Strain A in particular causes severe disease. There are subsets of each strain named after specific antigenic markers, for example H1N1, H3N1, etc.

There are many **modes of transmission** of the flu, largely linked to droplet infection. As well as droplet infection, it can be transmitted by direct contact with animal droppings or with virus-filled mucous from the nose. A third way of spreading is from surfaces which have been contaminated with the virus (formites) and by skin contact with viral-filled mucous.

Flu viruses affect a number of species including birds such as chickens and ducks, and pigs, dogs and cats. Pigs are vulnerable to human, bird and swine flu viruses. Sometimes a mutation occurs that produces a strain of flu that will cross the species barrier and infect humans (**zoonotic infection**). The most dangerous scenario is when a virus develops that not only can be passed from animal to human, but also from human to human.

fig A Pigs are a major source of the flu viruses that cause disease in humans.

Did you know?

1.5 million chickens die

In 1997, in Hong Kong, a few hundred people caught a form of bird flu (H5N1) from chickens. The virus spread straight from birds to humans – usually the viruses go through pigs first. There were about 1.5 million chickens in Hong Kong at the time and they were all slaughtered to prevent the development of a major outbreak of disease. The virus had not mutated to spread easily from one person to another, so once the birds were slaughtered there were no more outbreaks of the flu.

Mode of infection

The **mode of infection** of the flu virus is to infect the ciliated epithelial cells of the respiratory system. The viral RNA reaches the nucleus of the host cell and takes over the biochemistry, producing new virus particles. Eventually the cell lyses, releasing more viruses and dying in the process. The reaction of the body to the lysis of the cells causes many of the symptoms of disease. The death of the ciliated epithelial cells of the trachea and bronchi leaves the airways open to infection, and many of the deaths associated with flu are from severe secondary bacterial infections on top of the original viral infection. The people most likely to die of flu are the very young who have no prior immunity, the elderly who have weakened immune systems and are very vulnerable to secondary infections, and anyone who is prone to asthma or heart disease. The increase in deaths from respiratory diseases is particularly marked during flu epidemics. During the 1989/90 epidemic in England and Wales, there were about 26 000 more deaths from respiratory diseases than would normally have been expected during that time period, and between 80 and 90% of those deaths were in people aged 65 or over, the group most vulnerable to flu.

Each year the flu virus is subtly different as the antigenic structures on the surface of the virus change. The change is usually quite small, so having flu one year leaves you with some immunity against infection for the next. But every so often there is a major change in the surface antigens and this heralds a flu epidemic or pandemic as nobody has any 'almost right' antibodies ready.

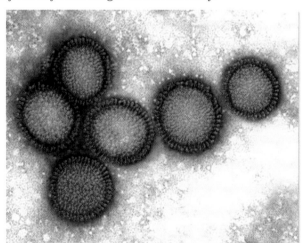

fig B These virus particles are only 80–120 nm in diameter, yet they can kill millions of people in a single outbreak of disease.

Pathogenic effect

The **pathogenic effect** of the flu virus on the body (the symptoms of disease) is quite widespread. It includes fever, often accompanied by shivering and sweating, feeling very unwell and unable to do anything, loss of appetite, aching muscles and painful joints. In addition, many people suffer from severe headaches, sore throats and shortness of breath. If no secondary infection sets in, the disease lasts for about 5–7 days before the fever goes down and convalescence begins. The lungs and the heart can both be severely infected. The exhaustion that follows a bout of flu can last from 6–12 weeks, even in patients without secondary infections.

Treatment and control

There is no cure for this disease. Rest, warmth, plenty of fluids to avoid dehydration and mild painkillers are the best advice. Anti-viral drugs may reduce the level of symptoms and slightly speed up the recovery but there is a lot of debate about the effectiveness of these drugs. If secondary bacterial infections set in, then antibiotics can be used to combat them.

The flu virus is an RNA virus, which makes it more prone to mutation. Vaccines can be used to protect people from becoming infected with the flu virus but, because of its changing nature, the vaccine has to be different each year. Scientists monitor the prevalent flu viruses and a vaccine cocktail is made up of the type A and B strains thought most likely to cause disease in any one year. This vaccine is made available to those in high risk groups.

One of the reasons flu spreads so rapidly is that people tend to carry on working and going about their daily routines as long as possible when they feel ill. This exposes many others to the virus via droplet infection. People would be better to stay at home when they feel ill. Modern transport links and global travel mean flu can be spread around the world very easily.

On the positive side, we also have modern surveillance methods, with scientists around the world constantly monitoring the emergence of new flu strains and sharing information through electronic media. Vaccines are regularly updated ready for mass vaccination programmes and stocks of anti-viral medicines prepared (although they are of limited use). Even with all this in place, a new strain of flu virus with the ability to pass easily from one person to another could cause devastation and death to many people before we could bring it under control.

Questions

1 Why does the viral nature of flu make it particularly dangerous as a disease?

2 Describe the transmission, mode of infection and pathogenic effect of the flu virus.

3 Compare the mortality rates of the four main flu pandemics of the twentieth and twenty-first centuries (see **table A**) and suggest reasons for the differences.

Key definitions

Modes of transmission are the different ways a pathogen is spread from one host to another.

A **zoonotic infection** is an infection in a person caused by a pathogen that can cross the species barrier from other animals.

The **mode of infection** is the way a pathogen causes infection.

The **pathogenic effect** of a microorganism describes the symptoms of disease it causes.

By the end of this section, you should be able to...

● explain the transmission, mode of infection and pathogenic effect of stem rust fungus on cereal crops

Fungi are a kingdom of eukaryotic organisms that have cell walls made of chitin and glucans. Yeasts are single-celled fungi, but many fungi grow as hyphae (long thread-like filaments) and reproduce using spores, usually produced in enormous quantities. Fungi cannot photosynthesise, they are all saprophytes or parasites. Many also live in mutualistic symbiotic relationships, for example the phylum Glomeromycota are mycorrhizal fungi, colonising plant roots and helping them absorb mineral ions from the soil. Relatively few human diseases are caused by fungi, but fungal pathogens have an enormous impact on human existence, because many of the plant diseases that threaten human food security around the world are caused by fungal infections. There are around 10 000 different fungal plant pathogens.

Recently, there have been increasing issues with fungal diseases, which emerge in the same way as healthcare-associated bacterial infections. An example is *Candida albicans*, the causative agent of thrush, which we all carry. It is a big problem for HIV patients or those with compromised immune systems. The occurrence of *C. albicans* can be useful as a disease progression marker in AIDS patients.

fig A Wheat stem rust has a devastating effect on crops around the world.

Puccinia graminis – the stem rust fungus of cereal crops

Wheat is the largest food crop in the world. It is a staple food in over 97% of countries. Barley is the sixth largest food crop, and between them they provide more than 25% of the world's food supply. The success of these two crops means the difference between having enough to eat and malnutrition for millions of people. Wheat and barley crops are threatened by the fungus ***Puccinia graminis***, commonly known as **stem rust fungus**. Over the years this disease has caused devastation to crops many times. The fungus can turn a field of wheat that is almost ready for harvest into a blackened mess and reduce the yield by 50–80%. There are different strains of stem rusts which affect almost all of the different cereal crops including oat, rye and barley, but it is the strain affecting wheat that is the most damaging to the human food supply.

Transmission

Puccinia graminis is a parasitic fungus – it feeds on the living tissues of its host. In fact it has two different hosts – cereal plants such as wheat and barley, and *Berberis* (barberry in the US), a genus of shrubs which grow freely in temperate and sub-tropical regions. It also has a very complex life cycle, and forms five different types of spores.

The disease is transmitted when spores from either infected wheat plants or infected *Berberis* are carried to young crop plants by the wind. They may become infected from parts of infected plants left in the soil after harvest, from *Berberis* bushes growing near the field or from infected crops growing hundreds or thousands of miles away. The pattern of infection helps determine whether the source of infection is a local bush (a fan of infection) or a distant wheat field (broader overall coverage).

Mode of infection

When spores land on the host plant they need water to germinate. Then a thread-like **hypha** emerges from the spore and penetrates one of the stomata of the leaves or stem. This gives the fungus a way into the internal tissues of the plant. The hypha secretes enzymes, such as cellulases, which digest the plant cells and the nutrients are then absorbed into the fungus. The hyphae branch to form a **mycelium** that feeds and grows, hidden in the stem or leaves of the healthy-looking wheat plant. The stem rust fungus grows best when it has hot days (25–30 °C), mild nights (15–20 °C) and wet leaves from either rain, dew or irrigation. The spore needs this water to germinate.

fig B Hundreds of thousands of rust spores are released from a single infected wheat plant, and each spore can infect another plant.

Pathogenic effects

The disease symptoms appear 7–15 days after the plant has become infected. Rusty-red pustules break through the epidermis of the stem or leaf. These contain up to 100 000 rusty-red spores, each of which can be blown by the wind to infect other wheat plants. This stage can be repeated many times during the growth of the wheat crop, until finally black spores, which can survive over the winter, are formed and the crop becomes blackened.

The parasitic fungus affects the crop in several ways:

- It absorbs nutrients from the plant, reducing the yield of grain.
- The pustules break the epidermis, making it more difficult for the plant to control transpiration. This results in less efficient metabolism, makes the plant more likely to dry out and gives entry to other pathogens.
- The mycelium grows into the vascular tissue, absorbing water and nutrients and interfering with the supply to the growing wheat grains.
- It weakens the stems so the plants are more likely to fall over in heavy winds and rain, so they cannot be harvested effectively.

Controlling stem rust in wheat

Stem rust is a disease that can ruin a crop in a very short time. Some modern farming practices actually encourage the spread of the disease. High nitrate levels favour the fungus, so fertilisers encourage rust. Many farmers in more economically developed countries now disturb the soil as little as possible so bits of infected plants are more likely to be at the top of the soil to infect the new crop. They also irrigate regularly, providing the water needed for the wheat to grow, but also for the stem rust spores to germinate.

There are, however, a number of ways in which this devastating disease can be controlled or reduced by good crop management. These include:

- bigger spaces between plants to reduce moisture and increase distance for spore to travel
- reducing the application of fertilisers
- use of earlier-maturing crops which avoid the time of maximum spread
- remove any wild *Berberis* so part of the life cycle is interrupted
- fungicides will control the growth of stem rust, but the cost of this control measure is prohibitive – the crop becomes uneconomic to produce.

Did you know?

Genetic resistance

The main way of controlling wheat stem rust has been the development of genetic resistance. In the mid-twentieth century scientists identified a number of genes that give wheat plants resistance to attack by rust, especially the gene Sr31. They bred strains of wheat containing Sr31 that were very resistant to *P. graminis* and by the 1970s the deadly fungus seemed under control.

Then, in 1999, a new strain of wheat stem rust fungus appeared in Uganda. Known as Ug99, this strain of rust fungus can overcome almost all of the known resistance genes in wheat. Sr31 has no protective effect. The spores have already been carried on the wind to affect other countries in East Africa, Yemen and Iran, and it continues to spread. It poses a new threat to worldwide wheat production as scientists have calculated that 80–90% of the wheat grown around the world is susceptible to this new strain of *P. graminis*.

Scientists are working hard to develop new rust-resistant strains of wheat to try and prevent Ug99 spreading into some of the most important wheat growing areas in the world. In research, they have developed a package of resistance genes that they can engineer into wheat varieties giving them resistance to all rust infections. However, a combination of cost and the environmental and ethical concerns in some countries about the use of genetically modified plants mean that this solution is not yet being widely adopted.

Questions

1 Why are all pathogenic fungi also parasites?

2 Explain how wheat rust fungus affects the yield of a crop.

3 Scientists are concerned that wheat stem rusts will become even more of a problem as the climate changes. Discuss this concern and investigate possible solutions.

Key definitions

Puccinia graminis/**stem rust fungus** is the fungus that causes wheat stem rust.

A **hypha** (plural hyphae) is a thread-like fungal structure that is a single unit of the mycelium.

A **mycelium** is a fungal body made up of a mass of thread-like hyphae.

By the end of this section, you should be able to...

● explain the transmission, mode of infection and pathogenic effect of the malaria parasite

The protozoa are a group of unicellular eukaryotic organisms that come in an enormous range of shapes and sizes and a wide variety of lifestyles and are part of the kingdom Protista. Many live singly but some exist in huge colonies. They feed in almost every available way and they reproduce both sexually and asexually. A number of protozoa live as parasites and these are the organisms that are pathogens of both plants and animals. They are responsible for diseases such as amoebic dysentery, trypanosomiasis (sleeping sickness in animals and humans) and malaria.

The malaria parasite (*Plasmodium* spp.)

Plasmodium **spp.** are parasites of the blood responsible for malaria. Malaria is a disease that affects around 200 million people each year worldwide, and kills up to a million of them. Many of the malarial deaths that occur in Africa are in children under 5 years old. Malaria also has a massive economic impact on the countries where it is present because many millions of adults are weakened by the disease, which affects their ability to work and support their families. The parasite that causes malaria has an extremely complex life cycle involving two separate hosts.

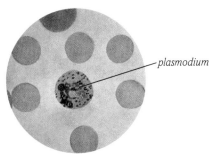

plasmodium

fig A The malaria parasite infects human red blood cells.

Transmission

Plasmodium spp., the malaria parasite, feeds on the living tissues of its host. The parasite is transmitted to a human host by a mosquito **vector** and the life cycle is split between two different hosts, the ***Anopheles* mosquito** and people. Transmission to the human host occurs when an infected mosquito takes a blood meal.

Female *Anopheles* mosquitoes need two blood meals to provide them with plenty of protein for their developing eggs. The female mosquito has piercing mouthparts that go through human skin and her saliva contains an anticoagulant that prevents the blood clotting as it passes up into her body. It is during the blood meal

that transmission occurs in both directions. If a person has malaria, the gametocytes (one stage of the parasite life cycle) pass into the mosquito ready for the next stage of their life cycle. When an infected mosquito pumps saliva into her host before the blood meal, the sporozoites (another stage of the parasite life cycle) pass directly into the human blood stream.

Mode of infection

When the malaria parasite enters the blood of the human host it travels to the liver. It remains in the liver for a time before releasing the next stage of its life cycle into the blood. At this stage the parasite invades the red blood cells and reproduces asexually. At regular intervals of about 48–72 hours, the parasites burst out of the red blood cells, destroying them before moving on to infect more red blood cells. This continues in cycles. Some of the parasites become gametocytes and are taken into a female mosquito during a blood meal, where they complete the life cycle.

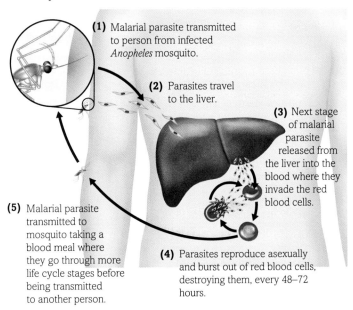

(1) Malarial parasite transmitted to person from infected *Anopheles* mosquito.

(2) Parasites travel to the liver.

(3) Next stage of malarial parasite released from the liver into the blood where they invade the red blood cells.

(5) Malarial parasite transmitted to mosquito taking a blood meal where they go through more life cycle stages before being transmitted to another person.

(4) Parasites reproduce asexually and burst out of red blood cells, destroying them, every 48–72 hours.

fig B A simplified version of the life cycle of the malaria parasite.

Pathogenic effect

When the malaria parasites burst out of the red blood cells, they cause the flu-like symptoms of the disease. The fever, sweating, shaking, muscle pains and headaches that result are linked to the response of the body to the lysis of the blood cells. There is long-term damage to the liver and a steady reduction in the numbers of blood cells and this leads to weakness, severe anaemia and eventually death. People who are already affected by other diseases such as HIV/AIDS are particularly badly affected by the malaria parasite.

The *Plasmodium* parasite's lifestyle is very well adapted to ensure its survival. It can take advantage of two different hosts and can survive in a dormant form for a long time in either without moving to a new host. When the parasite is inside the human body it spends most of the time inside either liver cells or red blood cells, so it is hidden from the immune system which therefore cannot destroy it. Although the parasite eventually kills the human host, this takes a long time, providing many opportunities for it to be passed on again into female mosquitoes.

Endemic disease

Malaria is an **endemic** disease. An endemic disease is one that is constantly present in a particular country or area. For example, influenza and chickenpox are endemic in the UK. Malaria is endemic in many areas of the world where the *Anopheles* mosquito survives and breeds (see **fig C**).

■ areas where malaria transmission occurs
■ no malaria

fig C Malaria is endemic in many areas of the world.

Treating any endemic disease raises particular problems:

- The disease is often widespread, so any eradication programme covers a large area.

- The pathogen is widespread in the environment, so it can be particularly difficult to track down and remove sources of infection.

- It requires the cooperation of large population numbers to eradicate a pathogen or vector, protect a community by vaccination, or deliver and use effective drugs.

- It costs a lot of money to treat and/or control endemic diseases, because they involve many people.

Many of the general issues of treating endemic disease can be recognised in the problems of treating and controlling endemic malaria.

Treatment of malaria

There are very limited treatments for malaria. Quinine, chloroquinine and artemisinin can all kill the parasite and prevent it recurring, but they are effective only if they are given very soon after infection. Anyone travelling to a region with endemic malaria is advised to take anti-malarial medicines so that if they are bitten by an infected mosquito, the parasite is destroyed before it can infect them. Unfortunately, the malarial parasite is increasingly developing resistance to the relatively cheap drugs such as chloroquinine, which used to be very effective. The advice now given by the World Health Organization (WHO), based on scientific evidence, is that malaria should always be treated by combinations of drugs which include artemisinin to avoid the problems of resistance as far as possible.

Prevention and control of malaria

So far there has not been an effective and economic vaccine produced against malaria. The malarial parasite spends most of its time hidden from the immune system inside body cells. The antigens on its surface change frequently, making it more difficult for it to be recognised by the immune system.

These adaptations also make it much more difficult to make a vaccine against the parasite. A vaccine which is effective against all the different stages of the life cycle of the malaria parasite, and all the different antigens it can display, and is also easy to administer and cheap to produce, still seems a long way off. However, in 2015 a new vaccine was introduced that seems to give children effective, although temporary, protection from the disease (see **fig D**).

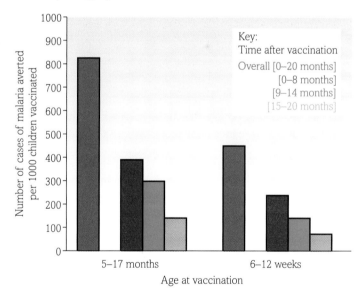

fig D Results from the trial of a new malaria vaccine for children.

In many of the countries affected by malaria most people live a long way from medical help and cannot afford expensive medicines. The simplest and most effective ways of controlling malaria involve controlling the mosquito that transmits the disease.

Avoiding contact with mosquitoes

This can be achieved in a number of ways:

- have mosquito screens on doors and windows
- sleep under mosquito nets – preferably nets impregnated with insecticide
- use insect repellents and insecticides in the home and on people
- wear clothes with long sleeves and legs to cover as much skin as possible.

Simple measures like these are very effective at reducing the rate at which people become infected with malaria.

Preventing mosquitoes breeding

Preventing mosquitoes from breeding halts the spread of malaria. Since the life cycle of the *Anopheles* mosquito takes place mainly in water, targeting these breeding sites can help reduce the number of vectors. Several control measures are employed:

- Mosquitos will lay eggs in any standing water, such as, garden ponds, old tyres, flower pots, old drink cans. If a community can be encouraged to remove possible breeding sites by removing rubbish and disposing of it properly, the incidence of disease can be reduced.
- Proper disposal of sewage – managing human waste so foul water is not left standing for mosquitoes to breed in.
- Biological control – seeding local water supplies with organisms that feed on mosquito larvae can reduce mosquito numbers.
- Chemical control – spraying local water sources with pesticides that kill the eggs and the larvae reduces mosquito numbers and so reduces the spread of disease.

Whichever methods are used to control endemic diseases such as malaria, they raise a number of issues:

- Social implications: people have to be persuaded to change their behaviour to reduce endemic diseases. For example, persuading families to sleep under mosquito nets in affected countries. Social changes are needed to reduce the spread of endemic diseases and they are not easy to bring about.
- Economic implications: the treatment, control and prevention of endemic diseases all involve substantial economic investment. People have to decide whether the cost of a particular course of action is worth the gain. Many of the countries affected by malaria are very poor. They have to decide where their money is best spent and aid agencies have to weigh up interventions such as providing mosquito nets against providing food or antibiotics because money is always finite.
- Ethical implications: treating endemic diseases often raises ethical issues, because the autonomy or choices of a few people may be sacrificed for the good of the many. Some people have reservations about vaccinations. There are always issues of informed consent in medical trials, especially in different countries and cultures. There are always risks involved in any medical treatment, yet it is possible that people may not

take enough notice of possible risks if they are trying to avoid a killer like malaria. The potential benefits and the potential risks must be clearly explained. Actions such as spraying insecticide on waterways to kill malaria mosquitoes will also affect other species of insects, and fish and birds. This needs to be discussed and determined by local people. Money spent on trying to develop an effective vaccine against malaria or a new treatment for certain types of cancer could be spent on educating a generation or developing new food plants. Money spent on saving the giant panda from extinction could be used to buy mosquito screens and nets for children at risk from malaria. Decisions such as these need careful consideration before projects are set up.

These are the types of issues that have to be considered whenever we look at trying to treat or eliminate the big endemic diseases that still plague the world.

The role of the scientific community

There are many different ways of tackling the problem of malaria. The spread of the disease is made worse because it is endemic in many of the poorest areas of the world, where there is little education, health infrastructure, or money and many compounding problems of malnutrition and other endemic diseases.

The scientific community plays an important role in the validation of different approaches to the control of this disease. The number of studies that have been done in the field is not large, and the sample sizes are often relatively small because of the difficulties in carrying out the studies in areas with high levels of malaria. Nevertheless, data on different aspects of malaria are used by the WHO to produce international guidelines for all countries. For example, in 2011 a major review of the data was carried out to produce a paper looking at the costs of different approaches to malarial treatments compared to their effectiveness in reducing incidents of malaria. This is one of many strands of evidence that are taken into account to produce the WHO guidelines.

Did you know?

If you want to find out more about the role of the scientific community in validating methods of control of malaria, this scientific paper is a good starting point:

Michael T. White, Lesong Conteh, Richard Cibulskis and Azra C. Ghani (2011) Costs and cost-effectiveness of malaria control interventions – a systematic review. *Malaria Journal* 2011 10: 337.
http://www.malariajournal.com/content/10/1/337
doi:10.1186/1475-2875-10-337

The international guidelines on aspects of malaria control, based on scientific evidence, include:

- Diagnosis: evidence shows that accurate diagnosis using microscopes to see the protozoa in the blood avoids treating diseases which cause similar symptoms with expensive anti-malarial drugs. This saves money and helps reduce the build-up of resistance in the population. Guidelines are produced to help people use microscopes accurately to make these vital diagnoses.

- Prevention: evidence shows that insecticide-impregnated mosquito nets can reduce malaria cases by up to 50% in areas where the disease is common. Untreated mosquito nets give about half the level of protection. As a result, the use of long-lasting insecticide treated nets (LLINs), which give protection for up to 3 years, is being introduced in as many malarial infected countries as is possible. Reducing the people available for the mosquitoes to feed on reduces the mosquito population too, also benefitting people who do not sleep under mosquito nets. Evidence shows that spraying homes with insecticide significantly reduces the risks of malarial infection, especially combined with the use of LLINs, so guidelines recommend the use of both strategies.

- Treatment: evidence shows increasing resistance of the malarial parasite to quinine and chloroquinine. However, when combined with artemisinin, these provide a very effective treatment and destroy the parasites in the body. This makes the treatment much more expensive, but it works. So the guidelines recommend, in line with the scientific evidence, that only cases of malaria which are diagnosed using microscopic examination of blood smears are treated with anti-malarial drugs, and that those cases are all treated with different combinations of anti-malarial drugs to overcome any problems of resistance.

All of the efforts of the scientific community in validating methods of diagnosis, control and treatment of malaria are, however, only of value if they are adopted by the governments and people of the countries where malaria is endemic. Persuading the people in affected regions to follow the WHO guidelines based on scientific evidence is key to removing the threat of this endemic but preventable and treatable disease.

Questions

1 Produce a flowchart of the life cycle of the malaria parasite.

2 How does the mode of infection and pathogenic effect of *Plasmodium* spp. compare with the mode of infection and pathogenic effect of the influenza virus?

3 Explain why endemic diseases are particularly difficult to control.

4 Give three examples of how malaria can be controlled and discuss the social, economic and ethical implications for each of them.

5 Investigate the scientific evidence for the effectiveness and the cost/benefit of the three methods you have chosen in your answer to question 4.

Key definitions

Plasmodium **spp.** are the parasitic protozoa that cause malaria. They have a life cycle split between two different hosts, female *Anopheles* mosquitoes and people.

Vectors are living organisms or environmental factors that transmit infection from one host to another.

Anopheles **mosquitoes** are the type of mosquitoes that carry the malaria parasite from one host to another.

An **endemic** disease is one in which the active disease, or the pathogen that causes the disease, is constantly present in a particular country or area.

[Note: In questions marked with an asterisk (*), marks will be awarded for your ability to structure your answer logically showing how the points that you make are related or follow on from each other.]

1   Which weather condition would quicken the spread and infection of *Puccinia graminis*?
    A   windy and wet
    B   no wind and dry
    C   windy and dry
    D   no wind and wet   [1]
                                                          [Total: 1]

2   One method of combating the influenza virus is by vaccination. Read the following statements and decide how many of them are true.
    (1) The virus cannot mutate.
    (2) The virus is not affected by antibiotics.
    (3) Vaccination must happen every year.
    (4) The virus has antigens.
    A   1
    B   2
    C   3
    D   4   [1]
                                                          [Total: 1]

3   This diagram shows the life cycle of the stem rust fungus *Puccinia graminis*. Straw is another name for the stem of a wheat plant, and *Berberis* is called Barberry in North America. Scientists can develop resistant breeds of wheat, but new strains of *Puccinia*, like Ug99, can arise and infect susceptible plants.

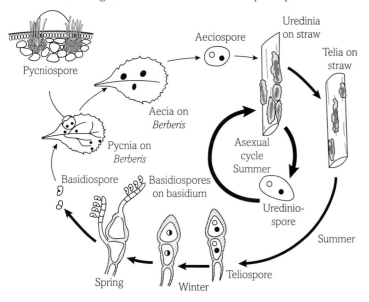

(a) State which **two** spores are able to infect wheat stems.   [2]
(b) State how a named spore is transmitted from *Berberis* to wheat.   [1]
(c) Scientists studying *Puccinia* noticed that urediniospores had two haploid nuclei, teliospores had one diploid nucleus, and basidiospores had one haploid nucleus.
     (i) Describe what happened to the urediniospore nuclei to form the teliospore nuclei.   [1]
    *(ii) Explain how the teliospore nuclei gave rise to the basidiospore nuclei.   [5]
(d) The *Puccinia* life cycle has a sexual stage in *Berberis*. The picture shows the aecia on the surface of *Berberis* leaves.

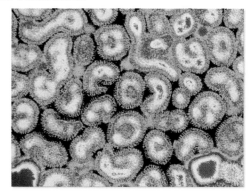

Countries such as the USA have banned the import of foreign *Berberis vulgaris* and have largely eradicated the native *B. canadensis*. Discuss how the removal of *Berberis* plants would affect the number of new strains of *Puccinia*.   [2]
                                                         [Total: 11]

4   Influenza is a disease caused by a virus. The virus's genome is made of 14 000 bases.

(a) State **three** ways the influenza virus can be transmitted between humans.   [1]
(b) State what type of nucleic acid contains the genome of the influenza virus particle as it infects a cell.   [1]

(c) This nucleic acid is converted into mRNA inside the cell's nucleus. This process causes an error rate of 1 base in 10 000. Calculate the average number of mutations that might be expected in a virus particle.
Give your answer to 1 decimal place. [2]

(d) Some of the proteins coded for by these genes form the antigenic proteins that the immune system recognises. Explain how a mutation in the genome could lead to the immune system no longer recognising the virus as 'foreign'. [4]

(e) (i) The influenza virus infects ciliated epithelial cells of the respiratory system. Explain why this might lead to a secondary infection by bacteria. [4]
(ii) Comment on how a mutation in the genome would lead to a **loss** of virulence of the virus. [2]

(f) Explain why antibiotics are ineffective against influenza virus. [2]

[Total: 16]

(b) Give **two** reasons why it is difficult for the human immune system to destroy the *Plasmodium* organisms. [2]

(c) (i) Explain how the gametocyte is transferred to the gut of the mosquito. [1]
(ii) Give **three** methods other than the use of insecticides that could be used to prevent the transmission of sporozoites to humans. [3]

(d) Malarone is the brand name of an antimalarial drug. It contains two drugs, one of which, atovaquone, inhibits cytochrome c, part of the electron transport chain of *Plasmodium*. This drugs needs to be taken 1–2 days before entering a malaria area and 7 days after leaving it.
(i) Explain how this drug stops an infection of *Plasmodium*. [2]
(ii) Give a reason why some people might have an allergic reaction to this drug. [1]

[Total: 10]

5 The *Plasmodium* parasite is a protozoan. Its life cycle is shown below.

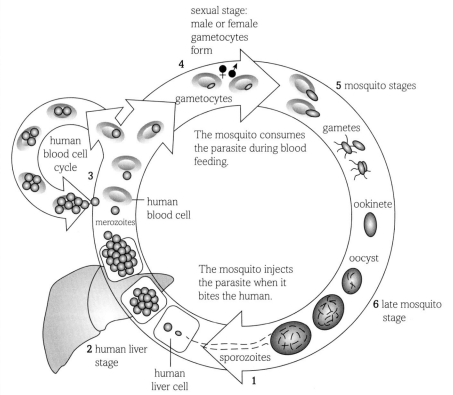

(a) Use information in the diagram to explain why a human will suffer the ill effects of *Plasmodium* but the mosquito will not. [1]

TOPIC 6
Microbiology and pathogens

6.3 > The response to infection

Introduction

In the 1980s, a new disease was observed – the disease we now know as HIV/AIDS. It spread like wildfire across the globe and still affects millions of lives. The human immunodeficiency virus (HIV) causes acquired immune deficiency syndrome (AIDS). Although we can now keep the worst of the disease at bay with a mixture of drugs, including anti-retrovirals, we still have no effective vaccine or cure for the disease.

Why is HIV so devastating? It attacks the immune system itself. The virus hides itself within key cells involved in the immune response. This has two major effects. Firstly, it reduces the normal immune response of the body to other invading pathogens. Secondly, because it is hidden within the immune cells, the virus cannot be detected and destroyed by the immune system. It remains safe inside the body to multiply and spread.

In this chapter you will discover how the body responds to invading pathogens. You will consider the non-specific responses to infection, including the macrophages, neutrophils and lymphocytes. You will also look at the complexities of the specific immune system and the elegant way in which the humoral and cell-mediated immune responses work together. You will investigate the roles of the T and B memory cells in the immune response and use this insight to help you understand how we can induce artificial immunity to protect ourselves against many potentially deadly diseases.

All the maths you need

- Translate information between graphical, numerical and algebraic forms (*e.g. considering mortality rates during pandemic disease*)

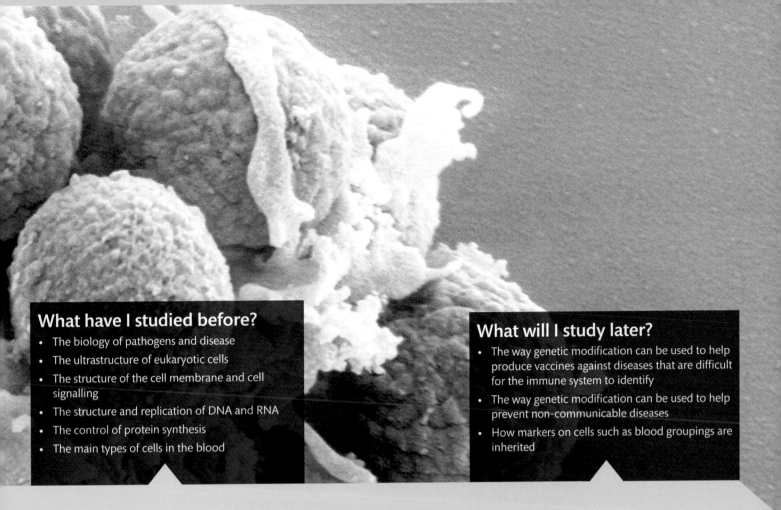

What have I studied before?

- The biology of pathogens and disease
- The ultrastructure of eukaryotic cells
- The structure of the cell membrane and cell signalling
- The structure and replication of DNA and RNA
- The control of protein synthesis
- The main types of cells in the blood

What will I study later?

- The way genetic modification can be used to help produce vaccines against diseases that are difficult for the immune system to identify
- The way genetic modification can be used to help prevent non-communicable diseases
- How markers on cells such as blood groupings are inherited

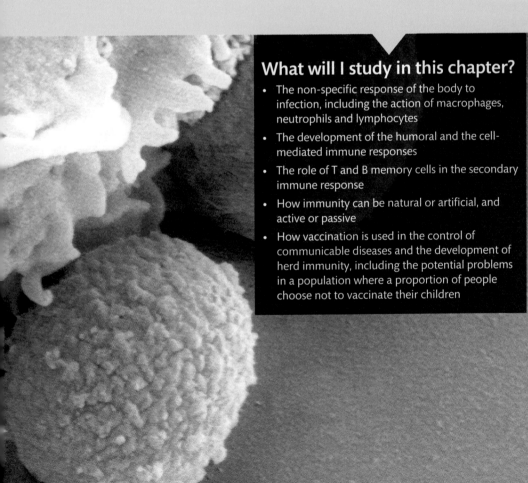

What will I study in this chapter?

- The non-specific response of the body to infection, including the action of macrophages, neutrophils and lymphocytes
- The development of the humoral and the cell-mediated immune responses
- The role of T and B memory cells in the secondary immune response
- How immunity can be natural or artificial, and active or passive
- How vaccination is used in the control of communicable diseases and the development of herd immunity, including the potential problems in a population where a proportion of people choose not to vaccinate their children

By the end of this section, you should be able to...

● describe the mode of action of macrophages and neutrophils

Even though the body has many barriers, millions of bacteria and viruses make their way in every day. But you do not constantly suffer from disease. This is because in most cases your body recognises that it has been invaded and destroys or inactivates the pathogens. This response is the result of a number of different defence systems in your body, many depending on a process of cell recognition.

Cell recognition

The ability of the body to distinguish between its own cells ('self') and foreign cells or organisms ('non-self') is key to the working of the immune system. The cells of different organisms have differing genetically-determined protein molecules on their surface membrane that seem to be key to cell recognition. These proteins include glycoproteins, which are protein molecules with a carbohydrate component (see **Book 1 Section 1.2.4**). These chains of sugar molecules can be varied, and are important in cell recognition in several ways.

The ability of the body to identify 'self' is vital when tissues and organs are forming in embryonic development. Similar sugar recognition sites may bind to each other, holding cells together. Equally important is the ability to identify pathogens and any other foreign cells that get into the body as 'non-self'. Non-self glycoproteins act as **antigens** and are recognised by white blood cells (leucocytes) during the specific immune responses. An antigen is any substance that stimulates an immune response in the body. Antigens are often chemicals on the surface of a cell such as proteins, glycoproteins or carbohydrates described here but they can also be toxins made by bacteria, or sometimes whole microorganisms such as bacteria or viruses.

Once pathogens get inside your body tissues they will be met by a number of different defences. Some of these responses are non-specific – they are triggered by any pathogen. Others are specific to particular pathogens (see **Section 6.3.2**).

Non-specific responses

Non-specific responses to infection are triggered by body cells breaking down and releasing chemicals, and by pathogens that have been labelled by the specific immune system. Many of the non-specific responses depend on the different types of leucocytes. For a reminder of these different types (see **Book 1 Section 4.3.2**).

Inflammation

fig A A skin rash – an example of inflammation of the skin.

Inflammation is a common, non-specific way in which our bodies respond to infection. It generally occurs when an infection is relatively localised, for example when you cut yourself and bacteria get into the wound. The inflammatory response involves a number of stages. Special cells called mast cells are found in the connective tissue below the skin and around blood vessels. When this tissue is damaged these mast cells and basophils (a type of leucocyte) release chemicals known as **histamines**. Histamines cause the blood vessels in the area, particularly the arterioles, to dilate, causing local heat and redness. The locally raised temperature reduces the effectiveness of pathogen reproduction in the area. The histamines also make the walls of the capillaries leaky as the cells forming the walls separate slightly.

As a result, fluid (plasma), containing leucocytes (mainly neutrophils) and **antibodies**, is forced out of the capillaries causing swelling and often pain. The antibodies disable the pathogens and the macrophages (more leucocytes) and neutrophils destroy them by phagocytosis. A fairly common symptom of a more widespread infection is a rash, which is a form of inflammation or tissue damage that particularly affects the skin, causing red spots or patches.

Fevers

One early, common, non-specific response to infection is a fever. Normal body temperature is maintained by the hypothalamus and follows a regular circadian (roughly 24 hour) rhythm, lowest in the early hours of the morning and highest at about 10 pm. When a pathogen infects the body it causes the hypothalamus to reset to a higher body temperature, so that we become aware of 'running a temperature' – in other words, we have a fever. This might seem a strange response, but a raised temperature seems to help the body combat infection in two ways:

- Many pathogens reproduce most quickly at 37 °C or lower. Therefore, a raised temperature will reduce the ability of many pathogens to reproduce effectively and so they will cause less damage.
- The specific response system (immune system) works better at higher temperatures and so will be more successful at combating the infection if the temperature is raised.

In a bacterial infection the temperature rises steadily and remains fairly high until treatment is successful or the body overcomes the infection. In viral infections the temperature tends to 'spike', shooting up high every time viruses burst out of the cells and then dropping down towards normal again. Although fevers can often be beneficial, if they get too high they can be damaging and even fatal. If your body temperature rises above 40 °C, the denaturation of some enzymes takes place and you may suffer permanent tissue damage. If the temperature is not lowered fairly quickly, death may result. Sweating is often associated with fever as the body sweats in response to the high temperature as a coding mechanism (see **Section 9.3.4**). If the fluid and electrolytes lost in the sweat are not replaced, dangerous dehydration and even death can result.

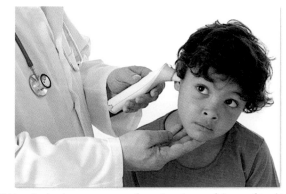

fig B Taking the body temperature of a patient is a useful way of tracking the course of many infections.

Phagocytosis

Phagocytosis is another non-specific response involving leucocytes, often seen in association with inflammation (see **Book 1 Section 4.1.4**). There are two main groups of leucocytes, the granulocytes, which have granules that can be stained in their cytoplasm, and the agranulocytes which have no granules. Phagocyte is a general term used to describe white blood cells that engulf and digest pathogens and any other foreign material in the blood and tissues. There are two main types of phagocytes:

- Neutrophils are granulocytes and make up 70% of the leucocytes in the blood. Each neutrophil can only ingest a few pathogens before it dies. They cannot renew their lysosomes so once the enzymes are used up the cell cannot break down any more pathogens.
- Macrophages are derived from monocytes, which are agranulocytes. They make up about 4% of the leucocytes in the blood. However, the monocytes migrate to the tissues and become macrophages, so there are large numbers of macrophages in the tissues. Macrophages have an enormous capacity for ingesting pathogens because, unlike neutrophils, they can renew their lysosomes so they last much longer. They accumulate at the site of an infection to attack the invading pathogens. You will learn more about the actions of the macrophages in **Section 6.3.2**.

The phagocytes engulf the pathogen. The pathogen is enclosed in a vesicle (membrane 'bag') called a **phagosome**. The phagosome then fuses with a lysosome. The enzymes in the lysosome break down the pathogen. The phagocytes can sometimes be seen as pus – a build-up of dead cells which are mainly neutrophils. The pus may ooze out of a wound or spot, or it may be reabsorbed by the body. The whitish/yellow colour of pus comes from the dead neutrophils.

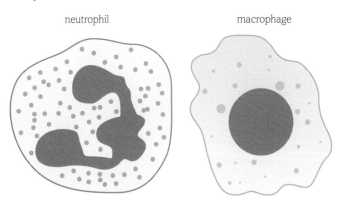

neutrophil   macrophage

fig C Neutrophils and macrophages are important in the non-specific responses of the body to invasion by pathogens. Macrophages also play a vital role in the specific immune responses.

When phagocytes have engulfed a pathogen they produce chemicals called **cytokines** in the surrounding tissues. Cytokines are effective cell signalling molecules that stimulate other phagocytes to move to the site of an infection. They also raise the body temperature and stimulate the specific immune response.

Some chemicals bind to pathogens making them more easily recognised by phagocytes. These chemicals are known as **opsonins**. Phagocytes have receptors on their cell membranes that bind to some of the most common opsonins, and the phagocyte then goes on to engulf the entire pathogen. The opsonins that have the strongest effect are antibodies such as immunoglobulin G (IgG) and immunoglobulin M (IgM), although other chemicals can also have a similar effect. Antibodies are glycoproteins that are produced in response to a specific antigen and are discussed in detail in **Section 6.3.2**.

Questions

1 The enzyme lysozyme is almost universally present in animals. Explain its role in the non-specific defence against disease.

2 Explain how inflammation protects the body against disease.

3 Inflammation and phagocytosis are used against pathogens that have invaded the body. Explain why they are referred to as part of the non-specific defences of the body.

4 If someone is ill with a fever, people often try to get their temperature down. Explain:

   (a) why this is not necessarily a good idea

   (b) why it can be very important to lower the temperature.

Key definitions

Antigens are glycoproteins, proteins or carbohydrates on the surface of cells, toxins produced by bacterial and fungal pathogens, and some whole viruses and bacteria that are recognised by white blood cells during the specific immune responses to infection. They stimulate the production of an antibody.

Inflammation is a common, non-specific response to infection involving the release of histamines from mast cells and basophils, causing the blood vessels to dilate, giving local heat, redness and swelling.

Histamines are chemicals released by the tissues in response to an allergic reaction.

An **antibody** is a glycoprotein that is produced in response to a specific antigen.

A **phagosome** is the vesicle in which a pathogen is enclosed in a phagocyte.

Cytokines are cell signalling molecules with several roles in the immune system, including stimulating other phagocytes to move to the infection site.

Opsonins are chemicals which bind to pathogens and label them so they are more easily recognised by phagocytes.

By the end of this section, you should be able to...

- describe the mode of action of lymphocytes
- describe the development of the humoral immune response
- describe the development of the cell-mediated immune response
- describe the role of the T and B memory cells in the secondary immune response

The **immune response** is the specific response of the body to invasion by pathogens. It enables the body to recognise anything that is 'non-self' and remove it from the body as efficiently as possible. Each organism carries its own unique set of markers, or antigens, on the cell surface membrane. Some are common to every member of a particular species, others to a particular individual. The more closely related two individuals are, the more antigens they have in common. Only genetically identical twins and clones have totally matching antigens.

The immune system of the body has four key characteristics:

- It can distinguish 'self' from 'non-self'.
- It is specific – it responds to specific foreign cells.
- It is diverse – it can recognise an estimated 10 million different antigens.
- It has immunological memory – once you have met and responded to a pathogen, you can respond rapidly if you meet it again.

Lymphocytes and macrophages are the two main types of white blood cells involved in the specific immune system. Lymphocytes are made in the white bone marrow of the long bones. They move around the body in the blood and lymph and are involved in recognising and responding to foreign antigens. They leave the bloodstream and move freely through the tissues.

Different kinds of lymphocytes

There are two main types of lymphocytes:

B cells

B cells are produced in the bone marrow. Once mature, they are found both in the lymph glands and free in the body. B cells have membrane-bound globular receptor proteins on their cell surface membrane that are identical to the antibodies they will later produce. All antibodies are known as **immunoglobulins**. Around 100 million B cells are formed as an embryo grows, each with a different membrane-bound antibody. Each then divides to form a clone of cells, giving a baby an immune system with the potential to recognise and tackle an enormous range of pathogens.

When a B cell binds to an antigen, the following types of B cell are produced:

- **B effector cells** – these cells divide to form the plasma cell clones.
- **Plasma cells** – these produce antibodies to particular antigens at a rate of around 2000 antibodies per second.
- **B memory cells** – these provide the immunological memory to a specific antigen, allowing the body to respond very rapidly if you encounter a pathogen carrying the same antigen again.

T cells

T cells are produced in the bone marrow but mature and become active in the thymus gland. The surface of each T cell displays thousands of identical T-cell receptors. T-cell receptors bind to antigens on infected body cells and then several further types of T cells are produced and play different roles in the immune response:

- **T killer cells** produce chemicals to destroy infected body cells.
- **T helper cells** activate the plasma cells to produce antibodies against the antigens on a particular pathogen and also secrete opsonins to 'label' the pathogen for phagocytosis by other white blood cells.
- **T memory cells** are very long-lived cells that make up part of the immunological memory. When they meet a pathogen for the second time, they divide rapidly. This forms a large clone of T killer cells which then quickly destroy the pathogen.

The working of many of these cells depends on special proteins known as **major histocompatibility complex (MHC) proteins**. These proteins display antigens on the cell surface membrane, as you will see later. The immune response to infection is extremely complex so we will look at it one stage at a time.

The humoral response

The humoral response of the immune system reacts to antigens found outside the body cells including antigens on pathogens such as bacteria and fungi, and to antigen-presenting body cells. The humoral response results in the production of antibodies, which are not attached to cells but are carried around the body in the blood and tissue fluid. Although B cells produce the antibodies, T cells are first involved in activating the B cells. The humoral response consists of two main stages: the T helper activation stage and the effector stage.

T helper activation

When a pathogen enters the body, chemicals are produced that attract the phagocytes, including macrophages and neutrophils. When a neutrophil engulfs a bacterium it destroys it in about 10 minutes. Macrophages take longer but do more, preparing the way for the specific immune system. The macrophage separates off the antigens from the digested pathogen and combines them with the major histocompatibility complex. The complexes move to the surface of the macrophage cell outer membrane. The macrophage with these antigen/MHC protein complexes displayed on the cell surface is now known as an **antigen-presenting cell (APC)**.

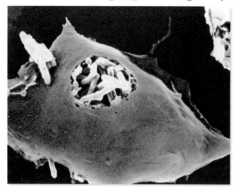

fig A Macrophages play an important part in the immune system by engulfing bacteria (small rod shapes).

The next step involves T cells, which have receptors on the outer membrane that bind to the specific antigen of the antigen/MHC complex on the APC. The binding of the T cell with the APC triggers the T cell to reproduce and form a clone of cells.

Most of these cloned cells become active T helper cells, which are then used in the rest of the immune system. The remainder of the cloned cells form inactive T memory cells, which remain in the body and rapidly become active if the same antigen is encountered again.

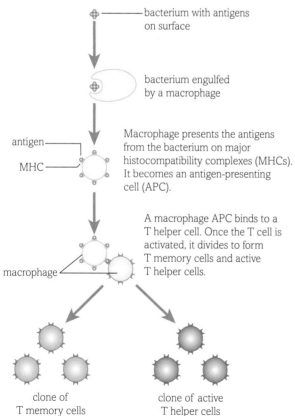

fig B T helper activation stage of the humoral response. The production of the antigen-presenting cell is an important step in the immune system of the body.

The effector stage

Some of the millions of different B cells will have immunoglobulins on their surfaces that are specific for the antigen presented by the pathogen and will bind to it. The B cell then engulfs the whole pathogen by endocytosis. As with the macrophage, the vesicle formed fuses with a lysosome and enzymes break down the pathogen to leave fragments of processed antigen. These fragments become attached to MHC proteins within the cell, and the MHC/antigen complex is transported to the cell surface membrane where the antigen is displayed.

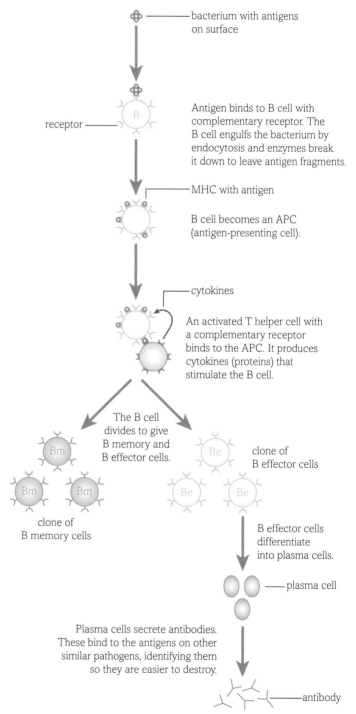

fig C Clonal selection – this is the process that results in millions of antibody molecules, which bind to pathogens and facilitate their destruction.

A T helper cell from the active clone produced in the T helper activation stage recognises the specific antigen displayed on the MHC complex on the B cell and binds to it. This triggers the release of cytokines from the T helper cell which stimulate the B cells to divide and form clones of identical cells. This is known as **clonal selection**. New clones of B effector and B memory cells are produced. The B effector cells differentiate to form plasma cells (**plasma cell clones**).

The plasma cells produce large amounts of antibodies that are identical to the immunoglobulin of the original parent B cell. An antibody is a special glycoprotein that is released into the circulation. It will bind to a specific antigen on the particular pathogen that has triggered the immune system, causing its destruction in one of several ways.

- **Agglutination**: When antibodies bind to the antigens on pathogens, the microorganisms agglutinate or clump together. This helps to prevent them spreading through the body and also makes it easier for them to be engulfed by phagocytes.

- **Opsonisation**: The antibody acts as an opsonin, a chemical which makes an antigen or pathogen more easily recognised by phagocytes.

- **Neutralisation**: Antibodies neutralise the effects of bacterial toxins by binding to them.

The ability of most pathogens to invade the host cells is dramatically reduced when they are combined with antibodies.

Plasma cells live for only a few days, but as they can produce up to 2000 antibody molecules per second this is long enough to be effective. They have extensive endoplasmic reticulum and many ribosomes, which are adaptations for their role in producing large quantities of protein antibodies. The antibodies remain in the blood for varying lengths of time and the memory cells may stay in the blood for years or even life.

The cell-mediated response

Sometimes, particularly in viral infections, the pathogen is inside the host cells and the humoral response is not very effective against it. This is when the cell-mediated response is important. The cell-mediated response involves T killer cells, which respond to specific antigens. T lymphocytes respond to cells that have been changed in some way, for example, to cells infected by a virus, to antigen-presenting cells, to cells changed by mutation to form cancer cells, or to the cells of a transplanted organ. When a body cell is infected with a bacterium or virus, the pathogen is digested and the surface antigens become bound to an MHC in a similar process to that seen in macrophages. As a result, the body cell effectively becomes an APC – but it is important to remember it is still infected by the pathogen.

T killer cells present in the blood have a wide range of complementary receptor proteins on the surface of their cell surface membrane. T killer cells bind to the matching antigen/MHC complex on the surface of the body cell. If the T cells are then exposed to cytokines from an active T helper cell, produced through T helper cell activation, they undergo a rapid series of cell divisions to produce a **clone** of identical active T killer cells

which can all bind to infected body cells. The T killer cells release enzymes that make pores form in the membrane of the infected cells. This allows the free entry of water and ions, so the cells swell and burst. Any pathogens that are released intact are labelled with antibodies produced by the plasma cells, and then destroyed. T killer memory cells are also produced.

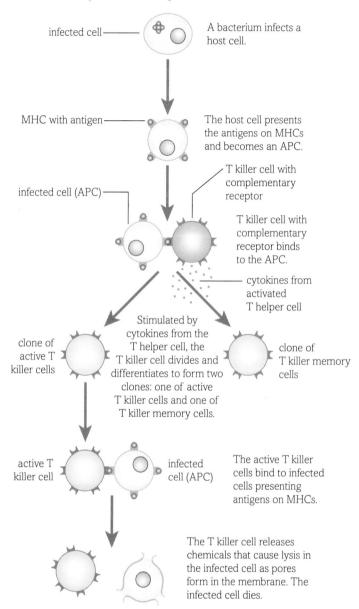

fig D The cell-mediated response.

Primary and secondary immune response

So why do we get ill? The primary immune response involves the production of antibodies by the plasma cells produced from the B effector cells and the activation of T killer cells, and it is extremely effective. However, it can take days or even weeks for the primary immune response to become fully active against a particular pathogen. This is why we get the symptoms of disease – we feel ill when pathogens are reproducing freely inside our bodies before the immune system has become fully operational against the pathogen concerned.

However, we have a secondary immune response which is quicker, greater and longer lasting. When the B-cell APC divides, it also produces B memory cells. Unlike the plasma cells, B memory cells are very long-lived. They are important in allowing the body to respond very rapidly to a second invasion by the same antigen. When you have had a disease once, you usually do not catch it again. This is not because you never come into contact with the disease-causing antigen again. Instead, when you do encounter it, the B memory cells help you produce the antibodies against it so rapidly that it is destroyed before the symptoms of the disease develop.

In addition, at the same time as a clone of active T killer cells are formed, some cloned T cells become T memory cells. These persist in the blood so that the body can produce a rapid response if the same pathogen invades again. These memory cells release a flood of active T killer cells to engulf and destroy infected cells. As yet no one is entirely sure exactly how memory cells provide immunological memory, but it is very effective.

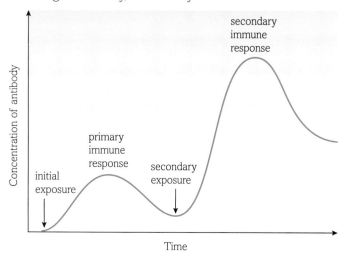

fig E The primary and secondary immune responses.

Questions

1 Produce a large diagram to explain in full how the immune system works.

2 Explain why the immune system does not attack the other cells of the body.

3 Explain the role of antibodies in the immune system.

4 Distinguish carefully between the roles of the B cells and the T cells in the immune response of the body.

5 Discuss the difference between the primary and secondary immune responses and explain the importance of the secondary immune response.

Key definitions

Immune response is the specific response of the body to invasion by pathogens.

Lymphocytes are granulocytes, made in the white bone marrow of the long bones, that make up the main cellular components of the immune system.

B cells are lymphocytes made in the bone marrow which are found both in the lymph glands and free in the body once they are mature.

Immunoglobulins are antibodies.

B effector cells divide to form the plasma cell clones.

Plasma cells produce antibodies to particular antigens at a rate of around 2000 antibodies per second.

B memory cells provide the immunological memory to a specific antigen, allowing the body to respond very rapidly if you encounter a pathogen carrying the same antigen again.

T cells are lymphocytes made in the bone marrow that mature and become active in the thymus gland.

T killer cells are lymphocytes that produce chemicals that destroy pathogens.

T helper cells are lymphocytes involved in the process that produces antibodies against the antigens on a particular pathogen.

T memory cells are very long-lived cells which make up part of the immunological memory.

Major histocompatibility complex (MHC) proteins are proteins that display antigens on the cell surface membrane.

Antigen-presenting cell (APC) is a cell displaying an antigen/MHC protein complex.

Clonal selection is the selection of the cells that carry the right antibody for a specific antigen.

Plasma cell clones are clones of identical cells that all produce the same antibody.

Agglutination is the clumping of cells caused when antibodies bind to the antigens on pathogens.

Opsonisation is a process that makes a pathogen more easily recognised, engulfed and digested by phagocytes.

Neutralisation is the action of antibodies in neutralising the effects of bacterial toxins on cells by binding to them.

A **clone** is a group of identical cells all produced from one cell.

By the end of this section, you should be able to...

- explain how immunity can be natural or artificial, and active or passive

- explain how vaccination can be used in the control of disease and the development of herd immunity

- explain the potential issues in populations where a proportion choose not to vaccinate

Once you have been exposed to a pathogen and your immune system has dealt with it, you are unlikely to suffer illness as a result of infection by that microorganism again. This is because your immune system has a 'memory', based on B cells, T cells and immunoglobulins (antibodies) (see **Section 6.3.2**).

Different types of immunity

Normally, when the body comes into contact with a foreign antigen, the immune system is activated, antibodies are formed and the pathogen is destroyed. This is known as **natural active immunity**, because your body actively makes the antibodies. During pregnancy in mammals, preformed antibodies are passed from the mother to the fetus through the placenta. The newborn baby gets additional protection from antibodies taken in through breast milk, particularly the milk made in the first few days after birth known as colostrum. This provides the baby with temporary immunity until its own system becomes active. This is **natural passive immunity** – passive because your body does not make the antibodies. It tends to be quite short-lived because the antibodies are not replaced.

fig A Young mammals are protected from a wide variety of diseases by immunoglobulins that they receive across the placenta before birth and through their mother's milk after birth.

Inducing immunity

An alternative approach to using drugs to treat an infection is to prevent it happening by using **immunisation**. Immunisation is the process of protecting people from infection by giving them passive or active artificial immunity. It enables you to develop immunity by exposing your immune system to antigens in a safe way that does not put you at risk of developing the disease, so that antibodies and memory cells can be prepared for meeting the real thing. **Vaccination** is the procedure by which you immunise people to produce immunity.

The aim of immunisation is to protect individuals against diseases that might kill or harm them. However, it has a wider role in society – to eradicate, eliminate or control diseases that cause large numbers of deaths, disabilities or illnesses within a population and which therefore place a strain on the structure of that society.

Artificial passive immunity occurs when antibodies formed in one individual are extracted and injected into another individual. The antibodies may be from a person who is already immune to a disease, or from a completely different species such as horses – they will still be effective. This does not confer long-term immunity, because the antibodies are gradually broken down and not replaced, but it can be enormously valuable if someone has been exposed to a rapidly acting antigen such as tetanus. Tetanus, also known as lockjaw, results from a toxin produced by the bacterium *Clostridium tetani*. The toxin affects striated muscles, causing them to go into spasm (or tetanus), making swallowing and breathing impossible and so causing death. If someone may have been exposed to tetanus, for example, from a bad cut while gardening or working with horses, they are injected with antibodies against tetanus. This prevents the development of the disease and death, but will not give prolonged immunity.

Artificial active immunity is the basis of most immunisation programmes. Here, small amounts of antigen (known as the vaccine) are used to produce immunity in a person. The pathogen is made non-infective without reducing its ability to act as an antigen. This can be done in a number of ways. If it is a toxin that causes the symptoms, a detoxified form with one or more chemical groups changed will be injected. Sometimes inactivated viruses or dead bacteria are used as vaccines and in other cases **attenuated pathogens** (viable but modified so they cannot produce disease) are used. Increasingly, fragments of the outer coats of viruses and bacteria, or even DNA segments, are used as vaccines. Your immune system will produce antibodies against the antigen, and appropriate memory cells will be formed without you becoming ill. If you subsequently come into contact with the active antigen, it will be destroyed rapidly without you experiencing the symptoms of the disease it causes (see **fig B**).

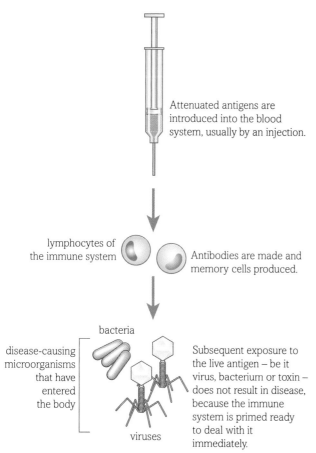

Attenuated antigens are introduced into the blood system, usually by an injection.

lymphocytes of the immune system

Antibodies are made and memory cells produced.

bacteria

disease-causing microorganisms that have entered the body

viruses

Subsequent exposure to the live antigen – be it virus, bacterium or toxin – does not result in disease, because the immune system is primed ready to deal with it immediately.

fig B The process of vaccination manipulates the immune system so that an individual is immune to a disease before they become infected.

Eradicating disease

For centuries, smallpox was feared. It killed millions and disfigured many more around the world. Now it has gone. Smallpox is the only disease that has been completely eradicated by a vaccination programme, although polio is close to being eradicated as well. Smallpox is no longer found either in people, animals or anywhere in the environment. The virus only exists sealed in two top-security labs.

Smallpox was very recognisable and had no non-human hosts, a long incubation period and a visible scar as evidence of immunisation. These features all made eradication possible.

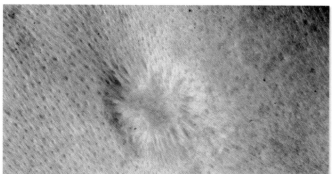

fig C Smallpox vaccination leaves a distinctive scar like this one. Since smallpox was officially eradicated across the world in 1980, immunisation against the disease is no longer necessary and scars like these are becoming a thing of the past.

There are many diseases where the pathogen survives in soil, water or animal hosts, so there is always a massive reservoir of potential infection. With these diseases, eradication is not realistically possible and elimination and control have to be the aims of immunisation. For example, the vaccination programme against polio has not entirely eradicated the disease, even in England and Wales. A live vaccine is used, so there is the very occasional instance where an infant, or an unvaccinated carer, develops the disease. Polio has almost been eradicated globally. It remains endemic only in Afghanistan, Nigeria and Pakistan and scientists and doctors are working hard to introduce universal vaccination programmes in these countries too.

Elimination is where the disease disappears but the pathogen remains in animals, the environment or in mild infections which are not recognised, and so immunisation must continue even when no clinical cases are being seen. When a disease is controlled it still occurs but not frequently enough to be a significant health problem. For some serious infectious diseases such as malaria, the development of a vaccine that is effective against all the different forms of a pathogen that evolves rapidly has so far proved impossible, but as you saw in **Section 6.2.3**, some progress is being made.

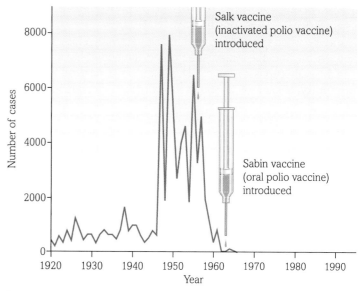

Salk vaccine (inactivated polio vaccine) introduced

Sabin vaccine (oral polio vaccine) introduced

fig D As a result of the polio vaccination programme, the impact of this disease on the population of England and Wales is now minimal. The days when swimming pools were closed and people had to sit in alternate seats in the cinema to avoid polio infection have long gone.

Herd immunity

If a disease is to be eradicated, eliminated or simply controlled, **herd immunity** is important. Herd immunity occurs when a significant proportion of the population is vaccinated against a disease. This makes it very difficult for the disease to spread because so few people are vulnerable to it. Herd immunity can effectively stop the spread of a disease through a community. It is important that everyone who can be vaccinated is vaccinated. This not only protects those individuals against the disease, but protects people who cannot be or have not been vaccinated. These include very young babies, very old people, people with compromised immune systems and people who are very ill with other diseases.

The percentage of the population that needs to be vaccinated to give herd immunity varies from one disease to another. It depends on factors such as how the disease is spread and how infectious it is. To give herd immunity against whooping cough (*pertussis*), 92–94% of the population need to be vaccinated. For measles it is 83–94%, and for polio only 80–86% vaccination is enough.

Some diseases are so serious that in the UK and many other countries all children are offered vaccines against them, giving herd immunity to the whole population. These include killer diseases such as diphtheria, tetanus, strains of meningitis and pertussis that killed and disabled thousands of people every year before the arrival of vaccines and better living conditions for all. In other cases, only vulnerable people are vaccinated as a general rule, with mass vaccination to provide herd immunity only introduced when there is a serious outbreak of disease (e.g. flu).

The pros and cons of vaccination

Immunisation is widely believed to be a good thing. However, there are two sides to every debate. The *pros* of vaccination programmes are very obvious and affect both individuals and society as a whole:

- The child, and the adult it grows into, is protected against diseases that could otherwise kill or disable it.

- Society benefits as the potential pool of infection is reduced with every vaccinated child through herd immunity, protecting children who cannot be vaccinated because of allergies or immune system diseases.

- The cost of treating serious diseases and caring for those left permanently damaged by them is kept to a minimum for a relatively small financial outlay (most common vaccines are relatively cheap).

The *cons* of vaccination have, until recent years, been less well known. They affect individuals but not society as a whole:

- Some of the live, attenuated vaccines are cultured in eggs, to which a small number of children may suffer a violent allergic reaction. To avoid this, children with egg allergies are not given vaccines grown in eggs and are protected by herd immunity.

- A tiny minority of children become extremely ill after a vaccination with what may be an extreme immune response. Some of these children may die and others have been left severely brain damaged. It is very difficult to prove a direct link between the vaccine and the damage to a child, because all children can get serious illnesses that affect them rapidly.

- Some scientists have suggested that mass vaccination programmes are linked to the recent rise in childhood asthma and allergies; however, this is one of many different explanations for this phenomenon.

- Some vaccines, such as rubella for boys, are given more for the benefit of society than for the direct benefit of the child. When girls are vaccinated to protect against cervical cancer, it benefits both individuals and society, but if boys are vaccinated the main benefit is to society at large.

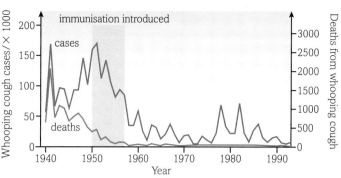

fig E In the UK during the 1970s, a link between whooping cough vaccination and serious brain damage in some children was suggested. Intense media coverage resulted in a sharp decrease in the numbers of children vaccinated. There followed, for the first time in years, an increase in the number of children affected by whooping infections and some died or were brain damaged. Levels of vaccine uptake are now back to around 96% of pre-scare levels, and herd immunity has been restored.

Questions

1 Explain and give examples of natural active immunity, natural passive immunity, artificial passive immunity and artificial active immunity.

2 What is herd immunity and why is it so important?

3 Some parents decide not to have their child vaccinated.
   (a) Explain how this puts their child at risk.
   (b) Explain how this puts other children at risk.
   (c) In the US children cannot start school unless they have been immunised. In the UK we do not insist on immunisation. Discuss the risks, benefits and ethics of both positions.

4 Using the data from **fig E**, explain how the story of the whooping cough scare illustrates the importance of establishing the difference between correlation and causation before taking action. Research the story of the MMR vaccine for more evidence.

Key definitions

Natural active immunity is when the body produces its own antibodies to an antigen encountered naturally.

Natural passive immunity is when antibodies made by the mother are passed to the baby via the placenta or breast milk.

Immunisation is the process of protecting people from infection by giving them passive or active artificial immunity.

Vaccination is the introduction of harmless forms of organisms or antigens by injection or mouth to produce artificial immunity.

Artificial passive immunity is when antibodies are extracted from one individual and injected into another e.g. the tetanus vaccine.

Artificial active immunity is when the body produces its own antibodies to an antigen acquired through vaccination.

Attenuated pathogens are viable pathogens that have been modified so that they do not cause disease but still cause an immune response that results in the production of antibodies and immunity.

Herd immunity is produced when a high proportion of a population is immune to a pathogen, usually by vaccination, lowering the risk of infection to all, including those not vaccinated, as they are less likely to encounter the pathogen.

THINKING BIGGER

THE BATTLE AGAINST RABIES

Due largely to Louis Pasteur, the idea that infectious diseases are spread by germs (microorganisms) became accepted by scientists, doctors and the public alike. This key realisation, so obvious today, has saved millions of lives.

LOUIS PASTEUR

In Louis Pasteur's time, as today, rabies was rare but the disease terrified people because once the symptoms started to show, death was the only possible result. Even today, without vaccination, rabies is still fatal. As a boy, Pasteur saw and was deeply affected by the attack of a rabid wolf on people in the village where he lived. When he became an eminent scientist, he decided he would attempt to find a rabies vaccine, both for the fame success would bring and to prove his germ theory once and for all.

Pasteur wanted to find the microbe that causes rabies and searched for it in the saliva, blood and brains of infected animals. He never found it, but he was sure a microbe was causing the disease and that he would be able to culture it, weaken it and produce a vaccine.

The veterinary surgeon Pierre-Victor Galtier discovered rabies could be transmitted to rabbits, where it has a much shorter incubation period than in dogs. Pasteur 'borrowed' these ideas and worked away secretly at the problem. In 1884 he announced he had discovered how to produce weakened rabies vaccines that could make dogs immune to rabies itself.

Then in July 1885 three terrified visitors arrived at Pasteur's laboratory. Two of them had been bitten by a rabid dog two days earlier. The skin of the owner was not punctured but little Joseph Meister, comforted by his frantic mother, had been badly bitten more than a dozen times. Two separate doctors agreed the child faced certain death from rabies. With the family's permission, Pasteur set out to treat him with a technique that had officially so far only been tried on dogs. Over 11 days the boy was given 13 injections made from the spinal cords of rabid rabbits, each injection more virulent than the last. After this the boy went home – and never developed rabies. Pasteur kept this secret from all but his closest family. Then three months later, he was asked to treat Jean-

Baptiste Jupille, a 15-year-old shepherd boy who had saved others from a rabid wolf but been bitten himself. Pasteur agreed to try his treatment, even though it was several days after the attack. Again the gruelling series of injections worked, and the boy was saved.

Pasteur's private notebooks later revealed that Joseph and Jean-Baptiste were not the first people he treated with his vaccine. Early in 1885 he treated a man who probably never had rabies, and a 6-year-old girl bitten on the lip when her puppy became rabid. She died two days into the treatment and Pasteur never mentioned her publically. In spite of this, the prevention of rabies was the final major discovery of Louis Pasteur's long and distinguished career and ensured his place in the history books of medicine.

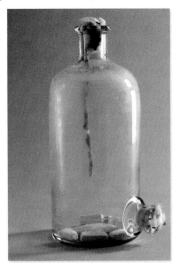

fig A One of Pasteur's original flasks, containing the dried spinal cord of a rabbit that died of rabies. Pasteur was convinced his new method of preparing a rabies vaccine would work, even though it was not properly tested before he tried it on desperate patients.

Where else will I encounter these themes?

Book 1   5.1   5.2   6.1   6.2   6.3   YOU ARE HERE 7.1   7.2

This extract is based on content from *A Biography of Louis Pasteur* by Ann Fullick.

1. What audience do you think this book is aimed at? Explain your answer.
2. Look at the account of how Pasteur developed and first used his rabies vaccines. Identify three aspects of his research that might appear unethical to modern eyes and discuss why this was not a problem in the 1800s.

Vaccinations have had a hugely positive impact on public health and have led to many diseases being eradicated or significantly reduced, for example, smallpox. However, as you have seen in this chapter, there are still ongoing ethical debates about vaccination programmes.

Now let us examine the biology. You already know about cell recognition, bacteria as pathogens, the response of the body to infection, artificial active immunity and the use of vaccination to control disease. This will help you answer the questions below.

3. How would developing a rabies vaccine prove Pasteur's germ theory of disease?
4. Suggest why Pasteur could not find the microorganism that causes rabies.
5. Discuss how the relatively short incubation time for rabies in rabbits helped Pasteur in his work developing a rabies vaccine.
6. Explain why Joseph Meister needed Pasteur's treatment, but the owner of the dog that bit him did not.

Activity

fig B Louis Pasteur with his son. Three of his beloved daughters died from infectious diseases in childhood. These losses drove him on in his work against the pathogens that cause disease.

Investigate the life and work of Louis Pasteur and produce one of the following resources:
- a brief biography of Louis Pasteur
- a timeline showing the main events in the life of Louis Pasteur.

● *A Biography of Louis Pasteur*, by Ann Fullick

1 Vaccination minimises the spread of disease in a population by developing herd immunity. What is the minimum percentage of a population that need to be vaccinated to ensure herd immunity for measles?

A 85%

B 90%

C 95%

D 97.5%   [1]

[Total: 1]

2 Which chemical is used by T helper cells to cause B cells to differentiate into active B cells?

A cytokinesis

B cytokines

C cytoplasm

D cytokinetics   [1]

[Total: 1]

3 A new technique for vaccinating people involves injecting them with DNA.

Viruses have proteins on their coats that are coded for by their DNA. The genes for producing viral proteins can be isolated and inserted into loops of DNA (plasmids). Plasmids can enter human cells which will then produce the viral proteins. The proteins will become part of the surface membrane of the human cell.

The immune system will recognise these proteins as foreign and respond by producing antibodies and T killer cells.

The process is summarised for one protein in the following diagram.

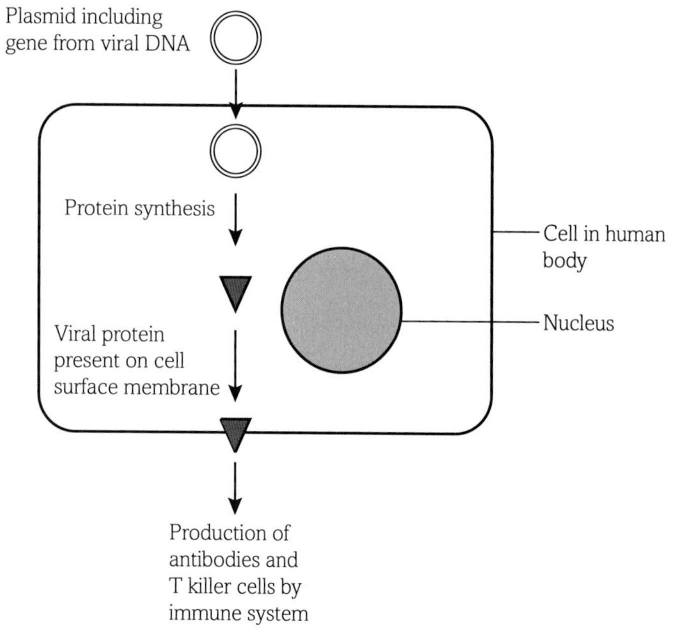

(a) Explain why the response of the immune system to the viral proteins is an example of active immunity.   [2]

(b) Explain how active immunity provides immunity against future infections by the virus.   [2]

(c) The table below compares the production and distribution of vaccines made using traditional methods with those made using DNA.

	DNA vaccine	Traditional vaccine
Time to develop vaccine against new strain of virus	2–3 weeks	4–6 months
Time to produce enough doses for effective protection of population	2–3 months	2–3 years
Treatment during distribution	No special treatment	Constant refrigeration

Use the information in the table to determine why the DNA vaccine is likely to be more effective at preventing the spread of a new strain of virus.   [3]

(d) A traditional vaccine involves the injection of viral protein into the body. This usually stimulates the production of antibodies but not T killer cells. Discuss how the use of viral DNA might be more effective than viral protein in producing immunity to a virus.   [4]

[Total: 11]

4 (a) The following list gives some examples of how immunity can develop in a mammal.

P Antibodies are transferred into the blood of a baby, from its mother, in the days before birth.

Q Killer T cells are produced by the body when it is infected by a virus.

R Polio virus, which has been made incapable of replicating, is given to babies to stimulate the production of memory cells.

S Anti-venom contains antibodies produced in an animal. This anti-venom can be injected to give protection against snake bite venom.

Complete a copy of the table below by writing a letter from the list above to match the type of immunity described in each case.

Immunity	Active	Passive
Natural		
Artificial		

[4]

(b) An investigation was carried out into the production of antibodies by lymphocytes.

Combinations of T cells and B cells were grown in four different cultures, W, X, Y and Z, as shown in the table below. The culture medium was checked for the presence of antibodies after several days.

Culture	T cells present	B cells present	Antigen present	Antibody production
W	✓	✓	✗	no
X	✓	✗	✓	no
Y	✗	✓	✓	no
Z	✓	✓	✓	yes

(i) Give **two** conclusions about antibody production that can be drawn from these results. [2]

(ii) Explain why no antibodies were produced in cultures X and Y. [2]

(c) In a rare genetic disorder called Bare Lymphocyte Syndrome (BLS), Major Histocompatibility Complex (MHC) proteins produced by cells are abnormal. As a result, T helper cells are unable to function normally and cannot respond to antigen.

Comment on what effect BLS is likely to have on the immune system of people with this disorder. [1]

(d) More than one gene is responsible for the production of MHC proteins. A mutation of any one of the genes will cause BLS in a person homozygous recessive for the faulty allele.

Following genetic testing, two people were each found to be heterozygous for the same two MHC genes. Their genotypes for these two genes can be shown as AaBb.

Draw a genetic diagram to show the probability that a child of this couple will have BLS. [3]

[Total: 12]

5 Immune responses involve communication between cells by means of chemicals called cytokines. A strain of mice unable to produce a cytokine called TNF (TNF deficient mice) was used to investigate the importance of chemical communication in preventing tuberculosis.

(a) Give **two** symptoms of tuberculosis. [2]

(b) TNF deficient mice and normal mice were both exposed to tuberculosis bacteria. After several weeks, the numbers of some types of white blood cells in the lungs were counted. The results are shown in the graph below.

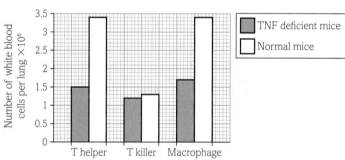

(i) Describe the effects of TNF on the numbers of these white blood cells in lungs. [2]

(ii) Describe the role of T killer cells in a normal immune response and give a reason for the results for T killer cells in this investigation. [3]

(iii) Using the information in the graph, explain why TNF deficient mice are more likely to die of tuberculosis. [3]

(c) Drugs that inhibit TNF production are commonly used to reduce inflammation in people with rheumatoid arthritis. Predict possible consequences of using TNF inhibitors for people who are infected with tuberculosis but have no symptoms. [2]

[Total: 12]

TOPIC 7
Modern genetics

7.1 ▶ Using gene sequencing

Introduction

In the twenty-first century we take gene sequencing – the ability to identify every DNA base that makes up the genome of an organism – almost for granted. The genetic sequence can be read in hours or days at the most. Genetic fingerprinting is another technique that is used on every TV detective programme and has become part of everyday language. Yet it was only in the 1970s that Frederick Sanger developed the technique that allowed the first genomes to be read, a process that took years to complete. His technique still underpins the modern machines that read genomes night and day in the Wellcome Trust Sanger Institute that carries his name. Sanger was also the first person to determine the order of amino acids in a protein. He determined the structure of insulin in the 1940s and 1950s. He is one of a very small number of people who have been awarded two Nobel prizes; one in 1958 and the second in 1980.

In this chapter you will discover what we mean by a genome. You will also look at the polymerase chain reaction (PCR). This technique enables us to amplify the tiniest sample of DNA and use it for gene sequencing and DNA profiling. You will learn about the very different techniques of gene sequencing and DNA profiling, each of them vital in their own way for our rapidly increasing understanding of the human genome.

All the maths you need

- Translate information between graphical, numerical and algebraic forms (*e.g. analysing DNA profiles*)

What have I studied before?

- The structure of DNA, including the structure of the nucleotides, base pairing, the two sugar-phosphate backbones, phosphodiester bonds and hydrogen bonds
- How DNA is replicated semi-conservatively in the nucleus of the cell
- The roles of the enzymes DNA helicase, DNA polymerase and DNA ligase in DNA replication
- The process of transcription of the DNA in the nucleus and translation at the ribosome
- The nature of the genetic code and that not all of the genome codes for proteins
- The sensitivity of enzymes to temperature
- How mutations in the DNA come about
- The process of gel electrophoresis
- How DNA sequencing and bioinformatics are used to distinguish between species and determine evolutionary relationships

What will I study later?

- Factors that affect gene expression, including transcription factors, post-transcription modification of mRNA and epigenetic modification
- The differences between totipotent, pluripotent and multipotent stem cells
- How epigenetic modifications can result in totipotent stem cells in an embryo developing into pluripotent cells in a blastocyst and finally into fully differentiated somatic cells
- How pluripotent stem cells from embryos produce opportunities to develop new medical advances and the ethical considerations
- How differentiated fibroblasts can be reprogrammed to form iPS cells that may give rise to fewer ethical issues
- The way recombinant DNA is produced and inserted into other cells using a variety of vectors
- How antibiotic resistance and replica plating can be used to identify recombinant cells
- The genetic modification of soya beans
- The use of 'knockout' mice as a tool to investigate gene function

What will I study in this chapter?

- The meaning of the term genome
- How PCR can be used to amplify DNA samples
- The way DNA amplified by PCR can be used to predict the amino acid sequence of proteins and possible links to genetically-determined conditions using gene sequencing
- The way DNA amplified by PCR can be used in forensic science and to test paternity using DNA profiling

By the end of this section, you should be able to...

● explain what is meant by the term genome

● explain how PCR can be used to amplify DNA samples

In the last 50 years, one of the fastest growing areas of biology has been the study of genetics, genomics and proteomics, highlighted by our ever-growing knowledge of the human genome.

What is the genome?

The genome is the total of all the genetic material in an organism. In prokaryotes the DNA is in the cytoplasm, both in the main chromosome and in the plasmids. In eukaryotes the DNA is in the nucleus of the cell and in the mitochondria, and in green plant cells DNA is also in the chloroplasts. The DNA that makes up the human chromosomes consists of many millions of base pairs. However, the genes themselves, that is the coding regions of DNA that determine the protein structures that underpin the entire organism, only make up around 2% of that DNA. These coding regions are known as the **exons**. The large, non-coding regions of DNA are removed from messenger RNA before it lines up on the ribosomes and is translated into proteins. They are known as the **introns**.

Analysing the DNA

In **Book 1 Chapter 1.3** you discovered the principles of the structure of DNA – the double helix made up of sugar-phosphate backbones with purine and pyrimidine bases, held together along the length of the molecule by hydrogen bonds. However, since Watson and Crick first proposed DNA as the molecule of inheritance, our ability to analyse the structure of the molecule has developed tremendously. We have analysed the entire human genome, not just once but thousands of times. In **DNA** or **gene sequencing** we analyse individual strands of DNA or individual genes, giving us the pattern of bases that codes for a particular protein in the cell. It can also be used to identify different species of living organisms. In DNA profiling we analyse the patterns in the non-coding areas of DNA (the introns) and use them to identify individuals, for example in a court of law or in paternity cases.

There is one reaction that is key to our successful analysis of DNA at both levels – the **polymerase chain reaction**, generally referred to as **PCR**.

The polymerase chain reaction (PCR)

Traditional DNA profiling needs at least 1 μg of DNA, which is the equivalent of the DNA from about 10 000 human cells. In a crime investigation, there may only be a minute DNA sample available. The polymerase chain reaction adapts the natural process that replicates DNA in the cell, enabling us to produce enough DNA to sequence a genome or produce a DNA profile from tiny traces of biological material. When a tiny sample of DNA is increased using PCR to produce a large enough sample for analysis we say it has been **amplified**.

Developing PCR

The problem for scientists when they were trying to develop a way of amplifying tiny amounts of genetic material was that the DNA sample needed to be heated to around 90–95 °C to separate the two strands and make them available for replication. This destroys DNA polymerase from most organisms.

Kary Mullis is the scientist who solved the problem. He decided to try using enzymes from a bacterium (*Thermus aquaticus*) that lives in hot springs to develop a technique for replicating DNA artificially in the laboratory. Because the enzymes in this bacterium have evolved to survive in extreme conditions, Mullis hypothesised that they would be robust enough to cope with the high temperatures and the frequent temperature changes needed to separate DNA strands and bring about replication.

fig A Kary Mullis, the developer of PCR and Nobel Prize winner, is a surfer, a scientist, and something of a maverick thinker.

How PCR works

The DNA sample that is to be amplified is mixed with the enzyme Taq (*Thermus aquaticus*) DNA polymerase, primers (small sequences of DNA that must join to the beginning of the separated DNA strands before copying can begin) and a good supply of the four nucleotide bases, as well as a suitable buffer for the reaction, and placed in a PCR machine. The mixture is heated to 90–95 °C, which causes the DNA strands to separate as the hydrogen bonds between them break down. The mixture is then cooled to 50–55 °C so that the primers bind (or anneal) to the single DNA strands. Finally, the mixture is heated to 72 °C, which is the optimum temperature for the Taq DNA polymerase enzyme to build the complementary strands of DNA.

These steps are repeated around 30 times, producing around 1 billion copies of the original DNA. On average, the whole process takes about 3 hours, and much of that time is taken up with heating and cooling the reaction mixture in the PCR machine.

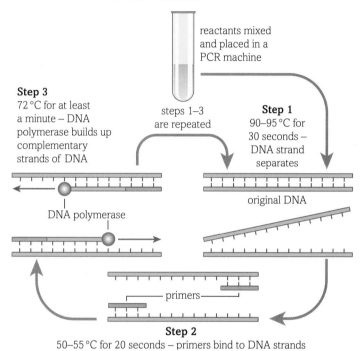

fig B The main stages of the polymerase chain reaction.

Questions

1 Describe how DNA can be amplified in the polymerase chain reaction.

2 Why was the use of enzymes from the bacterium *Thermus aquaticus* such a breakthrough?

3 Suggest why PCR has been an important development in the investigation of crime.

Key definitions

Exons are the coding regions of DNA (the genes).

Introns are the large, non-coding regions of DNA that are removed before messenger RNA is translated into proteins.

DNA or **gene sequencing** is the analysis of the individual base sequence along a DNA strand or an individual gene.

Polymerase chain reaction (PCR) is the reaction used to amplify a sample of DNA, to make more copies of it very rapidly.

When DNA is **amplified**, it is replicated repeatedly using the polymerase chain reaction to produce a much bigger sample.

By the end of this section, you should be able to...

- explain how DNA sequencing can be used to predict the amino acid sequence of proteins

- explain how DNA sequencing can be used to identify possible links to genetically-determined conditions

A technique that could successfully sequence all of the genome of an organism, revealing the order of bases in the DNA, was first developed by Frederick Sanger in the 1970s. The first organism to have its genome sequenced was phiX174, a bacteriophage – a virus that infects bacteria. Since those early days, when the processes were usually manual and took a very long time, sequencing techniques have been refined many times and automated. It took 10 years to produce the first rough draft of the human genome in the year 2000, but it can now be done in days. A project analysing 100 000 human genomes will take under 2 years to complete.

The process of sequencing

The process of sequencing entire genomes or individual genes has changed considerably over time, but there are some basic principles seen in all the different methods of DNA sequencing:

- The DNA strands are chopped into smaller pieces.

- The double strands are separated to give single strands.

- PCR is involved in replicating the DNA fragments to produce large quantities of material for analysis.

- Labelled **terminator bases** are added to the single strands of DNA. A terminator base is a modified version of one of the four nucleotide bases; adenine (A), thymine (T), cytosine (C) and guanine (G). When a terminator base is incorporated into a DNA molecule, the chain is halted as no more bases can be added. A T terminator stops DNA synthesis where a T base would be added, a G terminator stops it where a G base would go, and so on. Each type of terminator base has a fluorescent tag in a specific colour.

- The coloured tags enable the sequence of bases to be read very rapidly by an automated system.

Current sequencing technologies, for example **massively parallel sequencing**, work on millions of DNA fragments at a time and the processes are heavily dependent on state-of-the-art computer technology. The raw data is fed into computer systems that can reassemble the genomes. They compare all the fragments and find the areas of overlap between them.

There are 3 billion base pairs in the human genome and current technology can sequence them in days. The genome of a bacterium can be sequenced in less than 24 hours. The processes are constantly being developed and refined, so sequencing speeds are increasing and the costs of genome analysis are falling, making the technology more and more useful in an ever-increasing number of ways.

Using DNA sequencing

As DNA sequencing becomes faster and cheaper, the number of ways in which it can be used are also growing. In **Book 1 Chapter 3.1** you looked at how DNA barcoding is being used as a way of both identifying species and investigating the evolutionary relationships between different species. It has been a principle from the beginning that data on the human genome and other organisms should be available to all scientists, so each genome is made available online to anyone who wants to use it. However, the ability to interpret the data is a complex skill in itself.

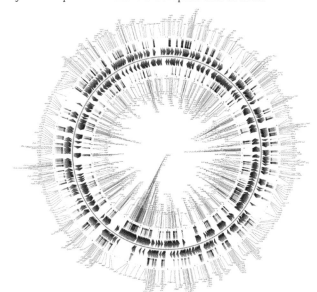

fig A A representation of the genome sequence of the bacterium *Erwinia carotovora*.

Predicting amino acid sequences

DNA sequencing can be used to determine the protein produced from any particular gene. Because of the universal genetic code, we recognise the start and stop codons in a gene. Analysing the base pairs enables us to work out which amino acids will be joined together to form a protein as a result of information contained in the DNA. In the past we thought one gene produced one protein. However, sequencing the human genome shows us that humans have 20 000–25 000 coding genes. There are many different estimates of the number of actual human proteins produced by the genome. Figures range from 17 000 to 1 000 000, with 500 000 being a realistic figure. It is certainly different from, and probably larger than, the number of genes. Biologists are becoming increasingly aware of the complexity of the relationship between the genome and the proteins produced. Some genes code for factors that affect the expression of other genes, changing the protein product. You will learn more about factors that affect gene expression in **Section 7.2.1**.

Links to disease management

One aspiration arising from the sequencing of the human genome is that the knowledge we gain will give us a better understanding of human disease. There are a small number of diseases that result directly from a mutation in a single gene. DNA sequencing makes it possible to identify a faulty gene, see which bases have changed and understand how the changes in the DNA affect the proteins produced. That in turn allows us to begin to work out how changes in the proteins result in the symptoms of the condition.

However, most non-communicable diseases are not the result of a change in a single gene. Many diseases are the result of the inheritance of a number of alleles or **gene variants** that increase the risk of an individual developing a specific disease such as diabetes, obesity or coronary heart disease. Environmental factors also play a part and in many cases it is the interaction between the genome and the environment that determines the incidence of disease.

Projects such as the 100 000 Genome Project, a UK initiative sequencing the genomes of 100 000 people with cancer and rare genetic diseases, is well under way, and should be completed by 2017. This will provide huge amounts of data showing how particular genotypes, combined with environmental conditions, raise the risk of or cause specific diseases. The genetic data from the patients and their families will be matched with clinical data held on DNA databases. This data will be used by scientists to identify mutations associated with different diseases and develop new, targeted therapies. Again computers play a key role, giving us the capacity to analyse and compare the genome sequences of thousands of people, revealing patterns in the DNA and linking these to disease. Gene sequencing has the potential to revolutionise epidemiology, completely changing the ways in which we understand and model patterns of disease, and having a long-term impact on the treatment of disease.

Sequencing in action

DNA sequencing has shown that certain genetic combinations greatly increase the risk of heart disease. Currently it is estimated that about 60% of all heart disease occurs in India. In a joint project with scientists in India, scientists at the Wellcome Trust Genome Campus identified a mutation in a gene coding for the heart protein MYPBC3. Around 25 bases are deleted from the gene, which causes major changes in the MYPBC3 protein. This in turn damages the structure of the heart muscle and makes the individual seven times more likely to suffer serious heart

disease than someone without the mutation. In young people the body seems to deal with the damaged protein. It causes health problems as people move into middle age, so the mutation is passed on because it does not have an effect until after most people have had their children. About 1 in 100 people have this mutation, but DNA sequencing has shown that it is four times as common in India where 4 in 100 people (1 in every 25 people) carry the potentially lethal mutation. This discovery opens up the possibilities of screening programmes, education and treatment to try and reduce the deaths from heart disease everywhere and especially in India and people of South Asian descent.

fig B Around 4% of these people will unknowingly have a mutation in their MYPBC3 gene, which makes them much more likely to be affected by heart disease as they get older.

Questions

1   What is a terminator base and why are they important in DNA sequencing?

2   The time taken to sequence human DNA has fallen from years to days. Suggest reasons for this.

3   Give three ways in which DNA sequencing is moving scientific knowledge and understanding forward.

Key definitions

Terminator bases are modified versions of the four nucleotide bases that halt the production of a DNA molecule as soon as they are incorporated as no more bases can be added.

Massively parallel sequencing is a very rapid method of sequencing millions of DNA fragments at the same time.

Gene variants are different versions of genes, an alternative term for allele.

By the end of this section, you should be able to...

● explain how DNA profiling is used in forensic science to identify criminals and to test paternity

One important discovery made is that you can identify individuals and species by patterns in their DNA. This has had enormous implications in many areas of science, including forensic science and paternity testing. The process is known as **DNA profiling**.

Introns and satellites

The human genome contains between 20 000 and 25 000 genes and the chromosomes are made up of hundreds of millions of base pairs. Less than 2% of the genome actually codes for proteins. Over 90% of the DNA is made up of the introns; repetitive coding regions between the genes. Their function is not yet fully understood, although we do know that some of these sequences can code for small interfering RNA molecules (siRNA) that interact with mRNA and prevent the production of certain proteins, and that they are inherited in the same way as the active genes. Introns are the regions of the chromosomes that are used in DNA profiling.

Within the introns there are short sequences of DNA that are repeated many times to form **micro-satellites** and **mini-satellites**. In a mini-satellite a 10–100 base sequence will be repeated 50 to several hundred times. A micro-satellite has 2–6 bases repeated between 5 and 100 times. The same mini- or micro-satellites appear in the same positions on each pair of homologous chromosomes. However, the number of repeats of each satellite will vary as different patterns may be inherited from your father and mother.

There are many different introns and a huge variation in the number of repeats so the likelihood of any two individuals having the same pattern of DNA is extremely remote, unless they are identical twins. However, the more closely related two individuals are, the more likely it is that similarities will be apparent in their DNA patterns.

How is a DNA profile produced?

The strands of DNA from a sample are cut into fragments using special enzymes known as **restriction endonucleases**. These enzymes cut the DNA at particular points in the intron sequences. There are many different restriction enzymes, each type cutting a DNA molecule into fragments at different specific base sequences known as **recognition sites**. Using restriction enzymes that cut either side of mini- and micro-satellite units leaves the repeated sequences intact, giving a mixture of different-sized DNA fragments, depending on the number of repeated sequences, made up largely of mini- and micro-satellite sequences.

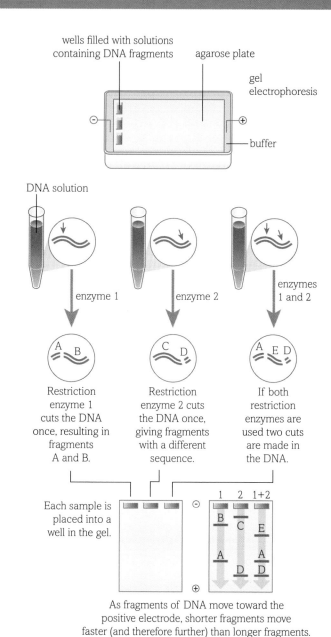

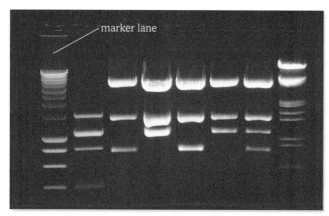

fig A A summary of the initial stages of DNA profiling.

Gel electrophoresis

The fragments need to be separated and identified. This process starts with gel electrophoresis, which is a variation of chromatography (see **Book 1 Section 3.1.5**). The DNA fragments are placed in wells in an agarose gel medium in a buffering solution (to maintain a constant pH) with known DNA fragments to aid identification. The gel contains a dye (e.g. EtBr, ethidium bromide) that binds to the DNA fragments in the gel. The dye will fluoresce when placed under short-wave ultraviolet (UV) light, revealing a band pattern of DNA. A visible dye is also added to the DNA samples. This does not bind with the DNA but moves through the gel slightly faster than the DNA, so that the current can be turned off before all the samples run off the end.

An electric current is passed through the apparatus and the DNA fragments move towards the positive anode because of the negative charge on the phosphate groups in the DNA backbone. The fragments move at different rates according to their size and dependent on charge. Once the electrophoresis is complete the plate is placed under short-wave UV light. The DNA fluoresces and shows up clearly so it can be identified.

This is the original method of DNA profiling, which needs a relatively large sample of DNA. It shows up large DNA fragments containing a minimum of 50 base pairs – in other words, mini-satellites. The resultant DNA profile (fingerprint) looks similar to a supermarket barcode. By using extensions of this technique we can now identify smaller regions of DNA and specific genes.

Southern blotting and gene probes

The next stage is **Southern blotting**, named after its British inventor, Edwin Southern. An alkaline buffer solution is added to the gel after electrophoresis and a nylon filter or nitrocellulose paper is placed over it. This dry absorbent material is used to draw the solution containing the DNA fragments from the gel to the filter, leaving the DNA fragments as 'blots' attached to the filter. The alkaline solution also denatures the DNA fragments so the strands separate and the base sequences are exposed. The DNA is then covalently bound by UV cross-linking.

Large quantities of **gene probes** are added to the filter after the Southern blotting process and bind with the complementary DNA strands in a process known as **hybridisation**. Gene probes are short DNA sequences that are complementary to specific sequences that are being sought. Each probe is labelled with a fluorescent molecule or a radioactive isotope.

Excess probes are washed away and the filter is placed under UV light to show up the DNA regions.

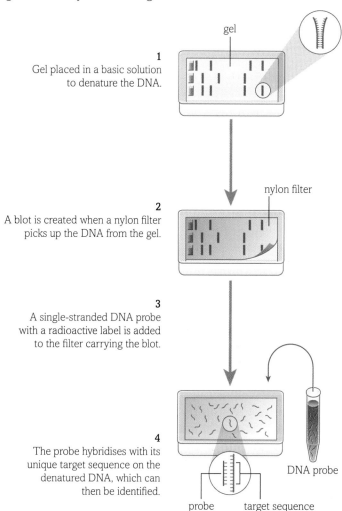

1 Gel placed in a basic solution to denature the DNA.

2 A blot is created when a nylon filter picks up the DNA from the gel.

3 A single-stranded DNA probe with a radioactive label is added to the filter carrying the blot.

4 The probe hybridises with its unique target sequence on the denatured DNA, which can then be identified.

fig B Southern blotting with DNA probes can give more accurate DNA profiles.

As a result of the different stages of the process, a DNA profile is produced as a graph (see **fig C**), with each peak representing the number of micro-satellite repeats in a fragment. It can be broken down to give a digital profile which provides numbers for the repeats at each micro-satellite. A single reading means that the same number of repeats was present on both chromosomes of a homologous pair. The more micro-satellites that are compared, the more accurate the DNA profile will be.

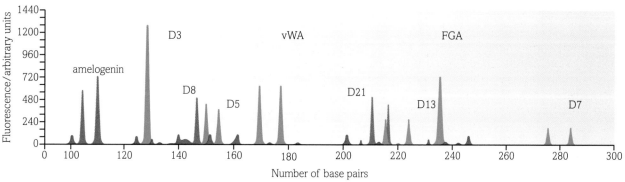

fig C This is a modern DNA profile – interpreting it is a job for the experts!

Did you know?

Inventing DNA profiling

In the 1970s Alec Jeffreys at Leicester University showed that when he used restriction endonuclease enzymes to cut the different regions of the chromosomes into smaller pieces, the repeating patterns in these units varied among individuals. Comparing seal and human myoglobin genes, he discovered groups of repeating DNA sequences coding for the same polypeptide in two different species. Then, working with DNA from their lab technician and her family, the team noticed the repeating units were more similar between related people than between complete strangers. DNA profiling – sometimes referred to as DNA fingerprinting – had been born.

Identifying individuals

In forensic science

Forensic science is the application of science to the processes of law. In forensic science, to develop a DNA profile in criminal investigations, gene probes are used to pick out **short tandem repeats**, which are micro-satellite regions that are widely used in DNA identification. The more micro-satellites used to make up a DNA profile, the more accurate it will be. Family members are more likely than unrelated individuals to have a number of micro-satellites in common, but the more sites that are examined the more likely it is that a different combination will have been inherited from the two different parents. Statistically, the chances of two people matching on 11 or more sites is so small that it is counted as reliable evidence in court.

In paternity testing

Sometimes there is doubt about who has fathered a child. This is almost always in the context of problems in human relationships. DNA profiling can be used to prove or disprove family relationships. The mother of the child is almost always known so it is paternity that has to be taken on trust, and therefore most DNA tests are to test paternity. Specific micro-satellite markers are used to make the matches.

Data such as **fig D** (provided by GFI Lab, a US DNA testing laboratory) is used in paternity cases. If the alleged father was the biological father, each of the child's micro-satellites would be from either the mother or the alleged father. In **fig D(b)**, the child has some micro-satellites in common with the mother (coloured green), some in common with the alleged father (coloured yellow) but some that do not come from either the mother or the alleged father (coloured pink).

Marker	Alleged father	Child		Mother
AMEL	X,Y	X	Y	X
D3S1358	15,16	15	16	15,16
D1S1656	16,17	15	16	15
D2S441	10	10	10	10
D10S1248	14	14	14	13,14
D13S317	9,14	9	14	9,10
Penta E	14,21	11	14	11,15
D16S539	10	10	12	12
D18S51	12	12	19	17,19
D2S1338	19,24	18	19	17,18
CSF1PO	11,12	11	12	12,13
Penta D	12,14	10	12	10
TH01	6,7	6	7	7
vWA	16	16	18	16,18
D21S11	29,30	30	30	30,31
D7S820	11,12	11	11	10,11
D5S818	11	11	11	11
TPOX	11,12	8	12	8,11
D8S1179	15	12	15	12,14
D12S391	17.3,19	18	19	18,19
D19S433	13.2,15	14	15	12,14
FGA	23,25	23	24	24,26
D22S1045	15,16	15	16	15,16

(a) Alleged father is the biological father

Marker	Alleged father	Child		Mother
AMEL	X,Y	X	Y	X
D3S1358	15,16	15	16	15,16
D1S1656	14,17	15	16	15
D2S441	12	10	10	10
D10S1248	14	14	14	13,14
D13S317	9,10	9	14	9,10
Penta E	14,21	11	14	11,15
D16S539	11	10	12	12
D18S51	16	12	19	17,19
D2S1338	20,24	18	19	17,18
CSF1PO	11,12	11	12	12,13
Penta D	12,14	10	12	10
TH01	7,8	6	7	7
vWA	16	16	18	16,18
D21S11	29,30	30	30	30,31
D7S820	11,12	11	11	10,11
D5S818	12	11	11	11
TPOX	11,12	8	12	8,11
D8S1179	10,12	12	15	12,14
D12S391	20	18	19	18,19
D19S433	13.2,15	14	15	12,14
FGA	23,25	23	24	24,26
D22S1045	15,16	15	16	15,16

(b) Alleged father is NOT the biological father

fig D The matching of micro-satellites is an accurate way to confirm or disprove paternity and produces data such as this, provided by GFI Lab, a US DNA testing laboratory.

Forensic science in action

Helena Greenwood was a talented British biochemist working on DNA analysis in the United States. In 1984 David Frediani broke into her home, held Helena at gunpoint for several hours and sexually assaulted her. She persuaded him not to kill her by saying that she would not tell anyone. Once he left she gave the police a description and they found Frediani. He was granted bail and a trial date was set for 1985.

Helena and her husband Roger were afraid Frediani might return so they moved to San Diego. But 3 weeks before Frediani's trial, Roger found Helena dead in their garden, strangled and beaten. Frediani was the prime subject. Evidence from his credit cards showed he had been in the area just before the murder, but police had no forensic evidence linking him to Helena's body. Frediani stood trial for the original assault and was imprisoned for 3 years. There was no proof he was involved in Helena's death – but no other suspects emerged either.

Then, in 1999, the San Diego police reopened a number of unsolved murders to see if new forensic techniques such as PCR and DNA profiling could help convict the perpetrators. Skin fragments from Helena's attacker were found beneath her fingernails when her body was examined. Using PCR to amplify the samples they got enough good material to produce a DNA profile – which matched David Frediani! More than 15 years after he killed Helena Greenwood, Frediani was tried, found guilty and sentenced to life imprisonment.

Another instance where DNA profiling is important is for inheritance claims. If someone claims to be a member of the family, with a claim on an estate, DNA evidence can prove things one way or the other.

Familial DNA testing is also important in immigration cases. If someone has immigrated to a country and wants their children to join them, they may have to prove that the children are really theirs. Sometimes birth certificates are not available, but DNA testing can quickly prove whether people are genuinely biologically related.

Questions

1 Explain the term DNA profiling.

2 Describe how gel electrophoresis is used to separate DNA fragments produced by restriction enzymes in the production of a DNA profile.

3 (a) Explain why DNA profiling is useful for identifying different individuals in forensic investigations.

   (b) Individuals have been wrongly accused of serious crimes partly because at some point another member of their family has committed a minor crime. Discuss how this might be possible and how it can be avoided.

Key definitions

DNA profiling is the identification of repeating patterns in the non-coding regions of DNA.

A **micro-satellite** is a section of DNA with a 2–6 base sequence repeated between 5 and 100 times.

A **mini-satellite** is a section of DNA with a 10–100 base sequence repeated 50 to several hundred times.

Restriction endonucleases are special enzymes used to chop up strands of DNA at particular points in the intron sequences.

Recognition sites are specific base sequences where restriction endonucleases cleave the DNA molecule.

Southern blotting is the name of a process in which DNA fragments are drawn from an electrophoresis gel to a filter, leaving the DNA fragments as 'blots' on the filter. The process also denatures the DNA fragments so the strands separate and the base sequences are exposed.

Gene probes are short DNA sequences labelled with a fluorescent molecule that are complementary to specific DNA sequences which are being sought.

Hybridisation is the binding of gene probes to the complementary DNA strands.

Short tandem repeats are micro-satellite regions that are now widely used in DNA identification.

THINKING BIGGER

TIMELINE OF GENOMICS

The processes used to sequence DNA are complex and developing fast. Since the original Human Genome Project was completed, the cost of sequencing genomes has decreased more than a thousand-fold, and the speed of sequencing a genome has increased by many orders of magnitude. Some of the main events in this rapid development of technology and expertise are listed below.

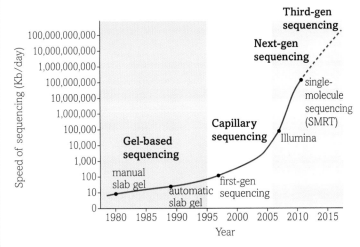

fig A A graph to show how the speed of DNA sequencing technologies has increased since the early techniques in the 1980s. (Image credit: Genome Research Limited).

- **1977** Frederick Sanger develops a DNA sequencing technique, which he and his team use to sequence the first full genome – that of a virus called phiX174.

- **1983** The polymerase chain reaction (PCR) is developed – a technique used for amplifying DNA – by Dr Kary Mullis at the Cetus Corporation in California, USA.

- **1985** Alec Jeffreys develops a method for DNA profiling. A DNA profile is produced by counting the number of short repeating sequences of DNA sequence found at ten specific regions of the genome.

- **1990** Human Genome Project is launched. The project aims to sequence all 3 billion letters of a human genome in 15 years.

- **1995** The first bacterial genome is completed (*Haemophilus influenzae*).

- **Mid-1990s** Capillary sequencing machines are introduced, speeding up the process of DNA sequencing.

- **1998** The genome of the nematode worm *C. elegans* is published.

- **1999** The first human chromosome (chromosome 22) is sequenced in the Human Genome Project (HGP).

- **2000** The full genome sequence of the model organism *Drosophila melanogaster* (fruit fly) is completed.

- **2001** First draft of the human genome sequence released.

- **2003** Human Genome Project is completed and confirms humans have approximately 20 000–25 000 genes. The human genome is sequenced to 99.99 per cent accuracy, 2 years ahead of schedule.

- **2007** A new DNA sequencing technology is introduced that increases DNA sequencing output 70-fold in one year!

- **2008** Next-generation DNA sequencing platforms result in a dramatic drop in sequencing costs. The 1000 Genomes Project launched – the first project that aims to sequence the whole genomes of a large number of people (2500).

- **2010** Wellcome Trust launches UK10K, which aims to compare the genomes of 4000 healthy people with those of 6000 people living with a disease of suspected genetic cause.

- **2010 onwards** Third-generation sequencing introduced including single-molecule sequencing in real time. These new methods are extremely efficient and sensitive.

Where else will I encounter these themes?

Book 1   5.1   5.2   6.1   6.2   6.3   7.1 YOU ARE HERE   7.2

This timeline is largely taken from the web resource developed by the public engagement team at the Wellcome Genome Campus near Cambridge. They have developed a wide range of materials to help school students and the general public understand the fast moving world of genomics.

1. Timelines are often used to help give a perspective on scientific events over time. Discuss how effective they are.

2. Compare the timeline of genomics given here with the original at www.yourgenome. org/facts/timeline-history-of-genomics. Summarise the differences between the short version in this book and the original and comment on how this affects the usefulness of the timeline for you personally.

3. Look at **fig A**. How many orders of magnitude faster is modern sequencing than the originally developed Sanger method?

> Using orders of magnitude can be a very useful way of making approximate comparisons. Remember, if two values differ by one order of magnitude, one value is approximately 10 (10^1) times larger than the other. If they differ by two orders of magnitude, they differ by a factor of around 100 (10^2).

Now let us examine the biology. You already know about the structure of DNA, how DNA sequencing is used to identify species and how PCR is used to amplify DNA samples. This will help you answer the questions below. You will be going on to look at factors affecting gene expression and gene technology, so you may like to revisit these questions later.

4. Discuss the value of sequencing the human genome and of projects such as the 1000 Genomes Project and the UK10K project.

5. In 2012 a study based on 30 research papers confirmed that the human genome contains 20 687 protein-coding genes. Suggest how this information might be useful in confirming the role of transcription factors and epigenetic modification in protein formation.

Activity

Choose one of the following activities:

- Investigate the different DNA sequencing techniques. Consider the technical details and how they have developed, as well as the impact they have had on genomics. Present your findings in whatever way you feel is most effective at communicating them to others.

- Investigate at least two of the people who have played major roles in the development of DNA sequencing and the unravelling of the human genome. Write a brief profile of each person you investigate, making sure that you present your findings in a way that is engaging and interesting for the reader.

> You can use online resources including www.yourgenome.org, which is developed by the public engagement team at the Wellcome Genome Campus. The campus is home to the biggest DNA sequencing facility in the world. It is sensible to always aim to use more than one source when you are carrying out research, to ensure that your information is reliable.

● Wellcome Genome Campus Public Engagement / www.yourgenome.org

8.1   8.2   9.1   9.2   9.3   10.1   10.2   10.3

1 The polymerase chain reaction (PCR) has a number of stages.
   (1) heat to 95 °C
   (2) add primers
   (3) repeat many times
   (4) cool to 55 °C
   (5) heat to 72 °C

Which is the correct sequence for this procedure?
A 1–4–3–5–2
B 1–5–4–2–3
C 1–4–2–5–3
D 5–4–2–1–3   [1]

[Total: 1]

2 Samples from PCR can be used in many ways:
   (1) identify the possible gene variants of a particular gene
   (2) identify a criminal
   (3) test for paternity
   (4) to see if there is a correlation between a gene variant and a disease

How many of these uses would require just one person's DNA sample?
A 1
B 2
C 3
D 4   [1]

[Total: 1]

3 The flow diagram below summarises some of the stages used to copy DNA in the polymerase chain reaction (PCR).

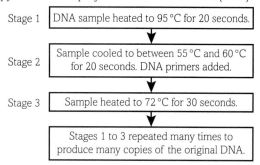

Stage 1   DNA sample heated to 95 °C for 20 seconds.

Stage 2   Sample cooled to between 55 °C and 60 °C for 20 seconds. DNA primers added.

Stage 3   Sample heated to 72 °C for 30 seconds.

Stages 1 to 3 repeated many times to produce many copies of the original DNA.

(a) Explain why the DNA is heated during Stage 1.   [2]

(b) Describe the structure of the primers used in Stage 2 and explain why they are used.   [3]

(c) Another method of producing many copies of a DNA sample is to introduce the DNA into bacteria and allow them to reproduce. Give one disadvantage of this technique compared with PCR.   [1]

[Total: 6]

4 Following a burglary, a DNA profile was created using a small sample of blood left behind on a broken window pane. This DNA profile was then compared with DNA profiles from four suspects, S1, S2, S3 and S4. These DNA profiles are shown in the diagram below.

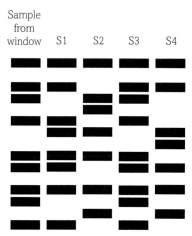

(a) (i) What is the name of the enzyme used in the process used to amplify the DNA in the small sample of blood taken from the crime scene?
    A endonuclease
    B invertase
    C polymerase
    D transcriptase   [1]

  (ii) What is the name of the process that could be used to separate DNA fragments to create the profiles shown in the diagram above?
    A amniocentesis
    B electrophoresis
    C endocytosis
    D chromatography   [1]

  (iii) Comment on which of the suspects is most likely to have left the blood sample on the broken window pane.

    With reference to the theory used in DNA profiling, explain how you came to this conclusion.   [5]

(b) Explain why evidence from DNA profiles may not be absolutely conclusive.   [2]

(c) Explain how DNA profiling could be useful to scientists who examine fossils of animals and plants.   [2]

[Total: 11]

5 In many of the hotter regions of the world, maize is grown as a crop plant. It is a good source of starch for humans and livestock such as cattle.

(a) Give **two** differences in the molecular structure of the two components of starch, **amylose** and **amylopectin**. [2]

(b) Explain the role of starch in the diet of humans. [2]

(c) Traditional varieties of maize can be severely damaged by insect pests. New varieties of maize have been developed to try to reduce the loss in yield caused by insect pests.

Hybrid varieties are produced by cross-breeding and selective breeding of maize varieties used in different parts of the world.

Genetically-modified varieties of maize have been produced which are resistant to attack by insects. These plants (Bt GM maize) have been genetically modified by inserting a gene for the production of Bt toxin, a natural insecticide. This gene is extracted from a species of bacterium, *Bacillus thuringiensis*.

In a survey of the yield of maize crops in six regions of southern Africa, data for three varieties of maize (traditional, hybrid and Bt GM) were compared. For each variety, the yield from one kilogramme of seed sown was measured.

The results of this survey are shown in the graph below.

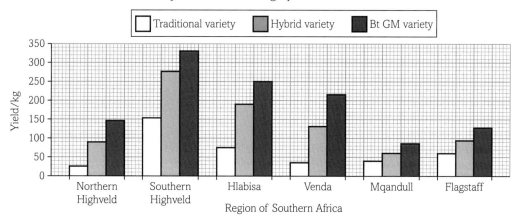

(i) Assess why it might be necessary to use the polymerase chain reaction (PCR) during the development of a Bt GM variety of maize. [2]

(ii) Explain how the data in the graph indicate that abiotic factors may vary considerably in the different regions. [3]

(iii) In all of the survey regions, using Bt GM variety maize gives the highest yield. With reference to the data in the graph, deduce why some regions might still choose to use the hybrid variety to increase the yield. [3]

[Total: 12]

CHAPTER

7.2 > Factors affecting gene expression

Introduction

If a racehorse damages a tendon, it can mean the end of a promising racing career. However, recently several racehorses have won prestigious races after having stem cell treatment to help repair and restore damaged tendons. The stem cells were taken from the horses themselves, grown on in the laboratory and then injected into the damaged area. In 2009 the winner of the Welsh Grand National had previously had stem cell treatment. In 2014 a trial of the same treatment began in human beings. Stem cells hold out the exciting prospect of being able to enable paralysed people to walk, blind people to see, and people threatened with degeneration of the brain to function normally.

In this chapter you will learn how gene expression is controlled to give many proteins from a single gene. You will look at transcription factors, spliceosomes, and the exciting new science of epigenetics, in which scientists are uncovering how the environment and genes interact. You will also learn about the different types of stem cells – totipotent, pluripotent and multipotent – and their roles in the body. You will discover how epigenetic modifications can result in totipotent stem cells becoming fully differentiated body cells. You will find out how pluripotent stem cells are derived from embryos and the opportunities they provide for us to develop new medical advances. You will consider the ethical considerations associated with the use of embryonic stem cells. You will find out more about ways in which differentiated fibroblasts can be reprogrammed to form induced pluripotent stem cells, and you will discuss reasons why these may be more acceptable to some people for medical and research use than stem cells derived from embryos.

All the maths you need

- Translate information between graphical, numerical and algebraic forms (*e.g. considering the rate of gene expression*)

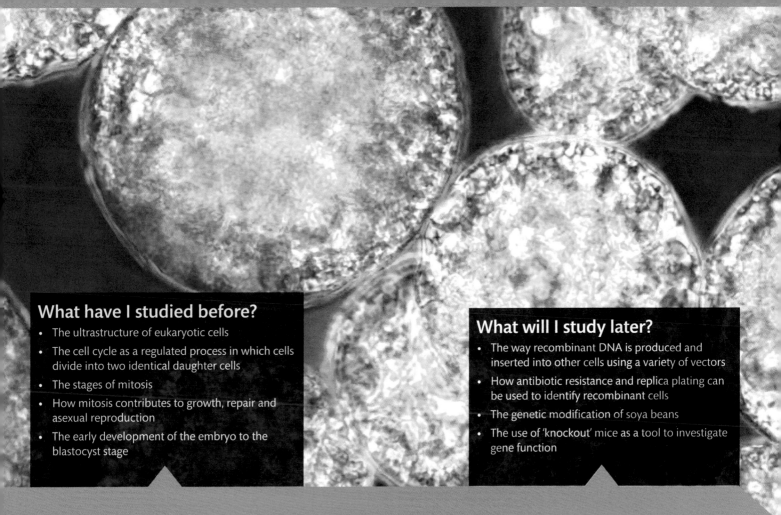

What have I studied before?

- The ultrastructure of eukaryotic cells
- The cell cycle as a regulated process in which cells divide into two identical daughter cells
- The stages of mitosis
- How mitosis contributes to growth, repair and asexual reproduction
- The early development of the embryo to the blastocyst stage

What will I study later?

- The way recombinant DNA is produced and inserted into other cells using a variety of vectors
- How antibiotic resistance and replica plating can be used to identify recombinant cells
- The genetic modification of soya beans
- The use of 'knockout' mice as a tool to investigate gene function

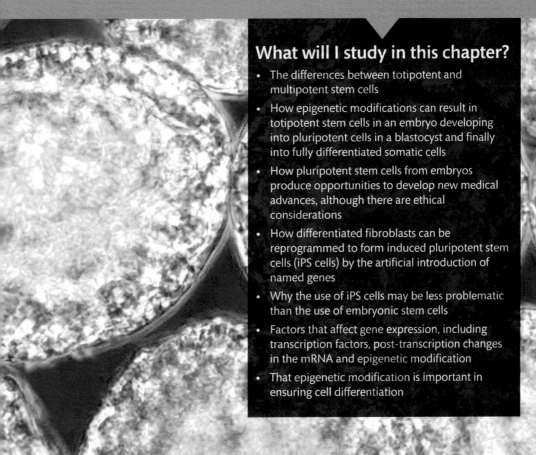

What will I study in this chapter?

- The differences between totipotent and multipotent stem cells
- How epigenetic modifications can result in totipotent stem cells in an embryo developing into pluripotent cells in a blastocyst and finally into fully differentiated somatic cells
- How pluripotent stem cells from embryos produce opportunities to develop new medical advances, although there are ethical considerations
- How differentiated fibroblasts can be reprogrammed to form induced pluripotent stem cells (iPS cells) by the artificial introduction of named genes
- Why the use of iPS cells may be less problematic than the use of embryonic stem cells
- Factors that affect gene expression, including transcription factors, post-transcription changes in the mRNA and epigenetic modification
- That epigenetic modification is important in ensuring cell differentiation

By the end of this section, you should be able to...

● describe transcription factors as proteins that bind to DNA

● explain the role of transcription factors in regulating gene expression

The genes in the DNA of a cell code for proteins. When a gene is expressed, the protein it codes for is synthesised. Many of these proteins are enzymes which in turn control everything that happens in the rest of the cell: the compounds that are synthesised, the compounds that are broken down, even which genes are subsequently expressed. So the factors that control gene expression control everything about a cell.

Gene expression in action

In a multicellular organism every cell contains the same genetic information but different cells perform different functions. They differentiate and develop into organs and tissues. As **cell differentiation** takes place to form tissues and organs, different types of cells produce more and more proteins specific to their cell type. The shape of the cell and the arrangement of the organelles will differ. One way in which scientists can measure the degree of differentiation between cells is to compare the proteins they contain using protein gel electrophoresis (see **Book 1 Section 3.1.5** and **Section 7.1.3**). As a result we know that almost all cells have a number of 'housekeeping' proteins in common. These are the proteins involved in the structures common to most cells, for example, the structural proteins of the membranes and the enzymes involved in cellular respiration.

But each different cell type also produces specific proteins that relate to the particular function of the cell. For example, the enzymes needed to produce insulin are only found in islet of Langerhans cells in the pancreas. This means that different genes must be expressed in different types of cells. By comparing the proteins found in different types of cells, scientists can measure the level of differentiation that has taken place and work out which genes have been expressed and which had been suppressed. Scientists have recently discovered that in human beings the testes have the biggest variety of extra proteins. They have 999 proteins on top of the normal housekeeping proteins. In comparison, the cerebral cortex of the brain has 318 extra proteins and the smooth muscle only expresses housekeeping proteins.

Gene probes illuminate gene expression

Much of our knowledge about genes and gene expression depends on the use of **gene probes**, which allow a particular section of DNA and mRNA in a cell to be identified.

To find a particular known gene out of the many present in the DNA of a human cell, you need a very specific probe. The gene probe finds the unique sequence of nucleotides on the DNA that make up the gene using a stretch of RNA that has the complementary sequence, in a process known as DNA–RNA hybridisation.

The DNA from the cells under investigation is isolated and heated gently. This breaks the weak hydrogen bonds holding together the two strands of DNA. Fluorescently-labelled mRNA for the required gene is added – this is the probe. Any DNA–RNA hybridisation that takes place shows that the required gene is present. This hybridisation is pinpointed using the fluorescent label on the mRNA. So, for example, by using gene probes you can show that both the red blood cells forming in the bone marrow (called reticulocytes) and the neurones in your brain have the gene for haemoglobin as part of their DNA. However, if you use a probe on the RNA from the cells rather than the DNA from the nuclei, the haemoglobin gene shows up in only the red blood cells. In other words, the gene for the production of red blood cells is present in both types of cell but it is only expressed in the red blood cells.

Controlling gene expression

The Human Genome Project found that the human genome has 20 000–25 000 individual genes. In a differentiated cell, between 10 000 and 20 000 of those genes are actively expressed. Different combinations are expressed in different cells, creating the variety of structure and function seen in cells of different tissues. The expression of a gene involves two key stages – transcription from DNA to mRNA and translation from mRNA to proteins (see **Book 1 Section 1.3.6**). Exerting controls at any of the stages of the process gives control over the expression of the genes. The different proteins present in a cell, and the quantities of the different proteins, determines the type of cell and its function in the body. In addition, the proteins can be changed once they have been synthesised, giving another level of control over the expression of a gene.

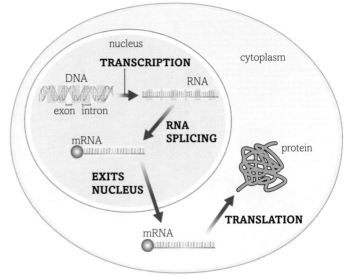

fig A Gene expression can be controlled at any stage in the process of protein synthesis.

Transcription factors and the control of gene expression

The most common way of controlling gene expression is by switching on and off the transcription of certain genes. Transcription describes the process by which the genetic code of the DNA is copied to a complementary strand of RNA (mRNA) before protein synthesis can take place. mRNA transcription is a very effective point at which to control gene expression, because a single mRNA molecule results in the production of many protein molecules at a ribosome or polysome.

Transcription factors are proteins that bind to the DNA in the nucleus and affect the process of transcribing the genetic material.

All transcription factors have DNA-binding regions that enable them to bind to specific regions on the DNA known as **promoter sequences**. Promoter sequences are usually found just above the starting point for transcription upstream of the gene – in other words, the 5′ end. Some transcription factors stimulate the transcription of a region of DNA simply by binding to a DNA promoter sequence, stimulating the start of transcription of that area of the DNA.

Other transcription factors bind to regions known as **enhancer sequences** and regulate the activity of the DNA by changing the structure of the chromatin, making it more or less open to RNA polymerase. An open chromatin structure is linked to active gene expression and closed chromatin structures are associated with gene inactivity. In this way these transcription factors can either stimulate or prevent the transcription of the gene. These regulatory sites can be at the site of the gene or they can be a considerable distance (thousands of base pairs) from the gene they are controlling, which makes it more difficult for scientists to work out exactly what is going on.

Often several different transcription factors will be involved in the expression of a single gene, giving many levels of control. Equally, a single transcription factor may control the activity of a number of different genes. It may stimulate the expression of one gene and suppress the expression of another. This is one way in which control over multiple genes is achieved. The control of transcription by transcription factors is the most common form of gene regulation. It means each gene can be expressed (switched on) or repressed (switched off) at different stages of development of the organism, in different cell types and under different circumstances in the body.

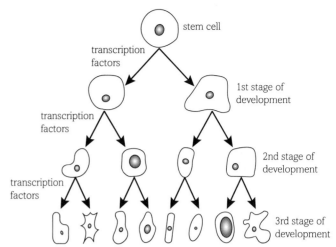

fig B The wide variety of different types of cells in an organism is a result of the activity of different transcription factors allowing different genes to be expressed or repressed in each cell type. This diagram gives you an idea of how many different types of cells can result from a single original as different transcription factors are switched on and off.

Questions

1 What is gene expression?

2 What is a transcription factor and why are they so important in gene expression?

3 Explain the difference between a promoter sequence and an enhancer sequence in the DNA strand.

4 Discuss the evidence that different genes are expressed in different cells of the same organism.

Key definitions

Cell differentiation is the process by which a cell becomes specialised for a particular function.

Transcription factors are proteins that bind to the DNA in the nucleus and affect the process of transcribing the genetic material.

Promoter sequences are specific regions on the DNA to which transcription factors bind to stimulate transcription.

Enhancer sequences are specific regions of the DNA to which transcription factors bind and regulate the activity of the DNA by changing the structure of the chromatin, making it more or less available to RNA polymerase, and so either stimulating or preventing the transcription of the gene.

By the end of this section, you should be able to...

● explain how post-transcriptional changes of mRNA in eukaryotic cells (RNA splicing) can result in different products from a single gene

● explain how gene expression can be changed by epigenetic modification including non-coding RNA, histone modification and DNA methylation

● describe the importance of epigenetic modification in ensuring cell differentiation

fig A The hearing mechanism of chicks is based on 576 proteins that are all produced by RNA splicing from a single gene.

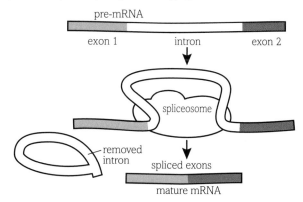

fig B RNA slicing to remove introns and sometimes rearrange exons enhances the expression of some genes.

One level of control of gene expression takes place in the nucleus where transcription factors control the transcription of regions of DNA into mRNA. However, there are many other ways in which the expression of a particular gene can be controlled in the cell.

RNA splicing

The RNA produced in the nucleus results from the transcription of all the DNA making up a gene, including the exons and introns (the non-coding DNA). We now know that this mRNA is not quite finished when it is first transcribed. A number of processes take place to modify it before it lines up on the ribosomes, so it is referred to as **pre-mRNA**. The modifications to the pre-mRNA always involve the removal of the introns and in some cases some of the exons are removed as well. The exons that are to be transcribed are then joined together by enzyme complexes known as **spliceosomes** to produce the mature functional mRNA.

The spliceosomes may join the same exons in a variety of ways in a process known as RNA splicing. As a result a single gene may produce several versions of functional mRNA transcribed from the same section of DNA. These different versions of mRNA code for different arrangements of amino acids, which in turn produce different polypeptide chains and so different proteins. Ultimately this can result in a single gene producing several different phenotypes. These post-transcriptional changes to mRNA lead to more variety in the phenotype than is coded for directly in the genotype. This is one of the ways in which the genotype can produce more proteins than there are genes.

One example of RNA splicing takes place in the cells which form the inner ear of a chick. As a result of spliceosome activity, a single gene gives rise to 576 different proteins that affect the sensitivity of the hairs in the inner ear. This enables chicks to hear a range of sounds from 50 to 5000 Hz.

Further modification of proteins may also take place after they have been synthesised. A protein that is coded for by a gene may remain intact or it may be shortened or lengthened by enzymes to give a variety of other proteins.

Epigenetics

Epigenetics is a relatively new area of research in biology. It studies genetic control by factors other than the base sequences on the DNA. So in a way RNA splicing is a form of epigenetic control, because it changes the mRNA and proteins produced from the original genetic code. Scientists are becoming increasingly aware both of the role of epigenetic control in the normal development of specialised cells, and of how environmental factors can influence the cell biochemistry and result in epigenetic changes that can be inherited by future generations. Three intracellular systems that can interact to control genes include:

• DNA methylation

• histone modification

• non-coding RNA.

DNA methylation

The methylation of DNA (addition of a methyl $-CH_3$ group) is a widely used mechanism in epigenetics. The addition of the methyl group always occurs at the site where cytosine occurs next to a guanine in the DNA chain with a phosphate bond between them (a CpG site). The methyl group is added by a DNA methyltransferase enzyme. **DNA methylation** can also modify the structure of the histones, so it can have an epigenetic effect in more than one way.

DNA methylation always silences a gene or a sequence of genes. The methyl group changes the arrangement of the DNA molecule and prevents transcription from taking place. DNA methylation has been found to be extremely important in controlling gene expression and has a big role in many processes including embryonic development and X chromosome inactivation. In many adult specialised cells, many genes are silenced by DNA methylation most or all of the time.

DNA demethylation is equally important. The removal of the methyl group enables genes to become active so they can be transcribed. Researchers are increasingly finding problems with DNA methylation or demethylation associated with diseases, including a number of human cancers.

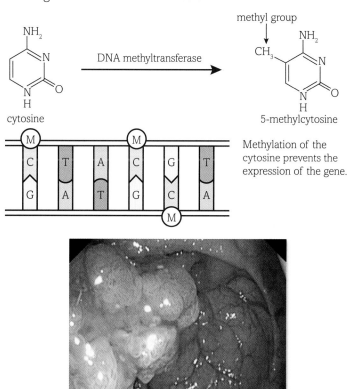

cytosine

methyl group

5-methylcytosine

DNA methyltransferase

Methylation of the cytosine prevents the expression of the gene.

fig C DNA methylation silences many genes. Scientists are finding demethylation of DNA may well be a factor in the growth of the tumour in some cancers, including ovarian and bowel cancers.

Histone modification

Histones can be modified in a number of ways to affect the transcription of DNA and therefore gene expression. As you saw in **Book 1 Section 2.3.1**, histones are positively charged proteins. DNA helices wind around the histones to form **chromatin**, the DNA/protein complex that makes up the chromosomes. The histones determine the structure of the chromatin. When the chromatin is densely supercoiled and condensed the genes are not available to be copied to make proteins and this is known as **heterochromatin**. Active chromatin is more loosely held together, with uncoiled regions of DNA opening up more genes for transcription so that new proteins are made. This is one way in which cells of different types are produced.

Many different factors affect the modification of the histones, including steroid hormones. Modification processes include:

- **Histone acetylation** – an acetyl group ($-COCH_3$) is added to one of the lysines in the histone structure. Adding an acetyl group usually opens up the structure and activates the chromatin, allowing genes in that area to be transcribed. Removing an acetyl group produces heterochromatin again.

- **Histone methylation** – a methyl group ($-CH_3$) is added to a lysine in the histone. Depending on the position of the lysine, methylation may cause inactivation of the DNA or activation of a region. Methylation is often linked to the silencing of a gene and even whole chromosomes. For example, histone methylation plays a role in the silencing of one of the X chromosomes in every cell in female mammals.

Did you know?

Histone modification in action

Moulting, the shedding of the exoskeleton of insects, is controlled by two hormones. Ecdysone, the 'moulting and metamorphosis' hormone, controls the events of the moult. It is a steroid hormone, which was first extracted from the pupae of silkworms. Juvenile hormone controls the kind of moult that occurs. As juvenile hormone levels decrease, more adult characteristics occur. When there is no juvenile hormone, the insect becomes an adult.

The way in which ecdysone has its effect has been studied using the larvae of *Drosophila* (fruit fly) and *Chironomus* (midge). These insects have giant chromosomes in their salivary glands. They are 100 times thicker and 10 times longer than normal chromosomes and easily visible with the light microscope.

Banding is visible on these chromosomes due to supercoiled areas of DNA. When an insect is undergoing a moult, or when ecdysone is injected artificially into an insect, 'puffs' appear on the chromosomes (see **fig D**). These puffs appear to be pieces of genetic material from supercoiled areas that have been opened up and made available for transcription. It is thought that they carry information about the new proteins needed in a more adult stage of the life cycle. This supports the view that steroid hormones have a direct effect on the DNA of a cell. Scientists think that many steroid hormones act in a similar way. Unfortunately not many organisms possess giant chromosomes, so the effect is not always easy to observe!

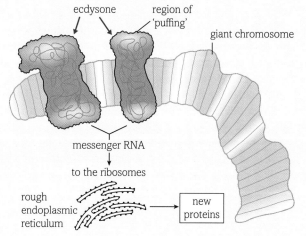

ecdysone

region of 'puffing'

giant chromosome

messenger RNA

to the ribosomes

rough endoplasmic reticulum

new proteins

fig D Chromosome puffs appear as the result of the hormone ecdysone. They are followed by the synthesis of much RNA, which in turn results in the formation of new proteins.

Non-coding RNA

About 90% of the human genome is transcribed into mRNA, but only about 2% of those RNA molecules actually code for proteins. Much of the rest of the non-coding RNA (ncRNA) seems to affect the transcription of the DNA code or modifies the products of transcription. Genes and even whole chromosomes have been shown to be silenced by ncRNAs. In female mammals, one of the X chromosomes in every cell is inactivated at random. This is largely due to the presence of an ncRNA called X-inactive specific transcript (Xist), which is produced by the active Xist gene on the inactive chromosome. The ncRNA coats one of the X chromosomes in female cells and deactivates it. The chromosome supercoils and condenses to form the stable, inactive Barr body. One X chromosome is inactivated in order to maintain the balance of gene products in males, XY, and females, XX. Another role for ncRNA in epigenetics seems to be in chromatin modification, where it acts on the histones to make areas of the DNA available or unavailable for transcription.

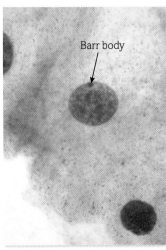

Barr body

fig E The random inactivation of one X chromosome in female mammals is responsible for the coat patterns of tortoiseshell cats and for the Barr body present in the nuclei of female cells. Female cats display the tortoiseshell patterns, while the male cats are ginger.

Cell differentiation

Cell differentiation takes place as unspecialised cells switch different genes on and off as needed to become specialised cells. Scientists are now aware that most of these changes are the result of epigenetic modification of the genetic material. Epigenetic modifications ensure that a wide range of very specific proteins are made within the cell as it differentiates for a specific function. The modification is often the result of DNA methylation or demethylation, or of histone modification. These epigenetic changes may be in response to internal stimuli from the cell itself or in response to changes outside the cell (an external stimulus) which affect the inside of the cell. For example, the sex hormones produced at puberty trigger changes in many cells in the body as a result of their effect on DNA methylation in the nuclei of the cells. You will learn more about cell differentiation in **Section 7.2.3**.

Learning tip

When you answer a question on epigenetics, be clear about the difference between histone methylation and DNA methylation.

Questions

1 How does RNA slicing enable a single gene to code for a number of proteins?

2 What is epigenetics?

3 Compare and contrast histone methylation and DNA methylation.

Key definitions

Pre-mRNA is the mRNA that is transcribed directly from the DNA before it has been modified.

Spliceosomes are enzyme complexes that act on pre-mRNA, joining exons together after the removal of the introns.

DNA methylation is the methylation of DNA (addition of a methyl $-CH_3$ group) to a cytosine in the DNA molecule next to a guanine in the DNA chain and prevents the transcription of a gene.

DNA demethylation is the removal of the methyl group from methylated DNA enabling genes to become active so they can be transcribed.

Chromatin is the DNA/protein complex that makes up the chromosomes.

Heterochromatin is the densely supercoiled and condensed chromatin where the genes are not available to be copied to make proteins.

Histone acetylation is the addition of an acetyl group ($-COCH_3$) to one of the lysines in the histone structure, which opens up the structure and activates the chromatin, allowing genes in that area to be transcribed.

Histone methylation is the addition of a methyl group ($-CH_3$) to a lysine in the histone. Depending on the position of the lysine, methylation may cause inactivation or activation of the region of DNA.

By the end of this section, you should be able to...

● explain what is meant by a stem cell, including the differences between totipotent, pluripotent and multipotent stem cells

● explain how epigenetic modifications can result in totipotent stem cells in the embryo developing into pluripotent cells in the blastocyst, and finally into fully differentiated somatic cells

Fertilisation starts a complex series of events that will eventually lead to the birth of a fully formed new individual. In humans the fertilised egg cell (zygote) has the potential to form all of the 216 different cell types needed for an entire new person. It is said to be **totipotent**. The future roles of individual cells are decided relatively early in the life of an embryo.

The early stages of development

As outlined in **Book 1 Section 2.4.5**, the first stage of embryonic development is known as cleavage. Cleavage involves a special kind of mitosis where cells divide repeatedly without the normal interphase for growth between the divisions. The result of cleavage is a mass of small, identical and undifferentiated cells forming a hollow sphere known as a blastocyst (**fig A**). In humans this process takes about 5–6 days, and it takes place as the zygote is moved along the oviduct towards the uterus. One large zygote cell forms a large number of small cells in the early embryo. The tiny cells of the early human embryo are known as **embryonic stem cells**. Stem cells are undifferentiated cells that have the potential to develop into many different types of specialised cells from the instructions in their DNA.

The very earliest cells in an embryo are totipotent like the zygote. By the time the blastocyst is formed the cells in the inner layer have already lost some of their ability to differentiate and the outer layer of cells goes on to form the placenta. The inner layer of cells can form almost all of the cell types needed in future, but not tissue such as the placenta. These cells are known as **pluripotent** embryonic stem cells.

Types of stem cells

Stem cells come in a number of different types, with different levels of potency.

Embryonic stem cells

The earliest embryonic cells are totipotent. By the blastocyst stage, when the embryo is implanted in its mother's uterus, the inner cells of this ball are pluripotent. Pluripotent stem cells become more specialised as the embryo develops, for example forming blood stem cells, which give rise to blood cells, and skin stem cells, which give rise to skin cells. By around three months of pregnancy the cells have become sufficiently specialised that when they divide they only form more of the same type of cell (see **Section 7.2.4**).

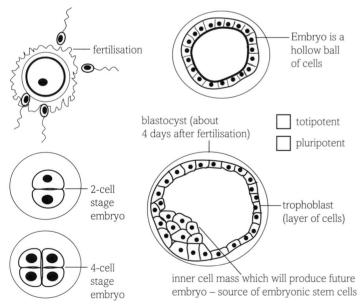

fig A In the very earliest stages of pregnancy any one of the cells has the potential to form an entire human being, but this ability is lost by the blastocyst stage.

(labels in fig A: fertilisation; Embryo is a hollow ball of cells; blastocyst (about 4 days after fertilisation); totipotent; pluripotent; 2-cell stage embryo; trophoblast (layer of cells); 4-cell stage embryo; inner cell mass which will produce future embryo – source of embryonic stem cells)

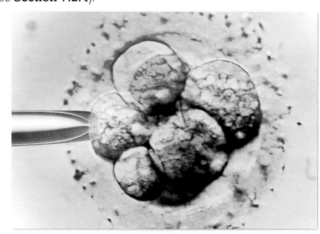

fig B Each of the 8 cells in this 3-day-old human embryo is totipotent. They have the potential to form every type of adult human cell. Within another day the cells will have divided more and become pluripotent as the result of some genes being switched off permanently.

Umbilical cord stem cells

The blood that drains from the placenta and umbilical cord after birth is a rich source of pluripotent stem cells. If this blood is frozen and stored, in theory those stem cells will be available throughout the life of the child should they or their family need them later for stem cell therapy (see **Section 7.2.4**). It may become possible to store stem cells from every newborn baby ready for when they might need them. However, it would take a lot of storage space to do this for everyone and would be expensive. Also, there is some evidence that the precursor cells of conditions like leukaemia are already present in the blood at birth. Currently there is very little evidence of cord blood being used successfully to treat anyone and many scientists are unconvinced of the value of the process. Because the benefits are as yet largely unproven, the only way parents in the UK can store umbilical cord blood for their children at the moment is to pay to do it privately.

Adult stem cells

An adult human is made up of many different types of highly specialised cells. However, some **adult stem cells** remain as undifferentiated cells found among the normal differentiated cells in a tissue or organ. They can differentiate when needed to produce any one of the major cell types found in that particular tissue or organ. For example, white bone marrow contains stem cells that can form white blood cells. Another term for adult stem cells is **somatic stem cells**.

There are only a very small number of adult stem cells in each different tissue. They are difficult to extract and most of them form a very limited range of differentiated cells. They are said to be **multipotent**. What is more, they are difficult to grow in the laboratory.

The formation of different cell types in development

As cells grow and divide in a multicellular organism, they become increasingly differentiated and specialised to carry out particular functions. Almost every cell in an organism contains the DNA instructions to make any other type of cell and, in the earliest stages of an embryo, each cell can produce any tissue. However, only days after conception, cells are already predestined or determined to become one type of tissue or another. This cell determination is closely linked to the position of the cells in the embryo. If the cells are surgically removed from the embryo and grown on, they will still produce the predetermined cell type, even if it is entirely inappropriate in the new setting (see **fig C**). No one is yet entirely sure of the mechanism of cell determination although epigenetics must play a part. Following determination, the cells differentiate and develop into organs and tissues. However, it is clear that, in almost all cases, differentiation is a one-way street. Once cells become specialised they do not become unspecialised again unless something goes wrong. For example, in many different types of cancer, scientists find evidence of epigenetic changes in the cells that make them less specialised and more rapid in their divisions.

Evidence for cell determination

Scientists working on cell determination and differentiation use a variety of organisms. Obviously such research on humans would be ethically unacceptable. However, amphibians such as frogs, and the fruit fly, *Drosophila*, have both proved very useful.

In a normal early amphibian embryo, cells undergo the first stages of differentiation and go on to become particular tissues as development continues. The hypothesis that this determination occurs irreversibly at an early stage was disproved by transplanting tiny patches of tissue from one area of an early embryo to another. The transplanted cells differentiated to form the tissue linked to their new position, not their original position (see **fig C**). When the same experiment was carried out with cells from a slightly older embryo, scientists saw that determination was complete and the cells differentiated to form the tissue determined by their original position rather than their new position.

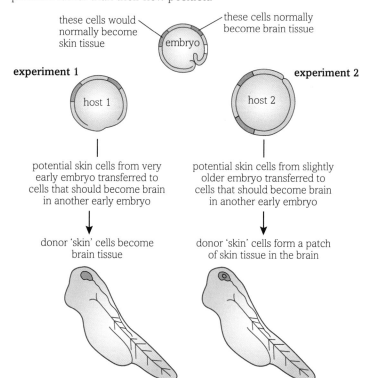

fig C Scientists use evidence from experiments like this to build up a model of how cell determination and differentiation works and when it takes place.

The development of an organism

Totipotent stem cells in the embryo become pluripotent stem cells in the blastocysts and finally become fully differentiated somatic cells in the body of the mature organism. How are these changes brought about?

The most common way of controlling gene expression is by switching on and off the transcription of certain genes. As you saw in **Sections 7.2.1** and **7.2.2**, this can be achieved in several ways including the action of transcription factors and epigenetic mechanisms such as DNA methylation, histone modification and ncRNAs. So, for example, during the process of differentiation some parts of the chromosomes undergo supercoiling to prevent certain genes being transcribed, and other areas uncoil, opening them up for transcription so that new proteins are made. Some genes are activated and others are silenced. As development progresses more genes are silenced in each cell. It is the combination of the particular genes that are activated or silenced that results in the different characteristics of fully differentiated mature cells.

An example of epigenetic control in human development

Haemoglobin is a vital molecule that carries oxygen around the body. Adult human haemoglobin contains two alpha and two beta globin chains. Fetal haemoglobin, which has a stronger affinity for oxygen than adult haemoglobin, contains two alpha and two gamma globin chains (see **Book 1 Section 4.3.3**).

During human development from embryo to fetus to baby, different versions of the globin genes are switched on and off. The levels of the different types of globin chains in the blood change through the 40 weeks of pregnancy and beyond. During this time, not only do genes for the different proteins get switched on and off, but also the genes are activated in different tissues as development progresses (see **fig D**). Globin production moves from the yolk sac in the embryo to the liver in the fetus and then the spleen, with the genes in the bone marrow taking over almost completely by the time of birth.

The genes controlling the production of alpha globin are needed in both the fetus and mature baby. The genes controlling fetal gamma globin are very important during fetal development but need to be silenced around the time of birth. The adult beta globin genes need to be activated just before birth in the bone marrow. This is a clear example of epigenetics at work in a number of tissues, changing the proteins which are made.

The mechanism by which the genes that code for the different chains are controlled is still the subject of a lot of active research. If scientists can work out how to reactivate the fetal haemoglobin gene in children and adults, they may be able to overcome the problems presented in genetic conditions such as sickle cell anaemia and thalassemia, which affect the structure of adult haemoglobin. So far scientists have found evidence of a number of epigenetic control mechanisms:

- Histone acetylation appears to activate the gamma globin gene in the fetus.
- DNA methylation appears to play a major role in silencing the fetal gamma globin genes just before and after birth.
- Histone methylation appears to play a complementary role to DNA methylation in silencing the fetal gamma globin.
- ncRNAs have been associated with the process, but scientists are not yet sure what they do.
- A number of transcription factors have been shown to be key in the switch to the production of beta globin in the spleen and bone marrow as the fetus approaches full term.

As we understand more about the control of cell differentiation, our ability to control stem cells will increase too.

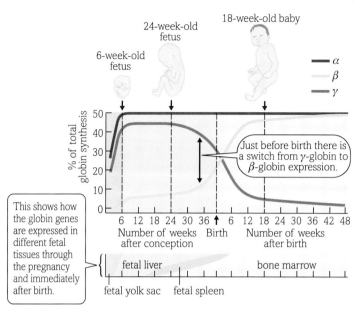

fig D Different genes for the production of globin molecules are switched on and off in different tissues during the development of a human being from an embryo to a fetus to a baby.

Questions

1 Compare embryonic and adult stem cells.

2 What are the differences between totipotent, pluripotent and multipotent cells?

3 Explain the difference between cell determination and cell differentiation.

4 Suggest reasons why animals such as fruit flies and frogs are often used as experimental organisms by scientists investigating cell differentiation.

5 Suggest reasons why scientists want to understand the control mechanisms behind the switching off and on of genes during the development of an organism.

Key definitions

Totipotent describes an undifferentiated cell that can form any one of the different cell types needed for an entire new organism.

Embryonic stem cells are the undifferentiated cells of the early human embryo with the potential to develop into many different types of specialised cells.

Pluripotent describes an undifferentiated cell that can form most of the cell types needed for an entire new organism.

Somatic stem cells/adult stem cells are undifferentiated cells found among the normal differentiated cells in a tissue or organ that can differentiate when needed to produce any one of the major cell types found in that particular tissue or organ.

Multipotent describes a cell that can form a very limited range of differentiated cells within a mature organism.

By the end of this section, you should be able to...

- explain how pluripotent stem cells provide opportunities to develop new medical advances, although there are ethical considerations

- explain how differentiated fibroblasts can be reprogrammed to be induced pluripotent stem cells (iPS cells) by the artificial introduction of named genes

- explain why the use of iPS cells may be less problematic than the use of embryonic stem cells

Scientists are working hard to find ways of using stem cells to treat medical conditions where the patient's own cells are damaged or faulty.

For about 30 years we have known about the stem cells in bone marrow, which are capable of forming all the different kinds of blood cells, and bone marrow transplants are used regularly in the treatment of certain cancers and immune system diseases. These transplants need to be taken from close relatives or from strangers who have matching immune systems, otherwise the body will reject and destroy the transplant.

Over time scientists have discovered stem cells in the bone marrow that can generate bone, fat, cartilage and fibrous tissue, and in the 1990s stem cells were discovered in the brain that can form the three main types of brain cells. Adult stem cells have been found in many different organs and tissues. In theory these cells can be extracted from a patient and treated so that they develop into the new cells that the patient needs.

Stem cell therapy

The use of embryonic stem cells to produce new tissues has not yet had the success that scientists hoped for when they were first cultured. It has been very difficult to control the differentiation of the cells. Some of the early treatments that were tried resulted in patients being cured of one condition but developing cancer. There have also been some ethical objections to the use of cells from embryos.

The problems with embryonic stem cells have driven research into adult stem cells and these have already been used successfully to produce new body parts such as the trachea. Stem cells from a patient are seeded onto a framework, which may be collagen based and from a donor, or completely synthetic. The stem cells grow to form the required cells and the new trachea can be returned to the patient with no risk of rejection. Hopefully other organs will be available soon.

Another area of active research using adult stem cells is in the repair of hearts damaged by heart attacks. Adult stem cells are injected into the heart and in some cases the improvement in

function has been marked. Some studies used adult stem cells from the bone marrow, others used them from the heart itself. Using the patient's own adult stem cells avoids the risk of rejection of new tissue. All of the studies so far have been on small numbers of patients and there is still much more research to be done.

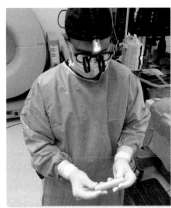

fig A At the moment organs grown using adult stem cells from a patient, like this new trachea, are rare. In future it may be the normal way to replace a damaged or non-functioning body part.

Therapeutic cloning

Somatic cell cloning or **therapeutic cloning** is an experimental technique that scientists hope can be used in the future to produce large quantities of healthy tissue to treat people with diseases caused by faulty cells, such as type 1 diabetes or Alzheimer's disease.

The first step is to produce healthy cloned cells from the patient. This is done by removing the nucleus from one of their normal body cells and transferring it to a human ovum which has had its original nucleus removed. A mild electric shock is used to fuse the nucleus with the new cell and trigger development. The newly formed pre-embryo cell starts to develop and divide, producing a collection of embryonic cells with the same genetic information as the patient. This is a cloned human embryo, but it has been formed with no intention to clone a whole human being. The embryo is simply a source of stem cells with genetic markers that match the patient perfectly. Stem cells are harvested from the embryo, which is destroyed in the process. The embryonic stem cells can then be cultured in a suitable environment so that they differentiate into the required tissue. These tissue cells can then be transferred to the patient, where they can do their job without the risk of the immune system rejecting them. This type of treatment is still very much at the experimental stage of development, not least because scientists are still trying to determine the exact triggers that control cell differentiation.

There is an additional factor to be considered in the treatment of genetic diseases. The adult stem cell nucleus would need to be genetically modified before being added to the empty ovum, otherwise the cultured stem cells would carry the genetic mutation that caused the problem in the first place.

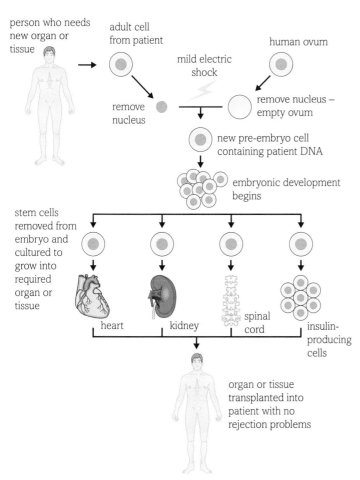

person who needs new organ or tissue

adult cell from patient

remove nucleus

mild electric shock

human ovum

remove nucleus – empty ovum

new pre-embryo cell containing patient DNA

embryonic development begins

stem cells removed from embryo and cultured to grow into required organ or tissue

heart   kidney   spinal cord   insulin-producing cells

organ or tissue transplanted into patient with no rejection problems

fig B Therapeutic cloning is still very experimental but this diagram shows how scientists and doctors hope it will be used in the future.

Pitfalls and potential benefits of stem cell therapy

There are problems that relate to the uses of all kinds of stem cells. For example, at the moment no one is quite sure how the genes in cells are switched on or off to form particular types of tissue. However, our understanding of transcription factors and epigenetic mechanisms is growing all the time, so the point when scientists can fully control the differentiation of embryonic stem cells is getting closer.

There are also risks associated with stem cell therapy. There are concerns that stem cells could cause the development of cancers in the body. There is some evidence that people who have been given bone marrow transplants (stem cell treatment) to help them overcome leukaemia are at higher risk of developing other cancers later. However, there are also great potential advantages. At the moment there are no cures for many of the conditions that stem cell therapy might solve. The ability to produce tailor-made cells to take over the function of damaged ones would revolutionise medicine.

Organ transplants create their own problems. Glycoproteins on the surface of your cell membranes act as part of your cell recognition system. Your immune system recognises your own cells ('self') and different cells ('non-self') and destroys the non-self cells. This works well if you pick up an infection, but is potentially lethal if your immune system attacks a transplanted organ. After a

transplant, people have to take immunosuppressant drugs for the rest of their lives, and this puts them at higher risk of infectious diseases. One advantage of using embryonic stem cell therapy is that it could avoid this risk of rejection. A mother's immune system does not attack and destroy her developing embryo, even though the markers on its cells are different from hers. Maybe new cells or organs created from embryonic stem cells would enjoy this same protection. And if the cells are cultured from the individual who needs them, rejection would be a thing of the past.

Induced pluripotent stem cells

In 2006 a team of researchers in Japan made an astonishing breakthrough. They took adult mouse cells and, using genetic engineering techniques, reprogrammed them to become pluripotent again. Effectively they produced stem cells without an embryo. What is more, the **induced pluripotent stem cells (iPS cells)** renew themselves. The next stage was to repeat the process in humans. The team used harmless, genetically-modified viruses to carry a group of four genes for specific transcription factors into skin cells taken from a 36-year-old and synovial fluid taken from a 69-year-old. The genes they used were Oct4, Sox2, Klf4 and c-Myc.

It has since been shown that Oct3 or Oct4 and one of the *Sox* family of genes are necessary to produce pluripotent stem cells. Other genes, including the *Myc* family, and the *Klf* family, increase the likelihood and efficiency of producing pluripotency in adult cells.

These induced pluripotent human stem cells appear to be very similar, although not identical, to human embryonic stem cells in their behaviour. The Japanese team even managed to make their iPS cells develop into brain cells and heart muscle cells.

One of the greatest potential benefits of this new technology is that it overcomes the ethical objections to using embryonic tissue, even embryos specifically created as a source of stem cells. There is also no risk of rejection if cells from an individual are used to provide their own stem cells. Persuading the cells to become pluripotent is not easy and making them differentiate into the tissue you want is even more difficult. How well, and for how long, they will behave as pluripotent stem cells remains to be seen. Perhaps most worryingly, the new stem cells show a tendency to become cancerous very quickly as Klf4 and c-Myc are oncogenes, genes that are particularly associated with cancer development.

In spite of the difficulties, many scientists think that iPS cells may well be the way forward in stem cell medicine. As there are still major problems to be overcome, there is a strong feeling in the scientific community that research into both embryonic stem cells and iPS cells should continue until it is clear which will produce the best treatments for patients in the long term. So far, most of the successful work on therapies has been done using embryonic stem cells (see **Section 7.2.3**).

Who could benefit from stem cell therapy?

Stem cell therapy has the potential to treat a wide range of diseases that are caused by faulty cells. Here are some examples of areas where scientists feel pluripotent stem cells could give real therapeutic value.

Parkinson's disease

Parkinson's disease is a brain disorder affecting about 2% of people over 65 years of age. Nerve cells in the brain that produce dopamine (dopamine neurones) stop working and are lost. As dopamine levels fall, people develop uncontrollable tremors in their hands and body. Their body becomes rigid and eventually they cannot move normally at all. Although drug treatments have improved a lot in recent years, there is still no long-term cure. Scientists hope that stem cell transplants will allow them to replace the lost brain cells and restore dopamine production, letting people return to a normal life.

Scientists have managed to get mouse embryonic stem cells to form dopamine neurones. These cells were transplanted into the brains of rats that had the symptoms of Parkinson's disease. The cells grew and released dopamine and the ability of the rats to control their movement improved. Scientists also showed that when dopamine-producing cells from relatively mature fetuses were transplanted into the brains of human adult patients with Parkinson's, the cells survived and produced dopamine for many years. It is, however, rarely possible to obtain mature fetal cells, so this is not a viable treatment. The hope is that pluripotent stem cells could be implanted into the brain and take over dopamine production from the damaged and dead cells. Most scientists agree that pluripotent stem cells offer the best current hope of an effective long-term treatment for Parkinson's disease.

Type 1 diabetes

Type 1 diabetes usually develops when people are young. The glucose-sensitive, insulin-secreting islet of Langerhans cells in the pancreas are destroyed or stop making insulin, so the blood glucose concentration is uncontrolled. This can be very serious or even fatal. Although insulin injections work well enough, people affected by diabetes have to monitor their food intake and blood glucose concentration and inject insulin regularly. Stem cell therapy could give them working pancreas cells again, restoring insulin production and therefore blood glucose control.

Scientists have succeeded in getting some mouse embryonic stem cells to form a group of cells that look and work just like insulin-producing tissue. Some of these cells were transplanted into mice with diabetes where they produced a rise in the blood concentration of insulin and improved control of blood glucose. Scientists have also successfully transplanted insulin-producing islet of Langerhans cells directly from one person to another.

In 2014 scientists at Harvard University reported a major breakthrough. Starting with human embryonic stem cells, they have developed mature human glucose-sensitive insulin-producing beta cells in the large quantities needed to use them in patients. The cells are now being trialled in animals with diabetes and if those trials are successful, the team hopes to use the new cells in human clinical trials within a few years.

Damaged nerves

So far there is no medical cure for damaged and destroyed nervous tissue in the brain and spine. These nerves do not usually regrow and so someone who suffers a major injury to their spine may be permanently paralysed below the location of the damage. Embryonic stem cells have been transplanted into mice and rats with damaged spines and the animals regained a certain amount of control and movement of limbs that had been paralysed. Recently, cells from the brain area linked to the sense of smell were transplanted into a paralysed man and resulted in the return of some control. These were not stem cells, but it showed that progress in people is possible. Examining the spinal cords of these rats and mice showed that the embryonic stem cells had grown into working adult nerve cells and the damaged spinal cords had at least partly been rejoined, offering hope for future human treatments.

Organs for transplants

Many people die because their organs no longer function properly. Hearts, kidneys, livers and many other organs can be replaced by transplant, but only if there is a suitable donor organ available. There is a desperate need for new organs, preferably ones that will not cause rejection problems when they have been transplanted. In 2013, a team of researchers in Australia produced embryonic stem cells from human skin cells. They then manipulated the differentiation of these stem cells into minute functioning kidney units. Although they were only millimetres across, the mini-kidneys functioned and produced urine. Also in 2013, a Japanese team produced functioning 3D liver buds from iPS cells, which connected with the circulatory systems of mice when they were transplanted into them. The hope is that eventually pluripotent stem cells will provide the huge supply of organs for transplantation we need globally each year.

Stem cell success

So far the number of successful therapies using pluripotent stem cells has been very small. Many ideas are being developed and trialled in animals, but so far not many have actually had any impact on human health although recent results with age-related muscular degeneration are very positive. Scientists and doctors expect that numbers of successful treatments will increase dramatically in the next 10 years.

Ethical questions

There have been some very powerful reactions to stem cell technology. As well as the many practical problems to be overcome before stem cell therapy becomes a standard treatment, society has many ethical issues to deal with.

The four ethical principles are:

- Respect for autonomy. Respect individuals, by not performing procedures without consent.
- Beneficence. The aim of doing good, by giving medicine to relieve suffering, etc.
- Non-maleficence. To do no harm.
- Justice. Treat everyone equally and share resources fairly, so as to avoid discrimination.

Embryonic stem cells

Perhaps the greatest problems with research into pluripotent cells from embryos are ethical ones. The cells used come either from aborted embryos or from spare embryos in fertility treatment.

Many people think the new work is a major breakthrough with the same potential to change healthcare as the discovery of antibiotics more than 70 years ago, offering hope of a cure to millions of people for whom there is, at the moment, no hope.

The vast majority of human embryos never make it beyond the early stages of development to form living babies, so the argument for using a small number of early embryos is considered acceptable in this context by many people. People who are in favour of stem cell research suggest that once tissue lines from a relatively small number of willingly donated embryos are established, the need to use further embryos will be reduced.

Also, many supporters of embryonic stem cell research feel that adult stem cells do not offer a good alternative, because they are much more limited in their scope for forming new and different tissues. They want research funding to be directed mainly at embryonic stem cell work.

However, other people feel that the use of embryonic tissue is wrong and an abuse of human rights. Some people, including many religious groups, think it is wrong to use a potential human in this way. Some objectors feel that every early human embryo has the potential to become a living human being and so should be afforded the same human rights as a fully grown adult. Others have strong religious convictions that using embryos is killing, and therefore wrong. They think that no medical advances are worth the moral evil of using embryonic tissue as a source of stem cells.

Many of these people feel that the use of adult stem cells offers an exciting and acceptable possible alternative to embryonic cells and they campaign for research funding to be directed to projects using these ethically less-sensitive cells. In addition, the use of embryonic stem cells from the umbilical cord of newborn babies may help to overcome many reservations. However, storing each baby's cord blood would be logistically and practically extremely difficult and very expensive, and there is little evidence yet that any therapies developed from the blood would work.

Therapeutic cloning

In therapeutic cloning, the embryos are not created from adults cells to develop into a new human being, they are simply there to provide embryonic stem cells. This is why many people are very optimistic about the future potential of this new technology and see no major ethical stumbling blocks to overcome. However, other people fear that if the cloning is allowed for therapeutic

purposes it could easily be taken further, with the cloned embryos implanted into a uterus to produce a cloned baby. What is more, even if the embryo used as a source of stem cells is produced in an unorthodox way, it is still an embryo. Many people have the same ethical or religious objections to using these for research as they do to the use of any other embryos. So therapeutic stem cell cloning can raise even more ethical problems than embryonic stem cell research.

iPS cells

iPS cells are pluripotent so they can be turned into most cell types by careful manipulation of transcription and epigenetic factors. They come from the individual patient so there are no issues of rejection. They do not come from embryos and they are not capable of forming a new embryo. The ethical issues are all answered by the use of this technology. The biggest problem is that these cells are not so easy to grow and manipulate as pluripotent embryonic stem cells.

Questions

1   What is the biggest obstacle scientists need to overcome to enable them to develop useful therapies with embryonic stem cells?

2   What are the main ethical objections to using embryonic stem cells and do you think they are issues that can be answered by scientific argument?

3   What are the advantages and disadvantages of umbilical blood and adult tissue as alternative sources of stem cells?

4   Choose one of the methods for producing embryonic stem cells. Prepare arguments for a debate on whether this treatment or the new method of producing iPS cells from skin cells should receive major funding in the next 10 years, or whether both should be funded.

Key definitions

Therapeutic cloning is an experimental technique used to produce embryonic stem cells from an adult cell donor.

Induced pluripotent stem cells (iPS cells) are adult cells that have been reprogrammed by the introduction of new genes to become pluripotent again.

1 Which of the following is the process by which one gene can produce many *different* proteins?
   A pre-translational modification
   B post-translational modification
   C pre-transcriptional modification
   D post-transcriptional modification   [1]
   [Total: 1]

2 Which of the following can DNA methylation **not** do?
   A silence a gene
   B induce a gene
   C affect embryo development
   D inactivate an X chromosome   [1]
   [Total: 1]

3 An experiment was carried out to study β-galactosidase activity in *E. coli* bacteria.

   β-galactosidase is an enzyme which normally breaks down the sugar lactose to galactose and glucose.

   Two test tubes were set up at 35 °C containing *E. coli* in nutrient broth, buffer solution, ONPG and methylbenzene (to break up the bacterial cells).

   ONPG is broken down by β-galactosidase to ONP and galactose. ONP is a yellow substance.

   (a) (i)   Explain how you would maintain the test tubes at 35 °C. [1]
        (ii) Explain why buffer solution is used in this experiment.   [2]
        (iii) In this experiment, lactose was added to one test tube but not the other. The experiment was repeated 5 times. The results are indicated below.

   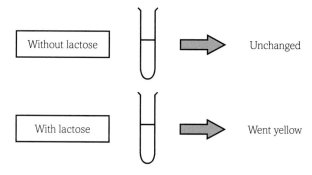

   Explain why a yellow colour appeared only in the tube with lactose.   [2]

   (b) *P. haloplanktis* is a species of bacterium which produces β-galactosidase in a similar way to *E. coli*. The graph below shows the rate at which β-galactosidase from each species causes ONPG to break down to (yellow) ONP, at a range of temperatures.

   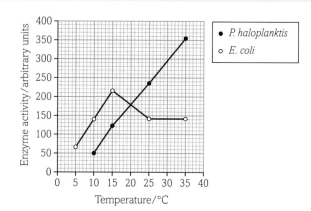

        (i)   Deduce a conclusion from the data shown in the graph.   [3]
        (ii) β-galactosidase is used to break down lactose in cheese manufacture. Explain why using β-galactosidase from *P. haloplanktis*, rather than from *E. coli*, might lead to increased efficiency.   [1]
   [Total: 9]

4 (a) A single stem cell can give rise to many genetically identical cells of different types. There are, for example, adult stem cells in the human brain which are capable of producing the different types of brain cells including nerve cells (neurones).

        Scientists are trying to find ways of growing such adult brain stem cells in the laboratory.
        (i)   Name the type of cell division by which a stem cell can give rise to many genetically identical cells.   [1]
        (ii) Explain how cells produced from stem cells can have the same genes yet be of different types.   [2]
        (iii) Give a reason why it might be useful to keep a supply of live stem cells from your brain in a laboratory.   [2]

   (b) Much research remains to be done and so it will be some time before adult stem cell treatments will be available to those who might benefit from them. However, research using embryonic stem cells is much further advanced.
        (i)   Explain why research with embryonic stem cells is further advanced than research with adult stem cells. [2]
        (ii) Some people hold the view that research into and the medical use of embryonic stem cells is unethical and that we should await the results of research on adult stem cells.

        Are you for or against embryonic stem cell research?

        Using your scientific knowledge and your understanding of the ethical issues associated with embryonic stem cell research, explain why you hold this view.   [4]
   [Total: 11]

5 Mesenchymal stem cells (MSC), normally found in bone marrow, can be implanted into damaged liver tissue enabling it to produce new cells and thus heal itself.

(a) Explain what is meant by **stem cell**. [2]

(b) MSC are adult stem cells. Describe how adult stem cells differ from embryonic stem cells. [2]

(c) State **one** advantage of using adult stem cells rather than embryonic stem cells. [1]

(d) Give **two** advantages of using stem cells, as an alternative to organ transplants. [2]

(e) Explain how the embryonic stem cells used in some types of medical research are produced and obtained. [2]

[Total: 9]

6 (a) The diagram below shows two different stem cells and the differentiated cells that they can form.

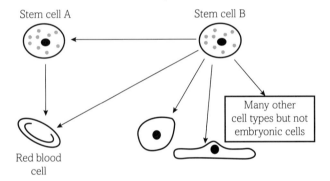

(i) Use the diagram to explain why stem cell B is described as **pluripotent**. [2]

(ii) Name **one** site where stem cell A may be found in an **adult** human. [1]

(iii) All the differentiated cells derived from stem cell B have the same genotype but have very different structures and functions. This is due to differential gene expression.

Explain how **differential gene expression** can enable cells which have the same genetic material to have very different structures and functions. [3]

(b) Three examples of how temperature affects organisms are given below. Which examples are due to differential gene expression? [1]

	Examples
A	The rate of protein synthesis within a plant is temperature dependent.
B	The gender of turtles is determined by the temperature of the ground in which the eggs are laid.
C	Asexual reproduction is more rapid in bacteria if the temperature is higher.

[Total: 7]

7 During an infection, some white blood cells make glycoproteins which become part of their cell surface membranes. To make glycoproteins, the white blood cells must first synthesise proteins on the surface of their rough endoplasmic reticulum.

(a) Explain how these newly-made proteins end up as glycoproteins on the cell surface membrane. [5]

(b) There are certain rare blood disorders in which there is a shortage of white blood cells. One potential treatment would be to inject totipotent stem cells into individuals with these disorders.

(i) Explain what is meant by the term **totipotent stem cell**. [2]

(ii) State why injecting totipotent stem cells may benefit a person with a shortage of white blood cells. [1]

(iii) Name **one** risk to the person receiving the stem cells. [1]

[Total: 9]

TOPIC 7
Modern genetics

CHAPTER

7.3 ▷ Gene technology

Introduction

Genetic modification (GM) is a widely used technique. Many scientists are convinced that producing GM plants with added nutritional benefits, or containing vaccines against serious diseases, is the key to the future of the human race. More than 50% of the soya beans used around the world are now genetically modified and almost 20% of the arable land is used to grow GM crops. However, some people are inserting genes into plants and animals that have less obvious uses. Starlight Avatar is a type of decorative, glow-in-the-dark houseplant made by introducing DNA from luminescent bacteria into the chloroplast genomes of a houseplant. The developers hope that eventually houseplants might become a form of lighting in every home!

In this chapter you will discover how genes from one organism are introduced into the genetic material of an organism of a different species. You will consider the types of changes that can be introduced to crops, using GM soya beans as an example. Originally these new genes were marked by linking them to genes for antibiotic resistance, and you will consider the potential problems associated with this and the alternatives now being developed. 'Knockout' mice are an increasingly valuable tool in biological and medical research and you will learn how they are produced and how they are used.

Finally, you will consider why the use of genetically-modified commercial crops and animals has caused so much public debate.

All the maths you need

- Translate information between graphical, numerical and algebraic forms (*e.g. use of GM crops globally*)

What have I studied before?

- The structure of DNA, including the structure of the nucleotides, base pairing, the two sugar-phosphate backbones, phosphodiester bonds and hydrogen bonds
- How DNA is replicated semi-conservatively in the nucleus of the cell
- The roles of the enzymes DNA helicase, DNA polymerase and DNA ligase in DNA replication
- The process of transcription of the DNA in the nucleus and translation at the ribosome
- The nature of the genetic code
- The aseptic techniques used in culturing organisms
- The principles and techniques involved in culturing microorganisms, including the use of different media
- Different methods for measuring the growth of a bacterial culture, including cell counts, dilution plating and mass and optical methods

What will I study later?

- How genetic modification of organisms by humans may have unforeseen effects on ecosystems

What will I study in this chapter?

- The way recombinant DNA is produced, including the role of restriction endonucleases and DNA ligase
- How recombinant DNA is inserted into other cells using a variety of vectors
- How antibiotic resistance markers and replica plating can be used to identify recombinant cells
- The genetic modification of soya beans and how it has been used to improve production, including altering the balance of fatty acids to prevent the oxidation of soya products
- The use of 'knockout' mice as an animal model to investigate gene function
- Why the widespread use of genetic modification of major commercial crops and other transgenic processes have caused public debate of their advantages and disadvantages

By the end of this section, you should be able to...

- explain how recombinant DNA can be produced, including the role of restriction endonucleases and DNA ligase

- explain how recombinant DNA can be inserted into other cells, and the use of various vectors such as viruses and gene guns

- explain how antibiotic resistance markers and replica plating are used to identify recombinant cells

Changing the genetic material of an organism, usually by inserting genes from one organism into the genetic material of another organism, is known as **genetic engineering** or **genetic modification**. Gene technology is already widely used, but the full potential benefits it can bring have yet to be explored.

Producing recombinant DNA

DNA that has been formed artificially by combining constituents from different organisms is known as **recombinant DNA**. Bacteria are the most widely used genetically engineered organisms, and the basic processes used to produce transformed bacteria are common to all genetic modification.

Artificial copies of a desired gene can be made by taking an mRNA molecule transcribed from the gene and using it to produce the correct DNA sequence. This uses the enzyme

reverse transcriptase and effectively reverses the transcription process to produce **complementary DNA** or **cDNA**. This can act as an artificial gene.

Alternatively, restriction endonucleases are used to cut up DNA strands into small pieces that can be handled more easily (see **Section 7.1.3**). Each type of endonuclease will only cut DNA at specific (restricted) sites within a particular DNA sequence, hence the name. Some restriction endonucleases can cut the DNA strands in a way that leaves a few base pairs longer on one strand than the other, forming a **sticky end**. Sticky ends make it easier to attach new pieces of DNA to them. Sticky ends will only join up to other compatible sticky ends so, for example, you can join together the sticky ends of *EcoR*1 cut DNA fragments.

The next step is to integrate the new gene into a vector. Plasmids, the circular strands of DNA found in bacteria, are frequently used as vectors to carry the DNA into a host bacterial cell. DNA ligases are used as 'genetic glue' to join pieces of DNA together (see **fig A**). This is how the desired DNA, cut from another organism or made artificially, is inserted into another piece of DNA that will carry it into the host cell. Once the plasmid is incorporated into the host nucleus, it forms part of the new recombinant DNA of the genetically engineered or transformed organism. However, in a bacterial host it most often remains as an autonomous plasmid, capable of independent replication, as bacteria do not have a true nucleus. Successfully transformed cells can be identified, isolated and cultivated on an industrial scale so that the proteins they make can be harvested for human use.

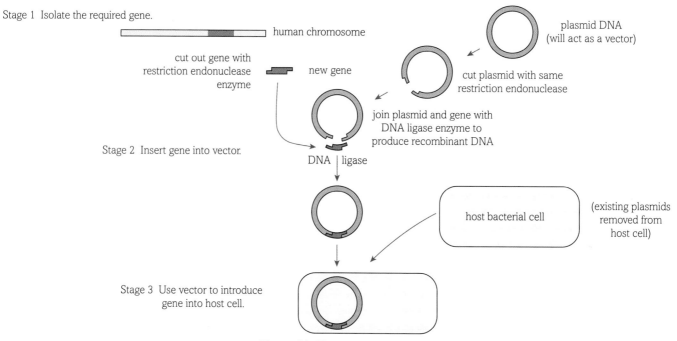

Stage 1 Isolate the required gene.

human chromosome

cut out gene with restriction endonuclease enzyme

new gene

plasmid DNA (will act as a vector)

cut plasmid with same restriction endonuclease

join plasmid and gene with DNA ligase enzyme to produce recombinant DNA

Stage 2 Insert gene into vector.

DNA | ligase

host bacterial cell

(existing plasmids removed from host cell)

Stage 3 Use vector to introduce gene into host cell.

The modified bacterium will now produce a different protein as the new gene is expressed and causes synthesis of the protein.

fig A The main stages involved in inserting a new gene into a bacterium.

Identifying transformed organisms

One bacterium looks very like another, therefore scientists transfer special marker genes along with the desired DNA so they can identify the microorganisms in which transformation has taken place. Originally, marker genes were often for antibiotic resistance. However, there is a major perceived risk in using antibiotic resistance to mark genetically engineered microorganisms. Some people are concerned that if they spread into the environment they would carry genes for antibiotic resistance with them, which could make it harder to treat bacterial infections. So increasingly, genes that make a bacterium dependent on a particular nutrient, or which cause the organism to fluoresce in UV light, are used as markers instead of antibiotic resistance.

A process known as **replica plating** can be used to identify recombinant cells. This involves growing identical patterns of bacterial colonies on plates with different media. It allows the identification of colonies that can survive in the presence of a particular antibiotic, or which cannot survive without a particular nutrient. These are the bacteria which have been genetically modified (GM). The main stages of the process are shown in **fig B**.

Vectors

Vectors play a key role in the formation of recombinant DNA, transferring the required gene, along with any marker genes, into the new cells. A successful vector targets the right cells, ensures that the desired gene is incorporated into the host genetic material so it can be activated (transcribed and translated) and does not have any adverse side effects. Vectors can of course recombine to the appropriate host chromosome, but others will remain as autonomous recombinant plasmids.

Did you know?

Microorganisms and human insulin – a GM success story

People with type 1 diabetes cannot make the insulin needed for their bodies to work properly. Everyone with type 1 diabetes (and some people who have type 2 diabetes) needs regular injections of insulin to keep them healthy.

Historically, the source of insulin for people with diabetes was the pancreases of slaughtered pigs, sheep and cattle. Although the insulin from these animals is similar to human insulin, it is not quite the same. This caused problems for some patients, because their immune systems reacted to the foreign protein of the animal insulin. Also, the supply of insulin was not always reliable, because it depended on how many animals were slaughtered for meat. Scientists developed a way of manufacturing human insulin using genetically modified microorganisms. The process was difficult because the insulin molecule is made up of two polypeptide chains. To tackle this problem, scientists in the 1980s managed to introduce a synthesised gene for each chain into different bacteria, which were then cultured in huge numbers. The microorganisms and the desired end products were separated from the rest of the mixture, producing two pure protein chains that were then oxidised to join them together.

The resulting chemical, often referred to as humulin, appears to function exactly like human insulin. Its purity is guaranteed, making it easier to calculate doses accurately. It has been a great advantage for the majority of people with diabetes. Even more recently, a synthetic gene has been developed that mimics the normal human gene for insulin and allows proinsulin to be made by just one type of engineered bacteria. At the end of the process, enzymes convert proinsulin to insulin. Using microorganisms in this way removes the problems of uncertain supply and provides a constant, convenient and pure source of a human hormone. This is a clear example of the way gene technology can have a positive effect in human medicine.

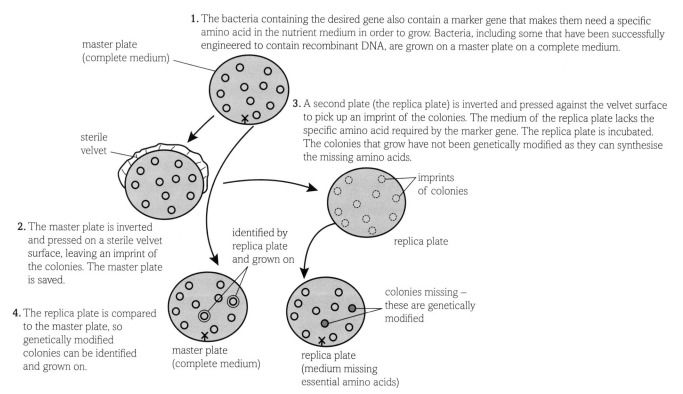

1. The bacteria containing the desired gene also contain a marker gene that makes them need a specific amino acid in the nutrient medium in order to grow. Bacteria, including some that have been successfully engineered to contain recombinant DNA, are grown on a master plate on a complete medium.

master plate (complete medium)

sterile velvet

3. A second plate (the replica plate) is inverted and pressed against the velvet surface to pick up an imprint of the colonies. The medium of the replica plate lacks the specific amino acid required by the marker gene. The replica plate is incubated. The colonies that grow have not been genetically modified as they can synthesise the missing amino acids.

imprints of colonies

replica plate

2. The master plate is inverted and pressed on a sterile velvet surface, leaving an imprint of the colonies. The master plate is saved.

identified by replica plate and grown on

4. The replica plate is compared to the master plate, so genetically modified colonies can be identified and grown on.

master plate (complete medium)

replica plate (medium missing essential amino acids)

colonies missing – these are genetically modified

fig B Replica plating to identify successfully GM bacteria.

Plasmids are particularly useful as vectors in the formation of GM bacteria and in the formation of GM plants (see **Section 7.3.2**). However, other vectors are needed to carry new DNA into some types of plant cells and into animal cells, especially human cells. There are a number of techniques being tried with varying degrees of success including:

- **Gene guns** – DNA is shot into the cell at high speed carried on minute gold or tungsten pellets. Some cells survive this treatment and accept the DNA as part of the genetic material.

fig C Gene gun in action on plant leaves.

- Using viruses – a harmless virus can be engineered to carry a desirable gene and then to infect the animal's cells, carrying the DNA with it.
- **Liposome wrapping** – the gene to be inserted is wrapped in liposomes (spheres formed from a lipid bilayer). These fuse with the cell membrane and can pass through it to deliver the DNA into the cytoplasm.
- **Microinjection** (DNA injection) – DNA is injected into a cell through a very fine micropipette. This is manipulated using a micromanipulator, because the steadiest hand would tremble enough to destroy the cell. The method is rather hit and miss, so many cells have to be injected before one takes up the DNA successfully, but it is the method that has resulted in most successful transgenic animals.

Once a vector is inside the cell the next challenge is to get the new DNA into the right place. This is proving very difficult, particularly when liposomes are used as the vectors. It has been estimated that only about 1 in every 1000 genes that enter a cell in a liposome get into the nucleus to be transcribed. Research is looking at modifying the vectors to get the new genes into the cells and then into the nuclei more effectively. So far viruses are much better at getting DNA into the nucleus than other methods, because inserting DNA into the genetic material of other organisms is part of how viruses work. However, just like pathogenic viruses, viruses used in genetic engineering can cause an immune response in some people.

Much of the research into genetic modification of human cells is focusing on non-viral vectors such as liposomes because of the lack of side effects and potential immune responses, even though they are not as effective.

Knockout organisms – silencing genes

Genetic modification does not always involve adding a new active gene to the genome of an organism. Knockout organisms are widely used and very important in genetic research. In a knockout organism one or more genes are silenced or 'knocked out', so they no longer function. This is done by inserting a new gene that is similar to the gene to be investigated, but which makes the original DNA sequence impossible to read so the gene is silenced and cannot make a protein. Knockout genes are usually accompanied by marker genes to show that they have been incorporated.

Knockout organisms can be used to identify the function of a gene. Many genes are identified in genome sequencing, but scientists often do not know what their function is in an organism. Knocking them out and observing the result can help to make their function clear.

They can also be used to investigate disease and test potential treatments. Genes that are known to be non-functioning in human diseases can be knocked out to create animal models of the disease. These animal models are invaluable for progressing our understanding of human disease and how it may be treated (see **Section 7.3.3**).

Questions

1 Produce a flow diagram showing the main stages of the formation of recombinant DNA.

2 Explain the importance of replica plating in the creation of a culture of GM bacteria.

3 Suggest advantages and disadvantages in the use of plasmids, viruses, gene guns and liposomes as vectors.

Key definitions

Genetic engineering/genetic modification is the insertion of genes from one organism into the genetic material of another organism or changing the genetic material of an organism.

Recombinant DNA is new DNA produced by genetic engineering technology that combines genes from the DNA of one organism with the DNA of another organism.

Reverse transcriptase is the enzyme used to make artificial copies of a desired gene by taking an mRNA molecule transcribed from the gene and using it to produce the correct DNA sequence.

Complementary DNA (cDNA) is DNA which can act as an artificial gene. It is made by reversing the transcription process from mRNA using reverse transcriptase.

A **sticky end** is the name given to the area of base pairs left longer on one strand of DNA than the other by certain restriction endonucleases, making it easier to attach new pieces of DNA.

Replica plating is the process used to identify recombinant cells that involves growing identical patterns of bacterial colonies on plates with different media.

Gene guns are used to produce recombinant DNA by shooting the desired DNA into the cell at high speed on minute gold or tungsten pellets.

Liposome wrapping is a technique for producing recombinant DNA that involves wrapping the gene to be inserted in liposomes. which fuse with the cell membrane and can pass through it to deliver the DNA into the cytoplasm.

Microinjection (DNA injection) is a technique for producing recombinant DNA that involves injecting DNA into a cell through a very fine micropipette.

Gene technology in plants

By the end of this section, you should be able to...

- explain the process of genetic modification of soya beans and how it has been used to improve production, including altering the balance of fatty acids to prevent oxidation of soya products

Genetically-modified (GM) bacteria are used to make a wide range of useful chemicals used in the pharmaceutical and food industries. However, prokaryotes simply do not possess the biochemistry to make some of the more complex human proteins. Therefore work has moved on to introduce desirable human genes into eukaryotic cells including yeast (a single-celled fungus), and more recently plants and even mammals. It is hoped that transgenic plants may become an important weapon in the worldwide fight against disease. Among scientists, genetic modification of plants is also seen as a key factor in providing enough food of the right nutritional value to feed the ever-growing world population.

Making transgenic plants

Introducing genes from one type of plant into another, or even from an animal into a plant, is usually achieved using the bacterium *Agrobacterium tumefaciens*. It causes tumours in plants that are known as crown galls. It contains a plasmid called the Ti plasmid, which transfers bacterial genetic information directly into the plant DNA. This is what normally causes the abnormal growth of the plant cells, but modified plasmids can be used to carry beneficial genes into the plant genome. Then, by the process of plant cloning, the modified transgenic cells can be used to produce whole new transgenic plants.

Genetic modification of plants often, although not always, involves the insertion of a gene from a closely related species of plant. For example, wild species often have good disease resistance or drought resistance but poor seed yield. Scientists can take those useful genes and add them to high-yielding crop varieties.

Cloning plants

Plant cells remain totipotent throughout life. As a result, GM plant tissue can be used to grow a mass of new GM plants by tissue culture. A small piece of GM plant tissue is grown on a gel medium impregnated with plant hormones. A mass of undifferentiated, genetically identical plant cells is produced by mitosis. Transferring the cells to another gel impregnated with different hormones encourages the development of roots and shoots and large numbers of new, identical, GM cloned plants result.

Stage 1
The Ti plasmid is extracted from *A. tumefaciens*.

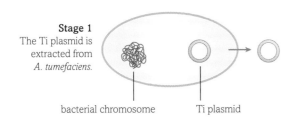

bacterial chromosome   Ti plasmid

Stage 2
The gene to be carried to the plant is inserted into the Ti plasmid which is then returned to the bacterium.

Ti plasmid

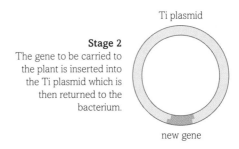

new gene

Stage 3
The plant is infected with the modified bacterium and part of the Ti plasmid with the engineered gene becomes part of the plant chromosomes.

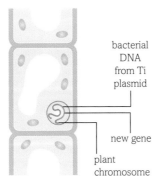

bacterial DNA from Ti plasmid

new gene

plant chromosome

Stage 4
A. tumefaciens causes a tumour to develop on the plant. These plant cells contain the new gene. If tumour cells are taken and cultured, whole new plants can be grown from them, containing the new genes. These are genetically engineered or transgenic plants.

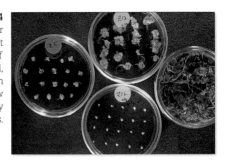

fig A One method of producing transgenic plants.

Genetic modification of crops

Plants, particularly cereal crops, are the main staple diet of most human beings. The problem is that these crops are vulnerable to disease, pests, adverse growing conditions and natural disasters such as floods and drought. Traditional plant breeding has resulted in many improvements but it is a slow process. Genetic modification can introduce many useful characteristics much faster. Many of the first commercially produced GM crops were modified to make life better for the producers and sellers of food, for example, giving longer shelf lives or herbicide resistance. The big second wave of GM plants delivered improvements for the consumer, such as improving the nutritional content of the food and making crops more resistant to adverse conditions. For example:

- Flood-resistant rice: complete immersion in water destroys rice crops, but global warming is causing more severe flooding. Strains of GM rice have been developed that can withstand total immersion in flood water for up to 3 weeks and still produce around 80% of the normal yields. These went from lab to field in 2 years and are already preventing starvation.

- Pesticide resistance: globally, 10–16% of all crops grown are lost to pests. If this could be reduced there would be a lot more food to go around. Genetic modification of crop plants can produce plants that make their own pesticide within their leaves (e.g. Bt toxin in cotton). This means farmers do not need to use chemical pesticides, which are expensive and can harm the environment. However, there are concerns that the pesticide genes will spread into wild plants, and insects and fungi will become resistant to the plant chemicals.

- Changing the nutrient values of plants: by genetically modifying crop plants, scientists have the potential to change the balance of different chemicals in the crop. If the vitamin or protein content of food can be increased (e.g. vitamin A in golden rice), or the balance of fats changed, the value of the food to the people who eat it can be greatly increased. It may even be possible to grow plants that contain vaccines against deadly diseases.

Genetic modification of soya beans

Soya beans, also known as soybeans, are one of the major food crops globally. They are grown as a source of protein-rich food for people and as a source of oils used commercially in food industries around the world. They are used in animal feed, as a biofuel, and have many uses in the cosmetics industry.

It is estimated that 90–95% of all the soya beans grown in the United States, a major producer, are now genetically modified. They have been modified in a variety of ways to make them more successful, both nutritionally and economically. The techniques used to add new genetic material to soya beans include the use of *A. tumefaciens* and gene guns. The main genetic modifications of soya beans include:

- Herbicide resistance (Roundup Ready® soybeans): the yield of soya beans is badly affected by competition with weeds, so farmers spray herbicides to destroy the weeds. However, soya plants are also killed by many broad-leafed plant herbicides, including Roundup®, a widely used commercial brand. So scientists genetically modified soya plants to be resistant to common weed killers. These GM soya beans are used very widely across the soya bean-growing world and yields have increased markedly as a result.

- Fatty acid balance: oil from soya beans is widely used in food production, but traditional soya oil contains relatively little oleic and stearic acid and relatively large amounts of linoleic acid. Linoleic acid oxidises easily, so soya oil tends to oxidise and go off. Also, linoleic acid is a polyunsaturated fatty acid, while oleic acid is a monounsaturated fatty acid, and studies have suggested that oleic acid is better for health than linoleic acid. Scientists have produced GM soya beans that have a radically different balance of fatty acids, with a lot of oleic acid and less linoleic acid. This has two benefits: the oil lasts longer and may be healthier.

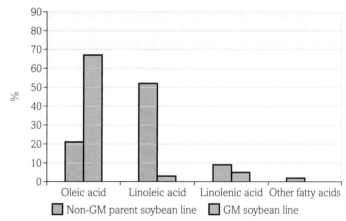

fig B The impact of genetic modification on the fatty acid balance in soya bean oil.

Questions

1. Compare and contrast the genetic modification of plants with the genetic modification of microorganisms.

2. Genetic modification of the fatty acid balance in soya beans offers advantages to both producers and consumers. Explain this statement.

3. Many scientists are convinced that GM crops will be the solution to the problem of feeding the growing world population. Suggest three reasons for their optimism.

By the end of this section, you should be able to...

- explain how 'knockout' mice can be used as a valuable animal model to investigate gene function

- explain why the widespread use of genetic modification of major commercial crops and other transgenic processes have caused public debate of their advantages and disadvantages

By considering some of the different uses of genetic modification in animals and people, along with the way genetically modified (GM) crops are being grown commercially around the world, we can begin to understand the advantages and disadvantages of the processes and consider the role of public debate in deciding where these techniques will lead.

Gene technology in animals and humans

As scientists have become better at genetic manipulation, the ways in which this technique can be used in animals and people have been expanding. No debate on the use of GM technology would be complete without an understanding of this area of the field.

Knockout mice in use

In **Section 7.3.1** you learned about the technique of knocking out genes. This is often done to produce animal models of the disease to make it easier for scientists to develop treatments or cures for conditions ranging from cancer and heart disease to cystic fibrosis and diabetes. To find a cure for a disease such as cystic fibrosis, it is important to understand exactly how it affects the body and what aspects of the disease can best be targeted for treatment. However, there is a limit to the investigations that scientists and doctors can carry out on people. Some experiments can be done using cells in culture, but sometimes there is no substitute for a living organism. This is where animal models play a vital role.

For example, in the early 1990s, David Porteous and Julia Dorin at the Medical Research Council Human Genetics Unit in Edinburgh used genetic manipulation techniques to develop mice with a faulty *CFTR* gene. These mice have similar chloride ion transport problems to humans affected by cystic fibrosis. They do not have all the same clinical symptoms, but they have been very useful both in helping understand the disease and in trialling new treatments, including some early gene therapy work using DNA and liposomes. The researchers demonstrated that this gene therapy was safe, but did not completely correct the chloride ion transport abnormality associated with the disease. The same results were later seen in human trials. Some people consider it unethical to manipulate and use animals like mice for scientific research. However, without this, our understanding of this disease and the development of potential new treatments would take much longer.

fig A The black mice in this picture have a homozygous defect in the *CFTR* gene. Coat colour is used as a marker for the gene. Mice like this give scientists an animal model for cystic fibrosis and other serious diseases.

GM animals

It is very difficult to transfer new DNA into eukaryotic cells, but there have already been some notable successes. Some exciting work was done on inserting human genes into the tissues of sheep and cattle to produce **transgenic animals**. In 1990 the first transgenic sheep was born: a ewe called Tracey who produced the human protein alpha-1-antitrypsin in her milk. This protein is missing in people who suffer from a genetic condition that affects their livers and lungs, causing emphysema to develop at a very early age. The early promise of this work has not yet been realised, at least in part because of the problems of producing large numbers of cloned transgenic animals.

The production of proteins using transgenic animals involves the introduction of a copy of a human gene that codes for the desired protein into the genetic material of an egg of a different animal species. As well as the gene for the specific protein, there is also a promoter sequence that makes sure the gene will be expressed only in the mammary gland of the lactating female. The fertilised and transgenic embryo is then replaced inside a surrogate mother, and the animal is born and grows to maturity. When the animal is mature and produces milk, that milk is harvested, purified and the human protein extracted.

So far transgenic sheep, cows, pigs, rabbits and mice have been produced, all capable of producing human proteins in their milk. Using large animals, such as cows and sheep, large volumes of milk, and therefore human proteins, can be produced, and the animal numbers can be increased by breeding or cloning.

More than 20 different human proteins have been produced so far from transgenic animals, showing that the technique can work, but none are in commercial use as therapies. They include factor VII and factor IX, which are important components of the human blood clotting cascade, alpha-1-antitrypsin needed for emphysema and cystic fibrosis sufferers, activated protein C for treating deep vein thrombosis, and modified milk for people with lactose and other intolerances. In 2011 scientists in China announced that they had developed a herd of around 300 GM cows producing milk similar to human breast milk.

The milk of transgenic animals is very expensive – it costs several thousand pounds per litre. The expectation was that as the numbers of transgenic animals increased, the price would eventually come down. However, there have been major problems, for example most of the transgenic sheep were destroyed when the company producing them hit financial difficulties.

fig B In the future, genetic modification may result in herds of cows that produce therapeutic proteins in their milk, or even produce a carefully matched substitute for human breast milk.

Human gene therapy

The prospect of being able to override the faulty genes that can cause some devastating diseases in people is a very exciting one, but the research path to a successful treatment is long and littered with disappointments. For example, scientists still hope cystic fibrosis might be relieved by inserting healthy versions of the *CTFR* gene into the cells of the lungs but progress has been slow.

Most gene therapy is carried out on normal body cells (somatic cells) and so it is known as somatic cell gene therapy. Even if successful, if the patient has a child they will pass on their faulty alleles. A potential solution to this problem is to alter the germ cells so that the faulty genes are no longer passed on. This could be done in the very early embryo immediately after *in vitro* fertilisation. The new individual would be free of the disease and would not risk passing it on to their own offspring.

This may sound like a good idea, and some people argue that when the technology is available it should be used, but many other people are very concerned. No one is yet sure of the effect on an early embryo of such an invasive intervention and the impact might not become clear until years into the life of the individual. Whilst attempting to remove the risk of genetic diseases seems a very positive aim, it could be difficult to know where to draw the line and some people would be prepared to pay a great deal of money to have the line moved a little. If it became possible to manipulate genes, not simply to remedy disease but to enhance longevity, change skin colour or increase adult height or intelligence, then some people would be tempted to try it.

At the moment most germ line research is banned in the UK, European countries and the US. But sooner or later, someone somewhere will try it and society needs to be prepared.

Pros and cons and public debate

There has been a great deal of public debate about the use of GM technology. In some parts of the world (e.g. America and Australia) people have few problems with the use of GM plants and foods. In other areas, including the UK and other parts of Europe, there is still a lot of uncertainty.

This area of science is very complex and many people do not understand it fully. Much of their knowledge comes from the media, and most journalists, even science correspondents, are not scientists. The evidence is not easy to gather or to access, nor is it easy to understand. Science is full of uncertainties, but people like things to be clear and certain. As a result many people, including some of the politicians who make laws about research, are arguing without all the evidence and in many cases with an incomplete grasp of the science. Interestingly, the use of GM bacteria to produce human medicines such as insulin raises relatively little debate. It is genetic modification in food plants, animals and humans that polarises opinions. Some of the issues that are often raised about GM technology include:

- Concern that if marker genes, which include antibiotic resistance, or other transgenic traits got into wild plant populations, there might be issues of the build up of antibiotic resistance. This issue is increasingly being addressed by the use of different marker genes, for example, fluorescence.

- Concern about infertile seeds. Scientists usually include marker genes in GM plants to make it easy to identify them. These marker genes may include alleles that ensure the plant is infertile, so that the plants cannot reproduce and cannot then spread into the environment. However, it does also mean that fresh supplies of transgenic plant seeds have to be bought each year for planting. This is more of a concern where food crops are involved than where plants are being modified to use as medicines, but some of the same issues arise. The big worry is that this will seriously disadvantage people in poorer countries, where the modified crops are most needed.

- Worries over ingesting alien DNA in GM food plants. They forget that we eat many different types of DNA every day and digest it, possibly including the organism that donated the new gene. For many people the risk/benefit balance shifts dramatically when GM food contains vital medicines or important vitamins.

fig C Genetic modification of crop plants offers hope of more food in many parts of the world where it is badly needed.

- Environmental concerns about gene transfer from GM plants and animals to wild species.

- Strong objections to the use of animals and other organisms for scientific research as some people feel that genetic modification violates the rights of the modified organism.

- A real concern that gene technology will be the property of just a few companies in developed countries. As a result it is feared that the new advances may be largely biased towards the needs of the richer countries. However, the work on plant-based vaccines and flood and drought resistant rice is being made available at very low cost where it is needed, which is helping to negate that fear.

At the moment the use of transgenic plants in medicine is relatively new and unknown, so concerns about risks in the general population are at a low level. Perhaps if the media coverage becomes negative, worries will grow. However, it seems likely that, as in the case of using GM microorganisms to produce human medicines, people will accept the benefits of this technology, seeing those benefits as outweighing any possible risks.

As the scientific evidence builds, scientists are increasingly convinced of the safety and value of GM crops. However, as with other GM food, there are concerns about the safety of these products both for the humans that use them and for the animals that are used to produce them. Despite this, for the great majority of the world, where food security is often an issue and infectious disease and malnutrition still kill huge numbers every year, GM crops and medicines may well offer the difference between survival and death.

This shows you some of the areas of concern about gene technology which are debated. In spite of these concerns, many people, including the majority of scientists, feel that the enormous benefits of developments resulting from GM technology in food production, medicine development and potential therapies far outweigh any risks or ethical stumbling blocks which may arise. When people express concerns about the risks of using GM organisms in medicines or food, it is wise to remember that risks can come from a wide variety of sources, and that GM organisms can also reduce some of these risks, making them part of the solution, not the problem.

Questions

1 Identify two ethical positions on the use of GM bacteria in the production of drugs.

2 Investigate the use of GM animals to produce human medicines. Select one drug and evaluate the success of this process so far.

3 Discuss the advantages and disadvantages of germ line therapy.

4 Produce a report or a poster to summarise the debate on the use of GM in global crop production.

5 Explain why the development of gene technology has caused so much public debate and explain some of the problems and advantages of such debates.

Key definition

Transgenic animals are animals that have had their DNA modified using gene technology, so that at least some of their cells contain recombinant DNA.

EXPLOSIVE PLANT SCIENCE

Genetically-modified (GM) plants are being used in an ever-increasing number of ways. Removing explosives from the soil is one of the more unusual ones.

EXPLOSIVE PLANT SCIENCE!

Although typically associated with developing enhanced food crops, GM technology has a wealth of potential applications. One of the most inspiring talks (to my mind at least) at the UK PlantSci conference was that of Elizabeth Rylott (University of York): "Plant Cillit Bang! And the dirt is gone! Using TNT to understand detoxification of organic pollutants by plants". Trinitrotoluene (TNT), a potent carcinogen, is a pollutant released from explosive materials and which contaminates approximately 10 million hectares in the USA (especially around military zones), poisoning groundwater sources. Plants are naturally able to detoxify low levels of TNT, using enzymes which "activate" the functional groups on TNT. This then allows enzymes called uridine diphosphate (UDP) glycosyltransferases (UGTs) to transfer the activated groups to an acceptor molecule, converting TNT into a less toxic compound. Transgenically over-expressing these enzymes can reduce the levels of TNT in plants grown on contaminated soil [...] TNT often occurs in nature alongside [another explosive,] cyclotrimethylenetrinitramine (RDX), which plants cannot detoxify. Nevertheless, RDX detoxifying enzymes have been identified in bacteria, including a cytochrome p450 monoxygenase in *Rhodoccocus*. When this is introduced into plants already engineered for enhanced TNT detoxification, the result is 'super plants' capable of restoring contaminated soil. Such 'bioremediation' strategies are becoming increasingly popular and could play a part in turning public opinion in favour of GM. Indeed, these plants were capable of removing all RDX from the surrounding soil within a week. Rather than just accumulating toxic RDX in the leaves, however, these plants convert it into a source of nitrogen – which they then use to fertilise their own growth! Hence, RDX does not simply move from the soil to the plant, but is converted into a completely harmless form.

So could this be a message of hope to those millions of devastated acres tainted with explosives? Excitingly, this research has now moved to the field trial stage in the USA, using transgenic switch grass. Amusingly, part of the preparations for these trials involved packaging the seed into soil plugs and dropping these off the roof of the lab... apparently, the military strategy for planting the transgenic plants will be to drop them from a helicopter, hence the team had to make sure the plants could stand up to this treatment!

fig A Genetically-modified switch grass could clear explosives from millions of hectares of contaminated land around the world.

Where else will I encounter these themes?

Book 1   5.1   5.2   6.1   6.2   6.3   7.1   7.2

This is an entry from a blog by Caroline Wood, a PhD student at the University of Sheffield, who is investigating the effect of parasitic plants on the plants they infect. In her blog, Science as a destiny, she enthuses about a talk she heard at the UK PlantSci conference of 2014, given by Dr Liz Rylott from York University.

1. How does a blog differ from the scientific writing you might find in a journal, a textbook or a science magazine?

2. There are a number of different sources online on this work. One is written by Dr Liz Rylott from York University, who gave the talk that inspired Caroline. Others are reports from bodies that have funded the research in the UK and the US. Visit all of these web resources, then compare and contrast the way the story is told.
   - https://sites.google.com/a/york.ac.uk/liz-rylott/home/wider-audience
   - www.bbsrc.ac.uk/documents/explosive-plants-pdf
   - www.serdp-estcp.org/Program-Areas/Environmental-Restoration/Contaminants-on-Ranges/Protecting-Groundwater-Resources/ER-1498

   Discuss the following.
   a. Which account did you most enjoy reading and why?
   b. Summarise the process of phytoremediation of land contaminated with explosives as you understand it from the blog, written from memory after a number of talks at a conference.
   c. Summarise the same process based on the account by Liz Rylott, one of the scientists involved in the research.
   d. Compare your summaries and highlight any inconsistencies between them, suggesting reasons for them.
   e. How does the content of the other websites compare with the blog and with Liz Rylott's account?

> Scientific conferences provide important opportunities for scientists to present and discuss their work. You could visit the website of the UK PlantSci conference to learn about the cutting edge plant science at this year's event.

Now let us examine the biology:

3. a. What does this piece of writing tell you about the way these genetically-modified plants have been developed?
   b. Outline the process by which bacterial DNA can be added to plant DNA.
4. a. Explain the mechanism by which the plants remove TNT and RDX from the soil and then make them safe.
   b. How is this process linked to the genetic modification of the plant DNA?
5. The blog refers to the transgenic plants as 'super plants'. Evaluate this statement – do you think it is justified?
6. Summarise the main arguments for and against GM used in public debates in the media.
7. The blog suggests bioremediation projects such as this could play a part in turning public opinion in favour of GM. How do you think work such as the development of bioremedial plants described in this blog might affect the debate, and why?

Activity

Discover more about the work on transgenic plants and their ability to remove explosives from the soil. Use the resources listed above and look online to see if any results are yet available from the US field trials of the genetically-modified plants.

Put all the information together to plan out either of the following:
- a big article for the science pages of a popular newspaper on this novel use of gene technology and the potential of GM plants for the future
- a short piece on local radio or on a national science programme aiming to get over the technology involved and the potential benefits of GM plants, starting with this work.

> Use a variety of resources in your work. In each case, judge the reliability of your source before you use it and make sure that you include a reference for each one.

● From a blog by Caroline Wood, http://scienceasadestiny.blogspot.co.uk/2014/04/explosive-plant-science.html

1 There are a number of ways in which recombinant bacteria can be identified. Which of the following techniques might cause ethical objection?
   (1) replica plating to test nutrient requirements
   (2) antibiotic resistance marker genes
   (3) ultraviolet fluorescence
   A only technique 1
   B only techniques 2 and 3
   C only technique 3
   D only technique 2   [1]
   [Total: 1]

2 Several restriction endonucleases cut after a guanine base. Using this information, look at these sequences and decide which group of sequences would result in the formation of sticky ends.
   (1) GGATCC
   (2) AGCT
   (3) GAATTC
   A sequences 1, 2, and 3
   B only sequences 1 and 3
   C only sequences 2 and 3
   D only sequences 1 and 2   [1]
   [Total: 1]

3 Which of the following is the process by which DNA ligase can join two sections of DNA?
   A repairing the peptide bond
   B repairing the phosphodipeptide bond
   C repairing the phosphodiester bond
   D repairing the phosphate bond   [1]
   [Total: 1]

4 The flow diagram below shows how a genetically modified organism may be produced by inserting a human gene into a bacterium.

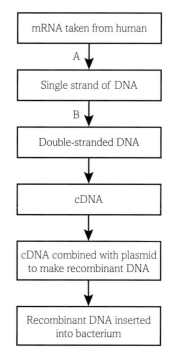

(a) Name the enzymes identified as A and B on the flow diagram. [2]

(b) Describe how the cDNA and the plasmid could be combined together. [3]

(c) Not all the bacteria would be successfully modified. Describe **one** method that could be used to identify the modified bacteria. [2]

[Total: 7]

5 One of the first stages in the production of cheese is the coagulation (clotting) of the milk protein, casein, which is brought about by the enzyme, chymosin.

The diagram below shows some of the main stages in the production of chymosin using yeast cells.

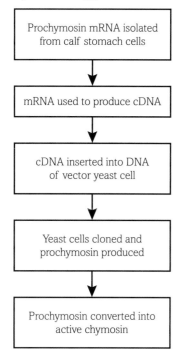

Prochymosin mRNA isolated from calf stomach cells
↓
mRNA used to produce cDNA
↓
cDNA inserted into DNA of vector yeast cell
↓
Yeast cells cloned and prochymosin produced
↓
Prochymosin converted into active chymosin

(a) Cheese produced using chymosin cultured from yeast cells is acceptable to most vegetarians. Give a reason why some consumers may still have concerns about eating cheese prepared in this way.   [1]

(b) With reference to the diagram above, explain how enzymes are used in the production of chymosin using yeast cells.   [4]

[Total: 5]

6 Cotton plants are used to produce the cotton from which cloth is made. They are grown in certain parts of the world, such as India and the USA, where cotton farming is an important way in which people earn their living. Normally, cotton plants need to be sprayed with chemical insecticides to kill insect pests. Recently, genetically-modified (GM) cotton plants have been developed which produce a natural insecticide of their own.

This insecticide kills the insect pests but is harmless to humans.

A field of cotton plants showing the 'bolls' of cotton fibres which are collected to make cloth. The bolls are the fruit of the plant and contain seeds.

(a) Describe advantages to cotton farmers and to the environment of growing genetically-modified cotton.   [3]

(b) Explain how the genetic modification of plants is similar to, yet distinct from, conventional plant breeding.   [4]

(c) For many people, genetic modification of plants remains controversial. In 2007, the European Union once again decided not to lift the ban on GM crops in Europe although they are widely grown in the USA and India.

State whether you are for or against the growing of GM crops.

Explain why you hold this view. Use scientific knowledge and make reference to social or ethical issues to support your explanation.   [4]

[Total: 11]

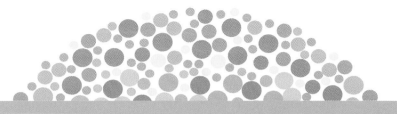

TOPIC 8
Origins of genetic variation

CHAPTER **8.1** ▶ **Genetic information**

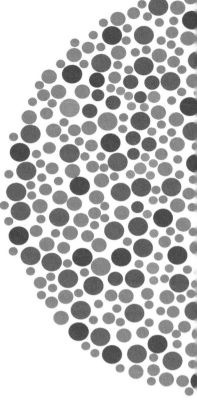

Introduction

With many experimental organisms, we can control the parents that are allowed to breed and manipulate the environment in which they grow and develop. However, we cannot choose human partners or deliberately manipulate human environments for genetic experiments.

Most human traits are polygenic, that is, the result of the interaction of a number of different genes. The environment within the uterus, during childhood and at adolescence also affects many aspects of appearance and behaviour. Identifying whether nature or nurture determines a certain phenotypic characteristic in humans is very difficult indeed. Twin studies are very useful to help us unravel some of these strands, with differences between identical, monozygotic twins studied and compared to the differences between non-identical twins and non-twin siblings. Identical twins that have been separated at birth and adopted by different families provide a very rich seam of evidence. Studies such as these play an important part in helping us to understand human genetics.

In this chapter you will consider the origins of genetic variation, including mutations, random fertilisation and crossing over during meiosis. You will learn how genetic information is transferred from parents to their offspring. You will be able to construct genetic crosses and pedigree diagrams for different organisms and be able to explain the inheritance of two non-interacting unlinked genes. In addition, you will learn about autosomal linkage due to the presence of genes on the same chromosome and you will see how this explains the results of certain genetic crosses. You will discover the importance of sex linkage on the X chromosome and how this explains the inheritance of haemophilia and other conditions in humans. You will learn how to use the chi squared test to test the significance of the difference between the predicted and observed results of any particular genetic cross.

All the maths you need

- Use ratios, fractions and percentages (*e.g. representing monohybrid and dihybrid crosses*)
- Estimate results (*e.g. estimate ratios for genetic crosses*)
- Understand simple probability (*e.g. probabilities associated with genetic inheritance*)
- Understand the terms mean, median and mode (*e.g. mean of a set of data referring to particular phenotypic characteristics*)
- Select and use a statistical test (*e.g. the chi squared test, to test the significance of the differences between observed and expected results*)

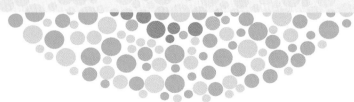

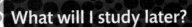

What have I studied before?

- Variation in a population
- Basic genetics and inheritance
- Simple Mendelian genetics
- The process of transcription of the DNA in the nucleus and translation at the ribosome to form proteins and affect the phenotype of an organism
- The nature of the genetic code

What will I study later?

- That selection pressures acting on the gene pool change allele frequencies in the population
- How stabilising selection maintains continuity in a population and disruptive selection leads to changes or speciation
- How changes in allele frequencies may be the result of chance and not selection, including genetic drift
- That allele frequencies can be influenced by population bottlenecks and the founder effect
- How the Hardy-Weinberg equation can be used to monitor changes in the allele frequencies in a population

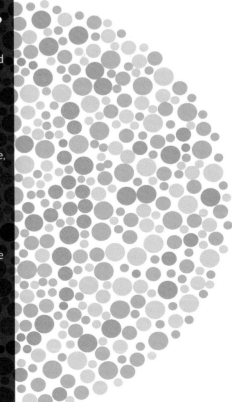

What will I study in this chapter?

- That mutations are the source of new variations and that the processes of random assortment and crossing over during meiosis give rise to new combinations of alleles in gametes
- How random fertilisation during sexual reproduction brings about genetic variation
- The meaning of the terms genotype, phenotype, homozygote, heterozygote, dominance, recessive, codominance and multiple alleles
- The construction of genetic crosses and pedigree diagrams
- The inheritance of two non-interacting unlinked genes
- How autosomal linkage results from the presence of genes on the same chromosome and that the results of crosses can be explained by the events of meiosis
- About sex linkage on the X chromosome, including the inheritance of haemophilia in humans
- How to use the chi squared test, to test the significance of the difference between observed and expected results

By the end of this section, you should be able to...

● explain how mutations are the source of new variations and that the process of random assortment and crossing over during meiosis gives rise to new combinations of alleles in gametes

● explain how random fertilisation during sexual reproduction brings about genetic variation

In all organisms that reproduce sexually there are both differences and similarities between the offspring and the parent organisms. These are the result of the genetic information inherited by the offspring from the parents. The variation that arises as a result of sexual reproduction is key to survival and evolution of individuals. In your biology studies so far you have met a number of different ways in which this genetic variation comes about. In this chapter, you are going to pull all the different strands together.

fig A Genetic variation in a family can lead to some strong family likenesses and some big differences.

Meiosis as a source of genetic variation

When the cells that are going to form gametes (sex cells) divide, they undergo meiosis (see **Book 1 Section 2.4.1**). Meiosis is a reduction division that occurs only in the sex organs. It halves the chromosome number in the cells from diploid to haploid. Meiosis introduces genetic variation in two main ways.

Independent assortment (random assortment)

In meiosis the components of the chromosome pairs from the father and the mother are distributed into the gametes randomly. So in humans, any number from none to all 23 chromosomes in your gametes could come from either your maternal or your paternal chromosomes. Independent assortment results in many new combinations of alleles and introduces considerable genetic variation.

Crossing over (recombination)

This process takes place during meiosis when large multi-enzyme complexes 'cut and join' bits of the maternal and

paternal chromatids together at the chiasmata. This exchange of genetic material leads to added genetic variation as many new combinations of alleles arise. Crossing over is also a potential source of mutation, which again introduces new combinations into the genetic make-up of a species (see **Book 1 Section 2.4.1**).

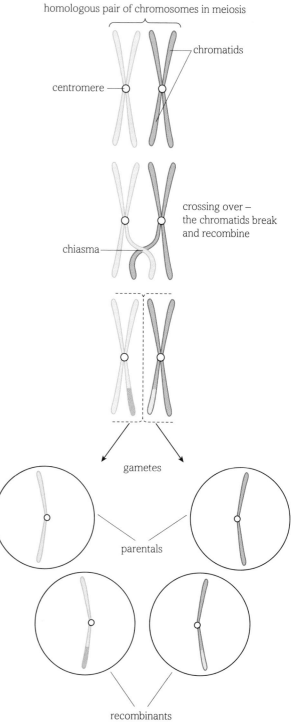

fig B Crossing over in meiosis gives rise to new combinations of alleles in gametes.

Mutation as a source of genetic variation

In **Book 1 Chapter 1.3** you looked at the structure of DNA, the genetic material of eukaryotic organisms, and saw how the genetic code carried on the DNA is translated into living cellular material through protein synthesis. If a single codon is changed the amino acid for which it codes may be different and this can affect the whole organism. A permanent change in the DNA of an organism is known as a mutation. Mutations can happen when the gametes are formed, although they also occur during the division of somatic (body) cells.

The majority of mutations result in slight changes which have little or no effect on the organism. However, when environmental conditions change, some of these variations may make an organism better adapted to survive and in this way they are the source of evolutionary change. Genetic disorders such as sickle cell anaemia, cystic fibrosis and Down's syndrome are also the result of random mutations in the genetic material of the gametes.

Different types of mutations

Although the body has its own DNA repair systems, with specific enzymes to cut out or repair any parts of the DNA strands that become broken or damaged, some mutations remain and are transcribed from the DNA into the mRNA and then translated when new proteins are made.

There are three main types of mutations:

- Point or gene mutations are caused by the miscopying of just one or a small number of nucleotides. They include substitutions, where one base is substituted for another, deletions, where a base is completely lost in the sequence, or insertions, when an extra base is added in that it is a repeat of one of the bases already there *or* a different base entirely (see **Book 1 Section 1.3.7**).

- Chromosomal mutations involve changes in the positions of genes within the chromosomes. These are more likely to make a measurable change in an organism than a gene mutation.

- Chromosome mutations occur when an entire chromosome is either lost during meiosis (cell division to form the sex cells) or duplicated in one cell by errors in the process. These usually have a major impact on the organism (see **Book 1 Section 2.4.2**).

Random fertilisation as a source of genetic variation

In sexual reproduction, the male and female gametes from two unrelated individuals fuse to form a new genetic individual. This introduces considerable genetic variation into the offspring. There are usually many more male gametes than female gametes, and the combination of the two which ultimately fuse to form a new individual is completely random.

For example, in humans several ova start to mature in the ovary each month. It is a random process that determines which one reaches the point of ovulation first. Each time a man ejaculates he produces on average 2–5 cm^3 of semen containing from 20–150 million sperm per cm^3. Each of those sperm carries a different combination of alleles and the process determining which of those sperm fertilises the ovum is completely random.

A combination of all these sources of genetic variation – meiosis, mutation and random fertilisation – ensures genetic variation in each new organism produced by sexual reproduction.

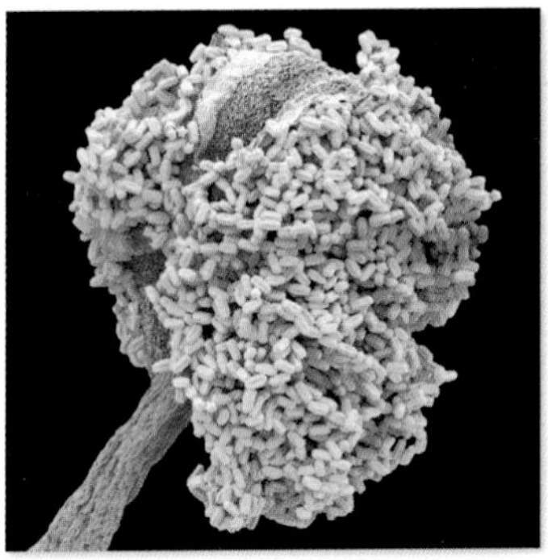

fig C The vast numbers of male gametes mean that random fertilisation adds another source of genetic variation.

Questions

1 Make a table to summarise the main ways in which genetic variation is introduced during the process of sexual reproduction.

2 'My sister and I look very alike – and just like our mum!' Explain how this statement and the concept of genetic variation are not mutually exclusive (one does not rule out the other).

By the end of this section, you should be able to...

- explain the terms genotype and phenotype, homozygous and heterozygous, dominance, recessive, codominance, and multiple alleles

- construct genetic crosses and pedigree diagrams

For centuries people bred the plants and animals we grow for food, our gardens and as companion animals without any real understanding of the breeding process. Now we understand many of the underlying mechanisms of inheritance, and the science of genetics has emerged.

Genetics: the basis of inheritance

The physical and chemical characteristics that make up the appearance of an organism are known as its phenotype, for example, the size of a cabbage, the colour of a flower or the shape of a nose. The phenotype is partly the result of the genotype (the genetic information) passed from parents to their offspring, and partly the effects of the environment in which the organism lives. For example, the size of a cabbage will depend on levels of soil nutrients and sunlight as well as on the genetic make-up of the individual plant.

As you know, differences in the genotype between individuals of a species are due to both the shuffling of genes during meiosis (see **Book 1 Section 2.4.1**) and the inheritance of genes from two different individuals in sexual reproduction.

The cells of any individual organism contain a particular number of chromosomes that is characteristic of the species. So, for example, humans have 46 chromosomes. Half the chromosomes are inherited from the female parent and the other half come from the male. The two sets can be arranged as matching homologous pairs in a karyotype (see **Book 1 Section 2.4.1**).

Along each chromosome are hundreds of genes, each gene being a different segment of DNA coding for a particular protein or polypeptide. The chromosomes in a homologous pair carry the same genes, with the only exception being the sex chromosomes. The gene for a particular characteristic is always found in the same position or locus, which means that you usually carry two genes for each characteristic.

Each gene exists in slightly different versions called alleles (variants). For example, at the locus for the gene for the height of a pea plant, the allele may code for a tall plant or for a dwarf plant. Since the pea plant has two homologous chromosomes carrying this gene, it may have two alleles for the tall characteristic, two for the dwarf, or one of each.

If both of the alleles coding for a particular characteristic are identical, then the individual is homozygous for that characteristic – it is a **homozygote** ('homo' means 'the same'). If the two alleles coding for a characteristic are different, the individual is heterozygous for that characteristic and is called a **heterozygote** ('hetero' means 'different').

Some phenotypes show **dominance**: their effect is expressed or shown whether the individual is homozygous or heterozygous for the allele. As long as one allele for the dominant phenotype is present it will be expressed, even in the presence of an allele for the **recessive** phenotype. Recessive phenotypes are only expressed when there are two alleles coding for the recessive feature, in other words, when the individual is homozygous recessive. In genetic diagrams, the alleles coding for dominant phenotypes are usually represented by a capital letter and those for recessive phenotypes by the lower-case version of the same letter.

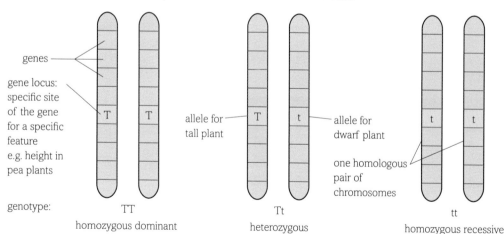

fig A In sexual reproduction the variants of the genes passed on determine the genotype and, therefore, eventually an aspect of the phenotype of the offspring.

Polygenic traits

In the next few pages you will be looking at genetic crosses involving one or two different loci. It is, however, important to remember that most traits in living organisms are determined not by a single gene but by several interacting genes. They are **polygenic**. Characteristics such as eye colour, weight and intelligence are determined by a number of different genes at different loci and in many cases interactions with the environment add further variety. So when looking at genetic crosses, remember this is a very simple model that helps us understand a much more complex reality.

Monogenic (monohybrid) crosses

Homozygotes are referred to as true breeding because if you cross two individuals that are homozygous for the same characteristic, all the offspring of all the generations that follow will show this same characteristic in their phenotype (unless a mutation occurs). Heterozygotes are not true breeding. If two heterozygotes are crossed, the offspring will include homozygous dominant, homozygous recessive and heterozygous types and at least two different phenotypes.

When one gene is considered at a time in a genetic cross, it is referred to as a **monogenic cross**. We can represent these crosses using simple diagrams called Punnett squares. A Punnett square shows you the potential alleles inherited from both parents, and the potential offspring that result. For example, **fig C** shows a cross between a pea plant homozygous for the dominant round pea seed shape and a pea plant homozygous for the recessive wrinkled pea (see **fig B** and **fig C**). The first generation of this cross is called the F_1 (first filial generation) and you can see that they all have the same genotype for the characteristic and they are heterozygous. They also all have the same phenotype – round pea shape because the round allele is dominant. There is no sign of the wrinkled pea allele. If we cross individuals from the F_1 generation we call the next generation the F_2 (second filial generation). In **fig C** you can see that theory predicts the genotypes to be 1 homozygous dominant : 2 heterozygous dominant : 1 homozygous recessive. In terms of the phenotypes that would result from these genotypes, you would expect to see three round peas for every wrinkled one. The recessive trait of wrinkled peas has become visible again, after being 'hidden' in the F_1 generation.

fig B Round or wrinkled peas may not seem very exciting, but they are genetic characteristics that have been studied since the earliest days of genetics, and they are easy to identify.

R = round
r = wrinkled

parental phenotype	round-seeded ×	wrinkle-seeded
parental genotype	RR	rr
gametes	R R	r r

	r	r
R	Rr	Rr
R	Rr	Rr

offspring genotype (F_1 generation)   Rr
offspring phenotype   round-seeded

if two F_1 offspring are crossed, we get the following results:

F_1 phenotype	round-seeded ×	round-seeded
F_1 genotype	Rr	Rr
gametes	R r	R r

	R	r
R	RR	Rr
r	Rr	rr

offspring genotypes (F_2 generation)   RR : 2Rr : rr
offspring phenotypes   3 round-seeded : 1 wrinkle-seeded

fig C Using a Punnett square makes it easy to predict the results of crosses between different parent pea plants.

Test cross

As you can see from **fig C**, individuals that are homozygous dominant or heterozygous have identical dominant phenotypes. For a plant or animal breeder this can present all sorts of difficulties. A breeder often needs to know that the stock will breed true, in other words, that it is homozygous for the desired feature. If the feature is a recessive phenotype, then any plant showing the feature in the phenotype must be homozygous. However, if the required feature is a dominant phenotype the physical appearance does not show whether the individual is homo- or heterozygous. To find out which it is, the individual must be crossed with a homozygous recessive individual (see **fig D**). This type of cross, known as a test cross, reveals the parental genotype.

Y = yellow
y = green

if a homozygous yellow parent is crossed:

| parental phenotype | yellow seeds | × | green seeds |
| parental genotype | YY | | yy |

gametes   Ⓨ Ⓨ   ⓨ ⓨ

	ⓨ	ⓨ
Ⓨ	Yy	Yy
Ⓨ	Yy	Yy

| offspring genotype (F₁ generation) | Yy |
| offspring phenotype | yellow seeds |

if a heterozygous yellow parent is crossed:

| parental phenotype | yellow seeds | × | green seeds |
| parental genotype | Yy | | yy |

gametes   Ⓨ ⓨ   ⓨ ⓨ

	ⓨ	ⓨ
Ⓨ	Yy	Yy
ⓨ	yy	yy

| offspring genotypes | Yy : yy |
| offspring phenotypes | 1 yellow : 1 green |

fig D A test cross to reveal the genotype for seed colour of a parent pea plant.

Did you know?

Carrying out genetic experiments

It is unethical to use people as experimental organisms in genetics. Also people usually have only one baby, after nine months gestation, families are small and many phenotypic characteristics do not show up until adulthood. Much of our understanding of human genetics comes from experiments on other organisms. A 'suitable' organism for a genetic experiment should, ideally, have the following features:

- be relatively easy and cheap to raise to maximise the chance of successful breeding and minimise experimental costs
- have a short life cycle so that the results of crosses and/or mutations can be seen quickly
- produce large numbers of offspring so that the results of any crosses are statistically relevant
- have clear, easily distinguished characteristics, such as tall or dwarf plants, or the wing length in insects.

Organisms commonly used by scientists for genetic investigations include the fruit fly *Drosophila melanogaster*, pea plants, fungi such as *Aspergillus nidulans* and a variety of bacteria, such as *Escherichia coli*. Using organisms such as plants, bacteria and fruit flies also raises fewer ethical concerns than experimenting with larger organisms such as mammals.

Sampling errors

The theoretical ratios predicted by a genetic cross are usually seen (approximately) in real genetic experiments, but the numbers are never precise. There are several reasons for this:

- Chance plays a large role in reproduction. The combination of alleles in each gamete is completely random as is the joining of particular gametes, unlike the theoretical diagrams that we draw.
- Some offspring die before they can be sampled. For example, some seeds do not germinate and some embryos miscarry.
- Inefficient sampling techniques. For example, it is very easy to let a few *Drosophila* escape.

So the sampling error must be taken into account when you look at a real genetic cross, and the smaller the sample, the larger the potential sampling error. If a $3:1$ ratio of phenotypes is expected in the offspring, it is unlikely to show itself if only four offspring are produced. Looking at 400 offspring increases the likelihood of the expected ratio emerging, and 4000 offspring is better still. This is why organisms such as fast-growing plants, *Drosophila*, certain fungi and bacteria are so useful – they all produce large numbers of offspring in a small space of time.

Questions

1 Define the following terms: homozygous; heterozygous; dominant; and recessive.

2 In pea plants the gene for plant height has two alleles. The tall phenotype is dominant to the dwarf phenotype. Choose suitable letters for the two alleles and draw genetic diagrams to show the following:
   (a) a cross between two homozygous tall parents
   (b) a cross between heterozygous parents
   (c) the F₁ and F₂ generations of a cross between a homozygous tall parent and homozygous dwarf parent.

3 Explain the importance to plant breeders of a test cross.

4 Why are organisms such as *Drosophila*, *Arabidopsis*, *Aspergillus* and *E. coli* bacteria suitable as experimental organisms in genetics?

Key definitions

A **homozygote** is an individual where both of the alleles coding for a particular characteristic are identical.

A **heterozygote** is an individual where the two alleles coding for a particular characteristic are different.

When a phenotype shows **dominance** it is expressed whether the individual is homozygous for the characteristic or not.

When a phenotype is **recessive** it is only expressed when both alleles code for the recessive feature, in other words the individual is homozygous recessive for that trait.

Polygenic phenotypic traits are determined by several interacting genes.

A **monogenic cross** is a genetic cross where only one gene is considered.

By the end of this section, you should be able to...

● explain the terms codominance and multiple alleles

The genetics we study today is built on the work of Gregor Mendel (1822–1884), an Austrian monk who taught botany, maths, physics and statistics. Mendel was fascinated by the variety of the plants in the monastery gardens and wanted to find out more. At a time when no one had seen chromosomes and DNA was unheard of, Mendel discovered some fundamental principles of genetics.

The work of Mendel

Mendel decided to investigate peas because he could see that they had characteristics (the colour and shape of the peas and pods, and the height of the plants) that varied in a clear-cut way. What is more, they were easy to grow and he could control which plants pollinated each other.

He spent the first two years of his research making sure that the plants he was using were true breeding for the characteristics he was investigating. He then tried crossing pure-breeding parents that had two different forms of a chosen characteristic and found the offspring were all the same. But when he crossed the first-generation offspring together he found that both characteristics reappeared in predictable ratios. **Fig A** shows some of the characteristics Mendel looked at, and **table A** shows some of his work.

Both the male and female parts of the pea flower are held within a hood-like petal so that self fertilisation frequently occurs. Mendel opened the bud of one flower before the pollen matured and fertilised the stigma with pollen from another chosen flower. In this way he could control the cross.

anther

stigma

Mendel used seven clearly differentiated, pure-breeding traits of the pea plant for his experiments. They are shown here in both their dominant and recessive forms.

	Stems		Pods		Seeds/flowers		
Dominant trait	Tall	Axial flowers	Green	Inflated	Round	Yellow	Red flowers
Recessive trait	Short	Terminal flowers	Yellow	Pinched	Wrinkled	Green	White flowers

fig A Mendel worked on clear-cut characteristics like these and built up a mass of experimental data.

Experiment	Characteristic	Results at second generation
1	round/wrinkled peas	336 round : 107 wrinkled
2	yellow/green peas	355 yellow : 123 green
3	purple/white flowers	705 purple : 224 white
4	fat/shrunken pods	882 fat : 299 shrunken
5	green/yellow unripe pods	428 green : 152 yellow

table A Some of Mendel's results.

The law of segregation: Mendel's first law

The first law that Mendel presented is known as the **law of segregation**. It was the result of his work with monohybrid crosses. The law states one unit or allele for each trait is inherited from each parent to give a total of two alleles for each trait.

The segregation (separation) of alleles in each pair takes place when the gametes are formed. This idea of independent units of inheritance, some dominant over others, that are maintained throughout the life of an individual and do not fuse to form a homogeneous mass, was Mendel's real breakthrough.

The law of independent assortment: Mendel's second law

The **law of independent assortment** states that different traits are inherited independently of each other. This means that the inheritance of a dominant or recessive phenotype for one characteristic, such as grey or ebony bodies in *Drosophila*, has nothing to do with the inheritance of alleles for other characteristics such as wing length or eye colour. See **Sections 8.1.4** and **8.1.5** to see where this law still stands up, and where it has been superseded as our knowledge of inheritance has grown.

Considering that Mendel formulated these laws so early in the history of our understanding of inheritance it is remarkable that they have relevance today, although we now recognise that the second law in particular has many exceptions as a result of gene linkage and polygenic inheritance (when several different genes interact to produce a given phenotype).

Multiple alleles

So far we have only looked at examples of traits that are inherited as genes with just two possible alleles, just as Mendel did. However, some features have **multiple alleles**. This means there are more than two possible variants. No matter how many possible alleles there are, any one diploid individual will only inherit two of them. These alleles are still inherited in a Mendelian fashion, although the patterns of dominance may be rather more complex. One clear example of multiple alleles is the human ABO blood group system.

Codominance

In the human ABO blood group system there are three possible alleles – A, B and O. These are usually shown as I^O, I^A and I^B. The different alleles code for the presence or absence of antigens on the surface of the erythrocytes (red blood cells). I^O codes for no antigens, I^A codes for antigen A and I^B codes for antigen B.

I^O is recessive. A homozygote with I^O will have no antigens on their erythrocytes and have blood group O. Both I^A and I^B are dominant to I^O so heterozygotes have the blood group of the dominant phenotype. An individual who inherits $I^O I^A$ or $I^A I^A$ will have A antigens on the erythrocytes and have blood group A, while someone who inherits $I^O I^B$ or $I^B I^B$ will have B antigens and be blood group B.

I^A and I^B are **codominant**. This means both alleles are expressed and produce their proteins, which act together without mixing. So an individual who inherits I^A and I^B ($I^A I^B$) will have both antigen A and antigen B on the surface of their erythrocytes and they will have the blood group AB. The A and B antigens will act in just the same way as if they were there individually – there is no blending in the phenotype. This is the key feature in codominance. Humans are not the only species that have the ABO multiple allele blood groups with codominance. They are seen in bonobos, chimpanzees and gorillas, and also in cows and sheep.

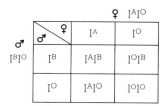

♂ \ ♀	I^A	I^A
I^O	$I^A I^O$	$I^A I^O$
I^O	$I^A I^O$	$I^A I^O$

♀ $I^A I^A$
♂ $I^O I^O$

parental phenotypes: mother group A
father group O

offspring genotypes: all $I^A I^O$
offspring phenotypes: all group A

♂ \ ♀	I^B	I^O
I^A	$I^B I^A$	$I^O I^A$
I^A	$I^B I^A$	$I^O I^A$

♀ $I^B I^O$
♂ $I^A I^A$

parental phenotypes: mother group B
father group A

offspring genotypes: $I^B I^A$: $I^O I^A$
offspring phenotypes: 1AB : 1A

♂ \ ♀	I^A	I^O
I^B	$I^A I^B$	$I^O I^B$
I^O	$I^A I^O$	$I^O I^O$

♀ $I^A I^O$
♂ $I^B I^O$

parental phenotypes: mother group A
father group B

offspring genotypes: $I^A I^B$: $I^O I^B$: $I^A I^O$: $I^O I^O$
offspring phenotypes: 1AB : 1B : 1A : 1O

fig B The genetics of the ABO blood group system.

Did you know?

Blood groups and blood transfusions

In the ABO blood groups system, the antigens on the erythrocytes are not the whole story. You have antigens on your erythrocytes plus corresponding antibodies in your blood plasma.

fig C Multiple alleles determine human blood groups.

Blood group	Antigen on erythrocytes	Antibody in the plasma
A	A	B
B	B	A
AB	AB	none
O	O	A and B

This affects what happens when blood transfusions are given. If the wrong blood groups are mixed, an antigen-antibody reaction results and the blood cells agglutinate (clump together). This can be fatal, so in blood transfusions the donor and the recipient must be carefully matched.

Blood group O has no antigens so it does not cause an antigen-antibody reaction and can be given to anyone. It is the universal donor. However, someone with blood group O can only receive blood group O. Their A and B antibodies would destroy any other blood group.

People with blood group AB can receive any type of blood as they have no A or B antibodies. They are universal recipients. However, AB blood can only be given to other people with AB blood as it would trigger agglutination in people with blood groups A, B and O.

Questions

1 **Table A** shows some of Mendel's published results.
   (a) Choose two of the crosses and draw suitable genetic diagrams to work out the theoretical results. Compare them with Mendel's actual results.
   (b) Collecting data as Mendel did was unusual at the time. Some people have suggested that Mendel 'tweaked' his experimental evidence to get the results he predicted mathematically. If this is true, does the end justify the means? Discuss.

2 A woman has blood group A and her partner has blood group B.
   (a) What are their possible genotypes?
   (b) What is the likelihood that one of their children will be a universal donor? Show how you work this out.

3 A true breeding chestnut coloured mare (C^HC^H) is crossed several times with a true breeding cream coloured stallion (C^hC^h). All of the foals are palomino (golden body, creamy-white mane and tail).
   (a) What does this suggest about the alleles C^H and C^h?
   (b) If a palomino mare and stallion are bred, what is the chance of them having a palomino foal? Show your workings.

4 Explain who people with blood groups A and B can donate blood to and receive blood from, and explain why.

Key definitions

The **law of segregation** describes Mendel's first law, which states that one unit or allele for each trait is inherited from each parent to give a total of two alleles for each trait. The segregation (separation) of alleles in each pair takes place when the gametes are formed and some alleles code for phenotypes that are dominant over others.

The **law of independent assortment** describes Mendel's second law, which states that different traits are inherited independently of each other. This means that the inheritance of alleles for one phenotype has nothing to do with the inheritance of alleles for another characteristic.

If a gene has **multiple alleles** it means there are more than two possible variants at a particular locus.

When alleles are **codominant**, both alleles are expressed and the proteins they code for act together without mixing to produce a given phenotype.

By the end of this section, you should be able to...

- explain the inheritance of two non-interacting unlinked genes
- use chi squared (χ^2) tests to test the significance of the difference between observed and expected results

A single trait inherited by a pair of alleles on the autosomes has been considered in all of the genetic crosses we have looked at so far. However, this is a very long way from the real mechanisms by which inheritance takes place. Hundreds and thousands of genes go to make up the genotype of any one individual and they are all passed on at the same time.

Digenic (dihybrid) crosses

Digenic crosses are breeding experiments involving the inheritance of two pairs of contrasting characteristics at the same time. Although still a very long way from the complexity of real events, this goes one step closer to the living cell.

Pea plants possess a variety of traits that are inherited through individual pairs of alleles. Mendel set up experiments to determine how two characters interact during inheritance, using the two traits of shape and colour of peas (see **fig A**). Round peas (R) are dominant to wrinkled peas (r). Yellow peas (Y) are dominant to green peas (y).

A dihybrid cross between a dominant homozygote (round and yellow, RRYY) with a recessive homozygote (wrinkled and green, rryy) results in F_1 offspring that are all heterozygotes (RrYy) and display the dominant phenotype characteristics (round and yellow).

A self-cross between two of these F_1 offspring or a test cross with a homozygous recessive gives rise to four different phenotypes: round and yellow, round and green, wrinkled and yellow, and wrinkled and green. Two of these – round and yellow, wrinkled and green – are the same as the original parents and are called **parental phenotypes**. The other two phenotypes – round and green, wrinkled and yellow – are different to the parents and are known as **recombinant phenotypes**. The ratios are different for the two types of cross. **Fig A** shows clearly the way in which a dihybrid cross works in peas, and also demonstrates the value of a Punnett square in dealing with these more complex crosses.

The results of these dihybrid crosses, which show the inheritance of two separate genes independently, follow Mendel's law of independent assortment. The two sets of alleles are on different chromosomes, separated into the gametes independently of

parental phenotypes:	round yellow	wrinkled green
parental genotypes:	RRYY	× rryy
gametes:	(RY) (RY) (RY) (RY)	(ry) (ry) (ry) (ry)

R represents the round allele   Y represents the yellow allele
r represents the wrinkled allele   y represents the green allele

offspring (F_1) genotypes:

Gametes	RY	RY	RY	RY
ry	RrYy	RrYy	RrYy	RrYy
ry	RrYy	RrYy	RrYy	RrYy
ry	RrYy	RrYy	RrYy	RrYy
ry	RrYy	RrYy	RrYy	RrYy

offspring (F_1) phenotypes: All round and yellow

round yellow	wrinkled yellow	round green	wrinkled green
○	(shape)	◐	(shape)

test cross on offspring (F_1)

offspring (F_1) genotypes:	RrYy	×	rryy
gametes:	(RY) (Ry) (rY) (ry)		(ry) (ry) (ry) (ry)

offspring (F_2) genotypes:

Gametes	RY	Ry	rY	ry
ry	RrYy	Rryy	rrYy	rryy
ry	RrYy	Rryy	rrYy	rryy
ry	RrYy	Rryy	rrYy	rryy
ry	RrYy	Rryy	rrYy	rryy

offspring (F_2) phenotypes:

round yellow	wrinkled yellow	round green	wrinkled green
1	1	1	1

selfing of offspring (F_1)

offspring (F_1) genotypes:	RrYy	×	RrYy
gametes:	(RY) (rY) (Ry) (ry)		(RY) (rY) (Ry) (ry)

offspring (F_2) genotypes:

Gametes	RY	rY	Ry	ry
RY	RRYY	RrYY	RRYy	RrYy
rY	RrYY	rrYY	RrYy	rrYy
Ry	RRYy	RrYy	RRyy	Rryy
ry	RrYy	rrYy	Rryy	rryy

offspring (F_2) phenotypes:

round yellow	wrinkled yellow	round green	wrinkled green
9	3	3	1

fig A In this dihybrid cross, each variant of one gene combines randomly with any of the variants of another gene. Due to this independent assortment there are four possible phenotypes that can result from a heterozygote cross, whether a self cross or test cross.

each other. The genes for the shape of the pea and the genes for the colour of the peas are on different autosomes. As a result, four different types of male gametes and of female gametes are produced. They segregate independently of each other and so can recombine in all possible ways. **Fig B** illustrates independent segregation in the gametes formed from two pairs of example chromosomes.

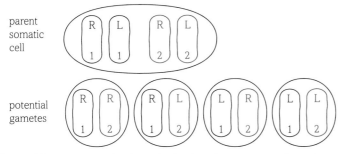

fig B Alleles are separated independently when the gametes are formed.

Another example of dihybrid inheritance occurs in the fruit fly *Drosophila*. Grey bodies are dominant to ebony bodies, and long wings are dominant to the very short vestigial wings. A cross between two heterozygotes, as seen in **fig C**, shows again the use of a Punnett square to determine the results of dihybrid inheritance. The crossing of two non-linked heterozygotes in a dihybrid cross always results in a 9 : 3 : 3 : 1 ratio of phenotypes.

L represents the allele for long wings.   G represents the allele for grey body.
l represents the allele for vestigial wings.   g represents the allele for ebony body.

parental phenotypes: long wings   long wings
                             grey body   grey body

parental genotypes: LlGg   ×   LlGg

gametes: LG   Lg   lG   lg   LG   Lg   lG   lg

offspring genotypes:

Gametes	LG	Lg	lG	lg
LG	LLGG	LLGg	LlGG	LlGg
Lg	LLGg	LLgg	LlGg	Llgg
lG	LlGG	LlGg	llGG	llGg
lg	LlGg	Llgg	llGg	llgg

offspring phenotypes: long wings grey bodies   long wings ebony bodies   vestigial wings grey bodies   vestigial wings ebony body

9 : 3 : 3 : 1

fig C The crossing of two heterozygotes in a dihybrid cross, whether plant or animal, always results in the typical 9 : 3 : 3 : 1 ratio of parental and recombinant offspring phenotypes.

There are some occasions when the ratios are not what you expect. There can be several explanations for this:

- small sample size
- experimental error – especially when working with organisms such as *Drosophila*, which can escape or die relatively easily
- the process is random and so sometimes the unexpected happens
- unexpected ratios can mean the genes being examined are both on the same chromosome (they are linked).

Using probability to predict the outcome of complex crosses

As crosses become more complex the use of simple diagrams to predict the possible offspring and their ratios becomes more difficult. You can use the idea of probability to work out very simply which genotypes might be expected for any given cross when the two characteristics are inherited independently of each other. In the cross between the heterozygous grey bodied, long-winged *Drosophila* shown in **fig C**, the probability of certain phenotypes occurring is worked out as follows:

- *Grey bodied, long-winged flies*: a grey body is dominant so $\frac{3}{4}$ of the F_2 offspring will be grey. Similarly, long wings are dominant so $\frac{3}{4}$ of the offspring will have long wings. Therefore, the probability of any fly inheriting a grey body and long wings is $\frac{3}{4} \times \frac{3}{4} = \frac{9}{16}$

- *Grey bodied, vestigial-winged flies*: as above, grey bodies have a $\frac{3}{4}$ chance of appearing in the offspring. The recessive character for vestigial wings has a $\frac{1}{4}$ chance of occurring in the phenotype. Therefore, the probability of grey bodied vestigial winged flies appearing is $\frac{3}{4} \times \frac{1}{4} = \frac{3}{16}$

- *Ebony bodied, long-winged flies*: the probability of ebony bodies appearing is $\frac{1}{4}$ as they are a recessive trait, and the probability of long wings appearing is $\frac{3}{4}$ as they are a dominant trait. Therefore, the probability of this phenotype being inherited is $\frac{1}{4} \times \frac{3}{4} = \frac{3}{16}$

- *Ebony bodied, vestigial-winged flies*: each of the recessive characters has a $\frac{1}{4}$ chance of occurring in the offspring so the probability of the double homozygous recessive phenotype being seen is $\frac{1}{4} \times \frac{1}{4} = \frac{1}{16}$

This gives the same 9 : 3 : 3 : 1 ratio of phenotypes seen in the Punnett square.

Please note that these calculations cannot be applied to linked genes (see **Section 8.1.5**).

The chi squared test – getting it right

In genetics, as in many other areas of biology, we rely heavily on the results of practical experiments to test the validity of our ideas and hypotheses. When we do a genetic experiment, for example the *Drosophila* cross shown in **fig C**, we have a good idea of the results we are going to get. Even if we are carrying out a completely new experiment we will have a hypothesis, which is a theory we are hoping to prove, or a null hypothesis, which is a theory we are hoping to disprove.

Drosophila, along with most other biological organisms, do not breed in neat multiples of 9, 3 and 1. Experimentally we are faced with several hundred fruit flies with different coloured bodies and wings. How do we know if the results we observe are close enough to our expected 9 : 3 : 3 : 1 ratio or whether they are so far out that some other unexplained factor such as the flies escaping before counting, getting stuck in the medium and dying, or gene linkage (see **Section 8.1.6**) has come into play? This is where the **chi squared (χ^2) test** is useful.

The chi squared test is a relatively simple statistical method that is used to establish whether any differences between the observed and expected results are significant or simply due to chance. The null hypothesis is where there is no significant difference between the expected and observed results. The test involves some fairly simple arithmetic followed by the use of a chi squared table to look up the related probability.

The formula for the chi squared (χ^2) test

$$\chi^2 = \sum \frac{(O - E)^2}{E}$$

In this equation, **O** *is your observed result*, **E** *is the expected result* and the symbol $\sum$ *means the sum of.*

What does this mean in practice?

Look back to simple monogenic crosses between round and wrinkled peas in **Section 8.1.2, fig C**. A cross between two round peas results in 880 new pea plants, 640 producing round peas and 240 producing wrinkled peas. We assume that the round pea phenotype is dominant to the wrinkled pea phenotype and, because both phenotypes have appeared in the offspring, we also assume that the parents are both heterozygotes. The expected ratio of dominant offspring to recessive is 3 : 1. Our null hypothesis is that any differences between our expected results and the observed results are purely due to chance.

If our hypothesis is correct and the ratio is 3 : 1, we would expect 660 round pea plants and 220 wrinkled pea plants $\left(\frac{880}{4} \times 3 \text{ and } \frac{880}{4}, \text{ respectively} \right)$.

Now we calculate χ^2 (chi squared):

Phenotype	Hypothesis	Observed	Expected	O – E	(O – E)²	$\frac{(O - E)^2}{E}$
round pea plants	3	640	660	640 – 660 = –20	–20² = 400	$\frac{400}{660} = 0.61$
wrinkled pea plants	1	240	220	240 – 220 = 20	20² = 400	$\frac{400}{220} = 1.82$

$\chi^2 = 0.61 + 1.82 = 2.43$

The value of χ^2 for this experiment is 2.43. On its own, the chi squared number does not tell us much. We also need to know the number of *degrees of freedom* in the system, which measures the spread of the data. The degree of freedom is always one less than the number of categories of observed information (different phenotypes in this case) you are working with. In this example there are two categories (two different phenotypes), so there is one degree of freedom.

Degrees of freedom = number of categories − 1

Once we have these bits of information we can use them with a χ^2 (chi squared) probability table to find out whether there is any significant difference between our predicted and observed results. In other words, whether the null hypothesis is correct and that the results are just chance occurrences, or the results are because another factor is affecting the inheritance of these particular alleles.

Degrees of freedom	Probability, p	0.99	0.98	0.95	0.90	0.80	0.50	0.20	0.10	0.05	0.02	0.01	0.001
1		0.000	0.001	0.004	0.016	0.064	0.455	1.64	2.11	3.84	5.41	6.64	10.83
2		0.020	0.040	0.103	0.211	0.446	1.386	3.22	4.61	5.99	7.82	9.21	13.82
3		0.115	0.185	0.352	0.534	1.005	2.366	4.64	6.25	7.82	9.84	11.35	16.27
4		0.297	0.429	0.711	1.064	1.649	3.357	5.99	7.78	9.49	11.67	13.28	18.47
5		0.554	0.752	1.145	1.610	1.610	4.351	7.29	9.24	11.07	13.39	15.09	20.52

fig D Part of a χ^2 (chi squared) table.

In most cases a probability (*p*) of 0.05 is used as the cut-off point to show that differences between observed and expected data are statistically significant. This means that 95 times out of 100 you will get the results you expected and the remaining 5 times any differences are the result of random chance.

Using the appropriate row of data for the number of degrees of freedom on a χ^2 table, look up the probability for your calculated value (see **fig D**). If the probability is greater than 0.05, there is no significant difference between what we expected and what we observed in the experiment so we can accept the null hypothesis. This means that the results match our prediction, supporting our hypothesis that the round pea phenotype is dominant to the wrinkled pea phenotype and that in this heterozygote cross the ratio of dominant offspring to recessive is 3 : 1. We can say with at least 95% confidence that the differences between the observed results and the theoretical predications are due to chance.

If the probability value for χ^2 had been less than 0.05, this would mean that there was a statistically significant difference between the observed numbers and the expected numbers and we reject the null hypothesis. This would indicate that the inheritance was not occurring as predicted; the results are not in the range we would have expected. We would need to look for an alternative explanation for them. This might include looking for errors in the experimental technique, or revising the entire hypothesis.

χ^2 for dihybrid crosses

The χ^2 test can also be used for dihybrid crosses. In the *Drosophila* cross described in **fig C**, two grey bodied flies with long wings reproduce. We assume that grey bodies are dominant to ebony bodies and that long wings are dominant to vestigial wings. We make the assumption that this is a heterozygote cross and that the observed results will show the expected ratio of 9 : 3 : 3 : 1. The null hypothesis is that any differences between the theoretical results and our observed results are due to chance. We use the χ^2 test to show if this hypothesis is correct, using results with 176 offspring.

Phenotype	Hypothesis	Observed	Expected	$O - E$	$(O - E)^2$	$\dfrac{(O - E)^2}{E}$
long wings, grey body	9	101	99	$101 - 99 = 2$	$2^2 = 4$	$\dfrac{4}{99} = 0.040$
long wings, ebony body	3	31	33	$31 - 33 = -2$	$-2^2 = 4$	$\dfrac{4}{33} = 0.121$
vestigial wings, grey body	3	34	33	$34 - 33 = 1$	$1^2 = 1$	$\dfrac{1}{33} = 0.030$
vestigial wings, ebony body	1	10	11	$10 - 11 = -1$	$-1^2 = 1$	$\dfrac{1}{11} = 0.091$

$\chi^2 = 0.040 + 0.121 + 0.030 + 0.091 = 0.282$

The value of χ^2 squared for this experiment is 0.282. There are four pieces of observed information (four different phenotypes) so subtracting 1 gives us 3 degrees of freedom.

By using the χ^2 squared table in **fig D** we can see that any number lower than 7.82 is not statistically significant. With our result of 0.282, we accept the hypothesis that there is no significant difference between the expected results and the observed results.

Questions

1. As any gardener knows, F_1 hybrid seeds do not breed true. Plants used from these seeds cannot be used as stock plants for the next year.
   Explain the biology behind this statement, including careful definitions of terms such as genotype, phenotype, homozygote, heterozygote, dominant, recessive, F_1 generation and true breeding.

2. *Drosophila* that are heterozygous for long wings and grey bodies are crossed with *Drosophila* that are homozygous recessive for both of these characteristics.
   (a) Show this cross using a Punnett square and explain the ratios of phenotypes that you expect to see.
   (b) Give two reasons why the data you collect may not fit the expected ratios exactly.

3. Look at the dihybrid cross between two heterozygous pea plants that produce yellow, round peas, as described in **fig A**. In an experimental cross, students achieved 219 round yellow pea plants, 81 round green pea plants, 69 wrinkled yellow pea plants and 31 wrinkled green pea plants. Use the χ^2 squared test to show that this fits the expected 9 : 3 : 3 : 1 ratio for a heterozygote dihybrid cross.

Key definition

Parental phenotypes describe offspring that have the same phenotypes as the parental organisms.

Recombinant phenotypes describe offspring that have different phenotypes to their parents as a result of recombination of the chromosomes during sexual reproduction.

Chi squared (χ^2) test is a statistical test used to compare differences between sets of data to evaluate whether they differ from each other significantly.

By the end of this section, you should be able to...

● explain how autosomal gene linkage results from the presence of alleles on the same chromosome, and that the results of the crosses can be explained by the events of meiosis

● explain sex linkage on the X chromosome

In the previous examples of dihybrid inheritance, you have seen how the alleles of the two genes segregate independently of each other. However, the situation is not always as predictable as that.

Gene linkage

When *Drosophila* with normal broad abdomens and long wings are crossed with flies displaying recessive narrow abdomens and vestigial wings, they produce the expected F₁ generation. All the offspring show the dominant phenotype, but possess the heterozygote genotype. However, when these F₁ flies are crossed, the normal 9 : 3 : 3 : 1 dihybrid ratio of parental and recombinant types does not occur (see **Section 8.1.4, fig C**). Instead there is a 3 : 1 ratio of dominant : recessive phenotypes, which is what we would expect in a monogenic cross. The explanation of this apparent discrepancy is that the genes are linked (**gene linkage**). This means they are sited on the same chromosome and inherited as if they were one unit, whereas the normal Mendelian ratios for a dihybrid cross only occur when the genes are on separate chromosomes.

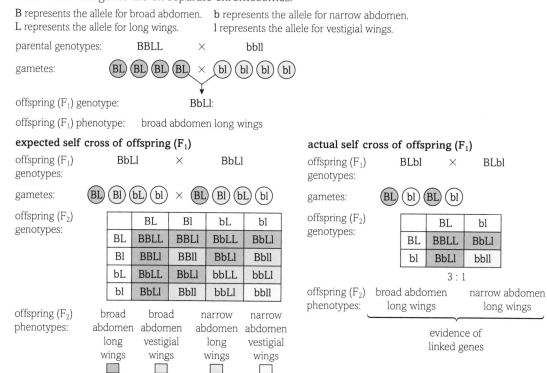

fig A These two genes are inherited as if they were one unit, because they are closely linked on the same chromosome.

Identifying linked genes

With completely unlinked genes, two genes are found on separate chromosomes and approximately equal numbers of gametes containing the parental alleles and recombinant alleles are formed. However, if genes are found on the same chromosome they are linked and inherited, to a greater or lesser degree, as if they were a single gene. When genes are closely linked, such as when they are located close to each other on the chromosome, recombination events during meiosis rarely occur in the gametes. If the genes are more loosely linked, such as when they are located further apart on the chromosome, then the number of recombination events in meiosis will be higher. Therefore, the tightness of the linkage is related to how close together the linked genes are on the chromosome. Genes that are very close together are less likely to be split during the crossing over stage of meiosis than genes that are further apart.

If two or more genes are positioned very close together on a chromosome they may be so tightly linked that they are never split up during meiosis, and so the gametes formed will always be of the parental types. If the genes are further apart, crossing over between them is more likely to occur. Although in the majority of cases they will be passed on as a parental unit, sometimes they will be mixed and recombinant gametes will be produced, which will in turn be reflected in the offspring.

The clue that linkage is involved in a dihybrid cross is when the expected $9:3:3:1$ ratio for a heterozygote cross does not occur. Small differences in numbers are usually due to experimental error. Larger discrepancies, especially when there is an unexpectedly high number of parental types, indicates that linkage between the genes is involved. A χ^2 squared test would establish this with the null hypothesis being rejected, showing that some interaction between the two genes exists.

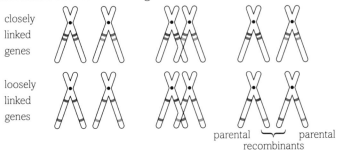

fig B During meiosis, closely linked genes will be passed into the gametes as a single unit. Genes on the same chromosome but further apart will form some recombinant gametes.

Chromosome mapping

Studying gene linkage on the autosomes gives us a way of working out how close together various genes are on a chromosome. Using linked genes we can build up a genetic map of a chromosome. To produce a genetic map, the crossover value (cov) has to be worked out:

$$\text{crossover value} = \frac{\text{number of recombinant offspring} \times 100}{\text{total number of offspring}}$$

Using this formula, closely linked genes producing only small numbers of recombinant offspring will result in low crossover values. Genes that are further apart produce larger numbers of recombinants and have higher crossover values. Crossover values can be used to build up a chromosome map, as shown in **fig C**.

Gene maps built up from crossover values of linked genes are still very valuable to scientists, although their production has been speeded up considerably by the development of ever more powerful computers and bioinformatics.

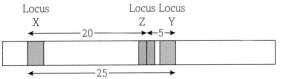

fig C To create a genetic map, the crossover values are converted into arbitrary map units. The first crossover value positions genes X and Y relative to each other. However, the position of gene Z can only be decided by reference to its crossover values in relation to both gene X and gene Y. This technique is vital when scientists want to alter individual genes or groups of genes in genetic engineering.

Sex linkage

The chromosomes in diploid organisms occur as homologous pairs. In organisms such as mammals, where there are clear differences between the males and females, sex is determined by the chromosomes. The autosomes make up all but one of the homologous pairs of chromosomes and they carry information about the general body cells and their biochemistry. The final pair of chromosomes carry information about the sex of the individual and are known as the sex chromosomes. In mammals the female has two large X chromosomes. Therefore all of her eggs contain an X chromosome and she is referred to as **homogametic**. The male has one X chromosome and one much smaller Y chromosome, so half of the sperm will contain X chromosomes and half will contain Y chromosomes and the male is **heterogametic**. In humans there are 22 pairs of autosomes and one pair of sex chromosomes. In females, one of the X chromosomes is inactivated at random in every cell to form a Barr body (see **Section 7.2.2**). In some organisms, such as birds and many reptiles, it is the male that is homogametic.

The small human Y chromosome has around 23 million base pairs and only 78 protein-coding genes. It seems to carry little but male sex information, including the crucial *SRY* gene (sex-determining region Y). This gene triggers the development of the testes and, through the hormones produced by the testes, the maleness of the fetus. The X chromosome is much bigger, with around over 150 million base pairs. Estimates for the number of protein coding genes on the X chromosome vary from around 800 to around 1200. The X chromosome codes for all the female characteristics and also carries a variety of other genes coding for traits including the clotting factors in the blood and the ability to distinguish between certain colours. However, there is enough similarity between X and Y to allow them to pair up during cell division.

Genes that are carried on the X chromosome are said to be sex-linked. Any recessive or mutant alleles passed on the X chromosome from a mother to her sons will be expressed in the phenotype as there is no corresponding allele on the homologous Y chromosome. Because of the importance of the X and Y chromosomes in **sex linkage**, they are shown on the genetic diagrams. Sex linkage was first discovered in *Drosophila* by Thomas Morgan in the early years of the twentieth century and we now know it occurs in many different organisms.

In *Drosophila*, eye colour is sex-linked. Their eye colour is the result of multiple alleles, with the most common eye colour being red and inherited through a dominant allele on the X chromosome. However, there is a wide range of mutant alleles giving eye colours which include white, apricot and purple (see **fig D**).

X^R represents the allele for red eyes.
X^r represents the allele for white eyes.

Female *Drosophila* have XX chromosomes, males have XY.

	♀		♂

parental genotypes: $X^R X^R$ × $X^r Y$

gametes:

♂ \ ♀	X^R	X^R
X^r	$X^R X^r$	$X^R X^r$
Y^O	$X^R Y$	$X^R Y$

offspring (F_1) genotypes: $X^R X^r$ $X^R Y$ $X^R X^r$ $X^R Y$

offspring (F_1) phenotypes: red-eyed female   red-eyed male   red-eyed female   red-eyed male

1 : 1
ratio of red-eyed males : red-eyed females

All of the F_1 generation resulting from a cross of a red-eyed female with a white-eyed male are red-eyed, showing that red is dominant to white.

offspring (F_1) selfcross:

	♀		♂

offspring (F_1) genotypes: $X^R X^r$ × $X^R Y$

gametes:

♂ \ ♀	X^R	X^r
X^R	$X^R X^R$	$X^r X^R$
Y	$X^R Y$	$X^r Y$

offspring (F_2) genotypes: $X^R X^R$ $X^R Y$ $X^r X^R$ $X^r Y$

offspring (F_2) phenotypes: red-eyed female   red-eyed male   red-eyed female   white-eyed male

All females are red-eyed.
Half of the males are red-eyed,
half white-eyed.

When a self cross is carried out on the F_1 generation, all of the female offspring have red eyes, but half of the male offspring have white eyes. This is a result of sex-linked genes on the X chromosomes.

wild type   white

garnet   apricot

vermillion   sepia

brown   purple

eosin female

fig D Sex-linked genes in *Drosophila* affect the eye colours seen in the insects.

Questions

1 (a) Show the resulting genotypes if one of the F_1 generation of flies in **fig A** with a phenotype of broad abdomen and long wings was test crossed with a homozygous recessive fly with a narrow body and vestigial wings.

   (b) Compare this to a similar test cross with a *Drosophila* that is a heterozygote for both grey body and long wings and explain the differences in the results.

2 In *Drosophila melanogaster*, kidney bean-shaped eyes are recessive to round-shaped eyes, and orange eye colour is recessive to red.

   A fly which is a homozygote for both round eyes and red eyes is crossed with a fly that has kidney bean-shaped, orange eyes. The offspring (F_1) were allowed to breed and the researcher expected to find a 9 : 3 : 3 : 1 segregation for eye shape and eye colour in the F_2 generation. Instead, the following progeny were produced.

Round, red eyes:	520
Kidney bean-shaped, orange eyes:	180
Kidney bean-shaped, red eyes:	54
Round, orange eyes:	45

   (a) Explain why the researcher expected a 9 : 3 : 3 : 1 segregation in the F_2 generation.

   (b) Explain the observed result.

Key definitions

Gene linkage is when genes for two characteristics are found on the same chromosome and are close together so they are linked and inherited as a single unit.

A **homogametic** individual produces gametes that only contain one type of sex chromosome.

A **heterogametic** individual produces gametes that contain two different types of sex chromosome.

Sex linkage refers to genes that are carried on the sex chromosomes.

Human genetics, sex linkage and pedigree diagrams

By the end of this section, you should be able to...

- explain sex linkage on the X chromosome, including haemophilia in humans
- construct genetic crosses and pedigree diagrams

Human genetics work in the same way as the genetics of the peas and fruit flies which you have been considering. In people, as in most organisms, few characteristics are the result of single genes. Almost every aspect of your phenotype is the result of the interactions of the variants of multiple genes, with the added impact of transcription factors and epigenetic influences. Looking at the inheritance of some single human genes can still be helpful, especially in understanding some of the more common inherited diseases that affect people around the world. For example, the structure of the haemoglobin molecule that carries oxygen in the blood (see **Book 1 Section 4.3.3**) is affected by multiple genes and a mutation in any one of them can affect the ultimate structure of the haemoglobin and therefore its ability to carry oxygen in the blood. Looking at one of the multiple alleles in isolation can help our understanding of inherited conditions.

The albino trait

The formation of the pigments that colour the hair and skin of animals is controlled by multiple genes. Albinism is a condition, seen in many different species, in which the natural melanin pigment of the skin, eyes and hair does not form. There are several different forms of albinism, as mutations in several different genes can give similar results in the phenotype. One of the most common forms of albinism is due to a mutant allele that prevents the formation of a normal enzyme in the cells. The enzyme tyrosinase, which is normally active in the melanocytes or pigment-forming cells, is not formed correctly and so the reactions that make melanin cannot take place. Albinism of this type is a recessive phenotype. The parents may appear normal, in which case they are both carriers of the albino allele, or one or more of the parents may be an albino themselves. In the general human population, about 1 person in 30 000 is affected by albinism. People who inherit this condition not only lack pigment in the cells of their skin, hair and eyes – their vision is often poor and they are at higher risk of developing skin cancers because they do not have the natural protective pigment melanin. However, albinism is not life threatening and both difficulties can normally be overcome.

Genetic pedigree diagrams

Genetic diagrams, such as those you have used in **Sections 8.1.2 to 8.1.5** and in **fig A**, show how a trait can be theoretically passed on and the probability of different offspring being produced. The use of a family tree or genetic pedigree diagram (see **fig B**) can show us what happens in reality in a human family over a long period of time. A pedigree diagram includes all the members of a family, indicating their sex and whether or not they have the disease. Because humans are not available as experimental animals (see **Section 8.1.2**), much of our direct understanding of human genes has come from the analysis of genetic pedigree diagrams. In families affected by conditions such as thalassaemia and cystic fibrosis, genetic pedigree diagrams can be useful in predicting which family members may be carriers of the genetic mutation, because they highlight carriers of a recessive phenotype. This allows people to consider their options before they conceive a child. Genetic pedigree diagrams are also extremely useful for identifying sex-linked traits, because a family tree indicates the males and females in the family.

Genetic pedigree diagrams are not only used in human genetics. They are widely used wherever people selectively breed animals. For example, pedigrees are extremely important in the breeding of thoroughbred race horses, and in specific breeds of animals such as dogs, cats, cattle and sheep.

Gametes	A	a
A	AA	Aa
a	Aa	aa

Gametes	a	a
A	Aa	Aa
a	aa	aa

fig A In albinism, a small change in the DNA leads to the production of a non-functional enzyme, which affects the pigment levels in the phenotype.

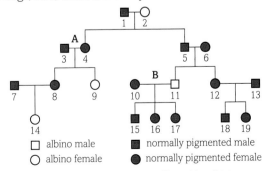

□ albino male   ■ normally pigmented male
○ albino female   ● normally pigmented female

fig B Genetic pedigree diagram for a family affected by albinism.

Sex-linked diseases in humans

Sex-linked genes occur in humans just as they do in other organisms. A mother always donates an X chromosome to her sons. The father always donates the Y chromosome. Because the Y chromosome is small and carries only genes which code for traits associated with maleness, any mutations in a gene on the X chromosome will affect the phenotype of the offspring, even if the characteristic it codes for is recessive. As a result, sex-linkage in humans leads to a variety of conditions known as **sex-linked diseases**. Some of these are relatively minor. Some are life-threatening or even fatal.

Red–green colour blindness

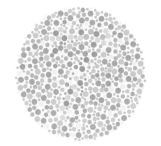

fig C Tests like this are used to demonstrate colour blindness.

Human colour vision is the result of three different types of light sensitive cells called cones found in the retina of the eye (you will learn more about how you see in **Section 9.2.5**). The ability to see in colour is the result of multiple genes coding for different aspects of the process. Many of these genes are found on the X chromosome and mutations in these genes can affect our ability to see in colour, causing different types of colour blindness. Red–green colour blindness is the result of one of these mutations and people affected still see red and green colours, but can have difficulty determining between certain tones of colour. For instance, a purple shirt may appear more blue to a red–green colour blind person as the red tone is not detected as much. Red–green colour blindness is usually an inconvenience but nothing more. It is due to a recessive mutation of a gene on the X chromosome. Because the condition is sex-linked, it is much more common in men than in women. However, because the gene does not markedly affect the chances of survival of an individual, and because the homozygous form is not lethal, colour blindness does occasionally occur in women (see **fig D**). In many populations, around 7–8% of males are affected by red–green colour blindness, but less than 1% of females. In the genetic pedigree in **fig D**, symptom-free carriers are shown as half-shaded circles.

Haemophilia

Colour blindness is usually an inconvenience but nothing more. **Haemophilia** is a much more severe sex-linked trait in which one of the proteins needed for the clotting of the blood is missing. The components of the blood clotting cascade (see **Book 1 Section 4.3.3**) are coded for by multiple genes. Many of these genes are carried on the X chromosome, so problems with the blood clotting cascade are often sex-linked diseases.

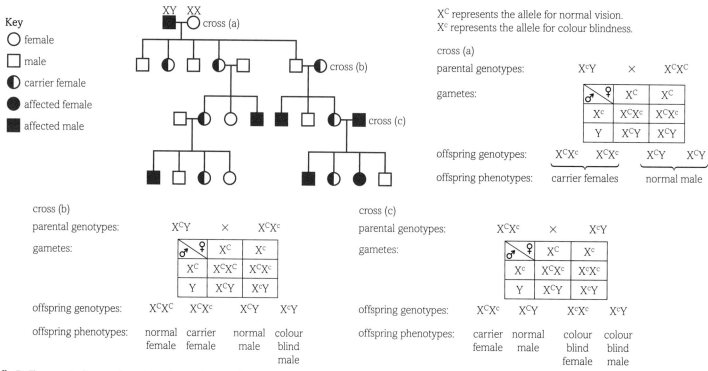

fig D The genetic diagrams here show the mechanism of transmission of the colour blindness genes through this family tree.

Did you know?

Haemophilia in history

Haemophilia A has been recorded since ancient times, and long before we had any understanding of genetics. The Talmud (books of Jewish law) specify that if a boy dies of bleeding after circumcision, his younger brothers should not undergo the ritual and his male cousins on the mother's side are also exempt. This shows an understanding that the problem of blood which does not clot runs in families and is inherited from the mother rather than the father. Haemophilia also plagued the royal families of Europe following a mutation in the earliest stages of the embryonic development of Queen Victoria, as **fig E** clearly shows.

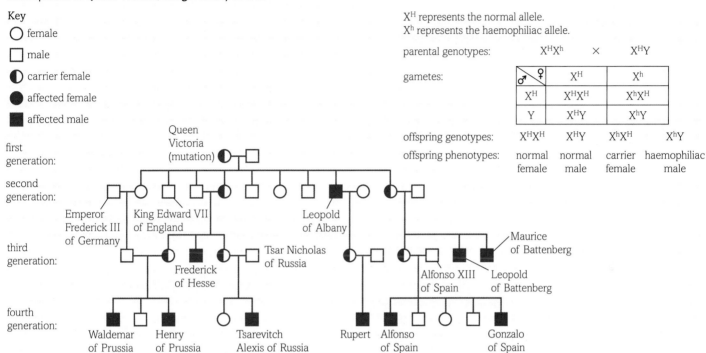

X^H represents the normal allele.
X^h represents the haemophiliac allele.

parental genotypes: $X^H X^h$ × $X^H Y$

gametes:

♂\♀	X^H	X^h
X^H	$X^H X^H$	$X^h X^H$
Y	$X^H Y$	$X^h Y$

offspring genotypes: $X^H X^H$ $X^H Y$ $X^h X^H$ $X^h Y$
offspring phenotypes: normal female — normal male — carrier female — haemophiliac male

fig E As a result of the wide-ranging intermarriages between various royal families, haemophilia spread from the original mutation in Queen Victoria to most of the royal houses of Europe.

One of the most common and best understood forms of haemophilia is haemophilia A. This is a sex-linked condition that involves the lack of clotting factor VIII. It is also known as factor VIII deficiency. Globally the condition affects between 1 in 4000 and 1 in 5000 live male births. The severity of the disease varies but it can be fatal if untreated. The homozygous form is rare and very severe and most affected female fetuses will not survive birth. In an untreated male haemophiliac, the slightest injury can lead to death through excessive bleeding – even exercise can result in internal bleeding of the joints. Historically haemophiliacs were treated by blood transfusions or with factor VIII extracted from donated blood. Now recombinant (genetically engineered) bacteria produce pure factor VIII in large quantities (see **Section 7.3.1**). Regular treatment with pure human factor VIII made by bacteria has given people with haemophilia A an almost normal life expectancy and life quality, as long as the condition is diagnosed and treated as soon after birth as possible – or even before birth.

Key definitions

A **sex-linked disease** is a genetic disease that results from a mutated gene carried on the sex chromosomes – in human beings, largely on the X chromosome.

Haemophilia is a sex-linked genetic disease in which one of the factors needed for the clotting of the blood is not made in the body.

Questions

1 In the United States, about 7% of men are red–green colour blind and less than 1% of women are affected. Using genetic diagrams to help your explanations, answer the following questions.
   (a) Why are there more red–green colour blind men than women?
   (b) Does every red–green colour blind man have a colour blind parent?
   (c) Does every red–green colour blind woman have a colour blind parent?

2 (a) Explain the difference between genetic diagrams and genetic pedigree diagrams.
   (b) Why are genetic pedigree diagrams so useful in human genetics?

3 (a) What is haemophilia A?
   (b) Why are there virtually no girls with haemophilia A?
   (c) A couple with no family history of ill health have four children. Their two daughters and one son appear healthy, but their eldest son has haemophilia A. Suggest two different ways in which this might have come about, using genetic diagrams or pedigree diagrams in your explanation.

155

THINKING BIGGER

FOREIGN GENES AND EVOLUTION

New evidence suggests that horizontal gene transfer – the movement of genes from organisms of one species to another – is more widespread than previously thought, and may be involved in evolutionary changes throughout the living world.

EXPRESSION OF MULTIPLE HORIZONTALLY ACQUIRED GENES

Many animals, including humans, acquired essential 'foreign' genes from microorganisms co-habiting their environment in ancient times, according to research published in the open access journal *Genome Biology*. The study challenges the conventional view that animal evolution relies solely on genes passed down through ancestral lines and suggests that, at least in some lineages, the process is still ongoing.

The transfer of genes between organisms living in the same environment is known as horizontal gene transfer. It is well known in single-celled organisms and thought to be an important process that explains how quickly bacteria develop resistance to antibiotics, for example. However, the idea that horizontal gene transfer occurs in more complex animals, such as humans, has been widely debated and contested.

The researchers studied the genomes of 12 species of fruit fly, four species of nematode worm, and ten species of primate, including humans. They calculated … how likely they were to be foreign in origin. By comparing with other groups of species, they were able to estimate how long ago the genes were likely to have been acquired.

In humans, they confirmed 17 previously-reported genes acquired from horizontal gene transfer, and identified 128 additional foreign genes in the human genome that have not previously been reported. A number of genes, including the ABO gene, which determines an individual's blood group, were also confirmed as having been acquired by vertebrates through horizontal gene transfer. The majority of the genes were related to enzymes involved in metabolism.

In humans, some of the genes were involved in lipid metabolism, including the breakdown of fatty acids and the formation of glycolipids. Others were involved in immune responses, including the inflammatory response, immune cell signalling, and antimicrobial responses, while further gene categories include amino-acid metabolism, protein modification and antioxidant activities…

Bacteria and protists, another class of microorganisms, were the most common donors in all species studied. They also identified horizontal gene transfer from viruses, which was responsible for up to 50 more foreign genes in primates. Some genes were identified as having originated from fungi.

… The majority of horizontal gene transfer in primates was found to be ancient, occurring sometime between the common ancestor of Chordata and the common ancestor of the primates.

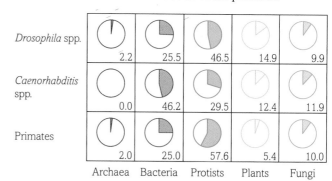

	Archaea	Bacteria	Protists	Plants	Fungi
Drosophila spp.	2.2	25.5	46.5	14.9	9.9
Caenorhabditis spp.	0.0	46.2	29.5	12.4	11.9
Primates	2.0	25.0	57.6	5.4	10.0

fig A Diagram to show the mean origin of foreign genes for each of the taxonomic groups studied, including primates like this squirrel monkey.

Where else will I encounter these themes?

Book 1   5.1   5.2   6.1   6.2   6.3   7.1   7.2

156

This is an extract from an article published by the University of Cambridge in 2015, in their research news for alumni (past students of the university). It discusses research published in full as an open access paper in *Genome Biology* in March 2015. The paper is titled 'Expression of multiple horizontally acquired genes is a hallmark of both vertebrate and invertebrate genomes.'

1. This excerpt is not a scientific paper, but neither is it a newspaper article for the general public. It is aimed at a relatively scientifically literate audience.

   a. Explain how we can tell that it is not a scientific paper – you can look at the original paper published in *Genome Biology* to see the difference.

   b. Discuss the clues that tell you this article is aimed at a scientifically literate readership rather than the general public.

2. Rewrite the article so that it could be published in the science pages of a popular newspaper, with an explanation of how you have changed the content and why you have made those changes.

> Open access (OA) journals publish all of their papers online, so that they are free for anyone to read. This enables scientists from all around the world to access the articles without financial barriers. Can you think of any advantages and disadvantages of this type of publishing?

Now let us examine the biology. You already know about biological molecules, enzymes, prokaryotic and eukaryotic cell structures, sexual reproduction, classification and the importance of DNA sequencing in determining how an organism is classified, evolution and natural selection, some of the major metabolic pathways, antibiotic resistance in bacteria, the immune response, gene sequencing and the transfer of genetic information. This will help you answer the question below.

3. Alastair Crisp, one of the authors of this research, is quoted as saying:

   This is the first study to show how widely horizontal gene transfer occurs in animals, including humans, giving rise to tens or hundreds of active 'foreign' genes. Surprisingly, far from being a rare occurrence, it appears that this has contributed to the evolution of many, perhaps all, animals and that the process is ongoing. We may need to re-evaluate how we think about evolution.

   Do you think he is right? Discuss the quote and justify your answer.

Activity

The normal process by which genes are passed from parents to offspring is called vertical gene transfer (VGT). This article discusses horizontal gene transfer (HGT) from one organism to another unrelated organism. Choose one of the following activities:

- Investigate the process of HGT in bacteria and produce a series of diagrams, a storyboard or simple animation to explain the process.
- Investigate the process of HGT and produce a resource to compare HGT with VGT that could be used to help explain the concepts to other A level students.

● Adapted from http://www.cam.ac.uk/research/news/human-genome-includes-foreign-genes-not-from-our-ancestors, which is adapted from a press release from BioMed Central.

1 The genetics of *Drosophila* fruit flies have been studied for many years. 'Goggle-eye' is a sex-linked trait whose allele is found on the X chromosome of *Drosophila* flies. This allele is recessive to that for 'standard eye'. In addition, the allele for white eye is recessive to red eyes, and the gene for eye colour is on an autosomal chromosome.

Using this information, answer these questions:

(a) If a heterozygous standard-eyed female mated with a goggle-eyed male, what would the proportion of standard-eyed male flies produced be?

    A 0%
    B 25%
    C 75%
    D 50%   [1]

(b) What proportion of *Drosophila* flies would be white eyed if two homozygous red eyed flies mated?

    A 0%
    B 25%
    C 75%
    D 50%   [1]

(c) Which female *Drosophila* flies could give rise to white goggle-eyed male *Drosophila* flies?

    A Standard eyes coloured red, homozygous for both genes.
    B Standard eyes coloured red, homozygous only for the colour genes.
    C Standard eyes coloured red, homozygous only for the eye shape.
    D Standard eyes coloured red, heterozygous for both genes.   [1]

                                        [Total: 3]

2 In maize, the seeds can be yellow or white in colour. In addition, the seeds may have a smooth surface or a wrinkled surface.

Each of these characteristics of maize seeds is an example of single-gene inheritance.

If a pure-breeding (homozygous) variety of maize with yellow, smooth seeds is crossed with a pure-breeding variety with white, wrinkled seeds, all of the F_1 generation have yellow, smooth seeds.

(a) (i) Draw suitable symbols that could be used in a genetic diagram for the alleles involved in these characteristics.
      Allele for yellow colour
      Allele for white colour
      Allele for smooth seed
      Allele for wrinkled seed.   [1]

    (ii) Using these symbols, give the genotype of both the pure-breeding varieties:
        Variety producing yellow, smooth seeds
        Variety producing white, wrinkled seeds.   [2]

(b) The F_1 seeds were sown and the resulting plants were allowed to self-pollinate. The numbers of seeds resulting from this cross are shown in the table below.

Phenotype	Observed frequency	Expected frequency
yellow, smooth	2046	
white, smooth	851	
yellow, wrinkled	750	
white, wrinkled	1953	
Total number of seeds	5600	5600

    (i) Copy and complete the table for the expected frequency of each phenotype in this cross.   [1]

    (ii) With reference to the events that occur during meiosis, explain why the results for the observed frequencies differ considerably from those for the expected frequencies.   [5]

                                          [Total: 9]

3 The ABO blood group system in humans is an example of multiple allele inheritance. Using this system, human blood can be classified into four possible blood groups: A, B, AB and O. The blood group of a particular individual is determined by a single gene pair.

(a) With reference to the inheritance of blood group in the ABO system, explain each of the following terms.
    (i) codominance   [2]
    (ii) multiple allele inheritance   [2]

(b) The family tree for a couple (P1 and P2) with three children is shown in the diagram below. The grandparents of the children and the blood group for each individual are also shown.

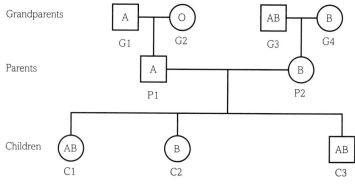

    (i) State the genotype of each of the children.   [1]
    (ii) There is a chance that the next child born to this couple will have group O blood.
        Use a genetic diagram to explain this.   [4]

                                          [Total: 9]

4 (a) Explain what is meant by the term **genetic carrier**. [2]

(b) Sickle cell anaemia is caused by the allele of a single gene. People who are homozygous for this allele have sickle cell anaemia and are severely anaemic.

The photograph below shows red blood cells taken from a person with sickle cell anaemia.

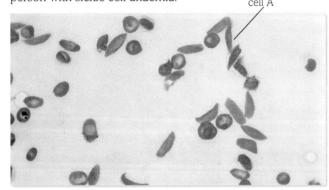

cell A

(i) Give **two** reasons why oxygen transport by cell A may be less efficient than a normal red blood cell. [2]

(ii) A couple have a number of children, only one of whom has sickle cell anaemia. Using a genetic diagram, show how this child inherited sickle cell anaemia. Use the symbols Hb^A to represent the normal allele and Hb^S to represent the sickle cell allele. [3]

(iii) Give the probability that this couple's next child is a genetic carrier of sickle cell anaemia. [1]

(c) One of the polypeptide chains in the haemoglobin of a person with sickle cell anaemia has a change in one of its amino acids. Explain how a change in DNA can lead to a change in a single amino acid in a polypeptide chain. [4]

[Total: 12]

5 The diagram below shows a pair of homologous chromosomes (a bivalent) during meiosis.

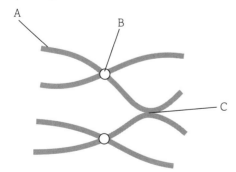

(a) (i) Name **one** stage of meiosis during which homologous chromosomes might look like the ones shown in the diagram. [1]

(ii) Name the structures labelled **A**, **B** and **C** on the diagram. [2]

(b) Guinea pigs are popular pets that have variations in both the colour and texture of their fur. The allele for black fur (**B**) is dominant to the allele for white fur (**b**). The allele for rough fur (**R**) is dominant to the allele for smooth fur (**r**). In a breeding experiment, a homozygous black, rough-furred guinea pig was crossed with a homozygous white, smooth-furred guinea pig. This cross was repeated several times to give offspring in the F$_1$ generation.

(i) State the genotype and the phenotype of the F$_1$ generation. [1]

(ii) Several of the offspring in the F$_1$ generation were interbred to produce an F$_2$ generation.

The phenotypes present in the F$_2$ generation are shown below. Give the expected ratio of the different phenotypes in the F$_2$ generation by writing appropriate numbers.

Black with rough fur =
Black with smooth fur =
White with rough fur =
White with smooth fur = [1]

(iii) The numbers of the different phenotypes in the offspring in the F$_2$ generation produced as a result of interbreeding are given in the table below.

Phenotypes	Number
Black with rough fur	31
Black with smooth fur	4
White with rough fur	2
White with smooth fur	11

With reference to the expected ratio of phenotypes in the F$_2$ generation, give an explanation for the results shown in the table above. [4]

[Total: 9]

TOPIC 8
Origins of genetic variation

8.2 > Gene pools

Introduction

European adders are distributed widely across Europe. However, their habitat has become rarer. Large-scale agriculture and the expansion of cities mean that many adder populations have become isolated. As a result, the frequency of different alleles has become fixed in each population. In this situation, numbers often fall as inbreeding affects fertility. If new individuals are introduced to an isolated population, the situation can improve rapidly. This was demonstrated with adder populations in Sweden, where the introduction of new males to isolated populations restored the numbers to above the original levels and changed the allele frequencies. If a true bottleneck occurs, with no introduction of new alleles, a population becomes very vulnerable, which is exactly the situation seen with cheetahs and rhinos in Africa.

In this chapter you will learn about the genetics of populations. You will consider the factors that affect the gene pool – the sum of all the genes in a population at a given time. You will look at the frequencies of different alleles within a population and use the Hardy-Weinberg equilibrium to measure changes in population genetics. You will consider the value of and limitations of this equation. You will consider the effects of stabilising selection, disruptive selection and chance on the allele frequencies in a population, and how this can sometimes result in speciation. You will look at the effects of genetic bottlenecks and the founder effect on the allele frequencies and genetic diversity of the subsequent populations.

All the maths you need

- Change the subject of an equation (*e.g. Hardy-Weinberg equation*)
- Substitute numerical values into algebraic equations (*e.g. Hardy-Weinberg equation*)

What have I studied before?

- Variation in a population
- Basic genetics and inheritance
- Simple Mendelian genetics
- That mutations are the source of new variations and that the processes of random assortment and crossing over during meiosis give rise to new combinations of alleles in gametes
- How random fertilisation during sexual reproduction brings about genetic variation
- The meanings of the terms genotype, phenotype, homozygote, heterozygote, dominance, recessive, codominance and multiple alleles
- The construction of genetic crosses and pedigree diagrams
- The inheritance of two non-interacting unlinked genes
- How autosomal linkage results from the presence of alleles on the same chromosome and that the results of crosses can be explained by the events of meiosis

What will I study later?

- Abiotic and biotic factors affecting the size and distribution of populations in an ecosystem
- Selection pressures in an ecosystem

What will I study in this chapter?

- How stabilising selection maintains continuity in a population and disruptive selection leads to changes or speciation
- How changes in allele frequencies may be the result of chance and not selection, including genetic drift
- That allele frequencies can be influenced by population bottlenecks and the founder effect
- How the Hardy-Weinberg equation can be used to monitor changes in the allele frequencies in a population

By the end of this section, you should be able to...

● explain how the Hardy-Weinberg equation can be used to monitor changes in the allele frequencies in a population

In biological terms, a **population** is a group of individuals of the same species occupying a particular habitat and a particular niche within that habitat. The habitat of an organism is the place where it lives and takes into account both the physical and biological elements of the surroundings. The niche of an organism is its place within the ecosystem, including its habitat and its effect on other organisms. You will learn more about these ecological concepts in **Topic 10**.

Population genetics

In population genetics we take the gene as the unit of evolution and look at how the genetic make-up of a population evolves over time. The sum total of all the alleles in a population at a given time is known as the **gene pool** and it will run to millions or even billions of genes. Fortunately, it is usually the gene pool for a particular trait that is considered. This is usually more manageable. At any point in time a population of organisms will have a particular gene pool, with different alleles occurring with varying frequencies. You studied the concept of allele frequencies in a population in **Book 1 Section 3.3.2**. Evolution can be considered as a permanent change in allele frequencies within a population. What does this mean and how is it measured?

Allele frequencies

The frequency at which a particular allele occurs in the individuals of a population is not fixed. As the environment changes, the frequency at which different alleles occur changes, due to the process of natural selection and adaptation. For example, warfarin is a chemical that prevents the blood from clotting and it has been used as rat poison since about the 1950s – the rats die of internal bleeding. When warfarin was introduced some rats already carried a mutation that, by chance, gave them resistance to the poison. The poison acted as a powerful selection pressure and resulted in a rapid increase in the frequency of the resistance allele. Soon the majority of rats were resistant to warfarin and new, more powerful poisons had to be developed. Evolution involves a change in the allele frequencies within a population.

The number of individuals carrying a certain allele in a population determines the allele frequency. It describes what proportion of individuals carry a certain allele and is usually expressed as a decimal fraction of 1. The frequency with which an allele occurs within the population has little to do with whether it codes for a dominant or recessive phenotype (see **fig B**).

fig A The presence or absence of the allele that gives rats resistance to warfarin is obvious only once they have been exposed to the poison.

fig B The frequency of a phenotype in a population is the result of the frequency of particular alleles, often independent of the dominant or recessive nature of the phenotype. In Scandinavia, blond hair and blue eyes, both of which involve a number of recessive alleles, are very common phenotypes.

Take an imaginary gene with two possible alleles X and x. If all of the individuals in a breeding population of 100 diploid organisms are heterozygous, then in theory the frequency of each allele is 100/200 or 0.5. It is very rare for this to be the case, but it gives us a way of building a model.

We have a general formula that can be used to represent the frequency with which the forms of an allele code for the dominant and recessive phenotypes occurring in the gene pool of a population. The frequency of the dominant allele is represented by the letter p and the frequency of the recessive allele is represented by the letter q. The frequency of the dominant allele plus the frequency of the recessive allele will always equal 1:

$$p + q = 1$$

This simple equation is of very limited use because it is usually almost impossible to distinguish between heterozygotes and dominant homozygotes based on their phenotype and so measuring the frequencies of heterozygotes and homozygotes in the population is not possible. However, we can readily observe the distribution of the recessive phenotypes in a population and from this we can calculate all the genotype frequencies using the Hardy-Weinberg equation. This is when the simple relationship $p + q = 1$ becomes very useful.

The Hardy-Weinberg equilibrium

The amount of change that takes place in the frequency of alleles in a population indicates whether the population is stable and unchanging, or evolving. In 1908 the British mathematician G.N. Hardy and the German physician W. Weinberg independently developed an equation that could be used to describe the mathematical relationship between the frequencies of alleles and genotypes within a stable theoretical population that is not evolving – these conditions are explained on the next page. The **Hardy-Weinberg equilibrium** theory states that in a population that is not evolving, the allele frequencies in the population will remain stable from one generation to the next in the absence of other evolutionary influences. If the population *is* evolving, the allele frequencies will change from generation to generation and so the population is not in equilibrium.

While the Hardy-Weinberg equation provides a simple model of a theoretical stable population, its main use is in calculating allele and gene frequencies in population genetics, providing a means of measuring and studying evolutionary changes.

The algebraic equation developed by Hardy and Weinberg is expressed as:

$$p^2 \quad + \quad 2pq \quad + \quad q^2 \quad = 1$$

p^2	$2pq$	q^2
frequency of homozygous dominant genotype in population	frequency of heterozygous genotype in population	frequency of homozygous recessive genotype in population

Using the Hardy-Weinberg equation

Since recessive phenotypes are easily observable, we can measure their frequency and calculate allele frequencies that we can then use in the Hardy-Weinberg equation to estimate genotype frequencies.

The frequency of homozygous recessive individuals is represented as q^2. From this, q is readily obtained by finding the square root of q^2. The result gives the frequency of the recessive allele and by substituting this figure into our initial formula of $p + q = 1$, the frequency of the dominant allele p can be found.

Let us take an example. In **Section 8.1.6** you looked at the inheritance of the albino trait. People who inherit the allele for the dominant pigmented trait may have the genotype AA or Aa. People who are albino are homozygous for the alleles determining the recessive phenotype, aa. Tests on a sample of North Americans showed that the frequency of albinos in the population was 1 in 20 000. This tells us that the frequency (q^2) of the homozygous recessive trait is 0.000 05, so we calculate the value of q:

$q^2 = 0.000\,05$

so $\quad q = \sqrt{0.000\,05} = 0.007$

we know $p + q = 1$

so $\quad p = 1 - 0.007 = 0.993$

By substituting these values into the expressions from the Hardy-Weinberg equation, the frequency of homozygous AA and heterozygous Aa genotypes can be calculated:

$p^2 + 2pq + q^2 = 1$

Frequency of homozygous AA = p^2

$p^2 = 0.993^2 = 0.986$

Frequency of heterozygous Aa = $2pq$

$2pq = 2(0.986 \times 0.007) = 0.014$

This gives us the frequencies for each of the three genotypes for albinism in the North American population. 98.6% of the population are homozygous for the the dominant phenotype, 1.4% are heterozygotes and 0.005% are homozygous for the recessive trait and are albinos. The allele frequencies must add up to 1, and the population percentages to 100%.

Conditions of the Hardy-Weinberg equilibrium

The Hardy-Weinberg equation describes the situation in a theoretical stable equilibrium, where the relative frequencies of the alleles and the genotypes stay the same over time. The implication is that in the absence of any factors that change the equilibrium, allele frequencies will remain constant within a population from generation to generation. In this theoretical population:

* there are no mutations
* there is random mating
* the population is large
* the population is isolated
* there is no selection pressure (i.e. all genotypes are equally fertile/successful).

In the real world these conditions are almost never met. Deviations from the Hardy-Weinberg equilibrium show that species are continuously changing. The factors that result in deviations from the hypothetical equilibrium state are the selection pressures that bring about a long-term change in the gene pool, changing the allele frequencies in the population and driving evolution forward. Upsetting the gene-pool equilibrium results in evolution.

Mutations

For the allele frequency to remain stable in a population, no mutations must occur. Mutations involve changes in the genetic material so the alleles are changed. As you know already, spontaneous mutations occur within a population all the time. Mutations in the somatic cells of animals will not be passed on to their offspring and may or may not affect the individual themselves. In animals, only mutations in the germ line cells – that form the eggs and sperm – will affect the alleles of the next generation. In plants, the germ line cells are not fixed in the embryo. A mutation that takes place in a single stem as it grows can therefore become part of the gametes of a flower that forms on that stem.

fig C Ruby red grapefruit arose as the result of two separate mutations in the growing cells of grapefruit trees. One produced pink-fleshed fruit. Several years later another mutation produced a redder and much sweeter flesh. In both cases the mutation produced a new dominant phenotype.

Although mutations occur continuously, they do not happen very rapidly. In a single generation each gene has between 1 in 10^4 and 1 in 10^9 chance of mutation. Recent work from the 1000 Genomes Project suggests that each of us has around 60 new mutations passed on from our parents. The vast majority of these will be recessive and will never be expressed, but occasionally mutations arise that will confer benefits to an individual and so will become entrenched within the gene pool.

Non-random mating

One of the most important requirements for a gene pool to remain in equilibrium is for random mating to occur. Random mating means that the likelihood of any two individuals in a population will mate is independent of their genetic make-up. If mating occurs randomly, the frequency of the alleles in the population will stay the same.

Non-random mating occurs when some feature of the phenotype affects the probability of two organisms mating. For example, when a male animal displays in some way to attract the female it is not random mating. The male peacock with the most impressive tail, the stag with the largest antlers and most effectively aggressive nature, the male stickleback with the brightest belly and most available nest – all of these will appear to be more attractive than average to the females of the species. As a result they will be more likely to have the opportunity to mate and pass on their genes, ensuring that their offspring in turn are likely to carry the alleles for these attractive characteristics.

fig D Male magnificent frigate birds display their red throat to attract a mate. The female selects the most impressive one, so mating is not random.

Within human populations, non-random mating is the normal situation. In every human culture, value judgements of one sort or another are used in the selection of a partner, either by the family or social groups to which the individuals belong, or by the individuals concerned. In different populations, different traits are regarded as desirable and so the gene pool shifts. For example, in the Kuna people of Panama, albinism is regarded as a very desirable trait and the incidence is approximately 1 in 200 people – the average frequency globally is 1 in 17 000. This is a remarkable shift in the allele frequency within the population and shows the effect non-random mating can have on the gene pool equilibrium (see **fig E**).

fig E Albinism is highly esteemed in the Kuna people of Panama and as a result they are reported as having the highest incidence of albinism in the world.

Populations of varying sizes

The Hardy-Weinberg equation is only valid if it is applied to a large population, with a minimum of several thousand individuals. This is because the maintenance of genetic equilibrium depends on a random assortment of the alleles. Large populations containing many individuals usually have large gene pools – the chance of losing an allele by random events is reduced in a large population. For example, if allele Z occurs in 10% of the population and that population consists of 10 individuals, then only one individual will carry the allele. If that individual is lost through predation or disease, the allele is also lost, even if it codes for an advantageous phenotype. However, in a population of 5000 individuals, 500 will carry the advantageous allele and the likelihood of all of those organisms being destroyed is remote. So there is a bigger chance of a potentially useful allele being maintained in the larger population. This is one reason why large, genetically diverse populations are needed to maintain biodiversity.

Isolation

If the Hardy-Weinberg genetic equilibrium is to be maintained, the population must exist in isolation. There should be no migration of organisms either into or out of the population. Of course this is very rarely the case in the living world. Insects carry pollen from one population of flowers to another and the wind can carry it

for miles. Male animals frequently leave their familial groups and go in search of other populations to find a mate. Many simple organisms release their gametes directly into the water to be carried great distances before fertilisation occurs. In all of these cases, migration of genetic material into or out of the population takes place. As a result of this **gene flow** occurs, tending to make the different populations more alike, but changing the allele frequencies within each individual population all the time.

Selection pressure

For Hardy-Weinberg equilibrium to apply, all alleles would have the same level of reproductive advantage or disadvantage, but we know this is not the case. Many alleles are neutral in their effect, but some alleles code for an advantageous or disadvantageous phenotype. If the environmental conditions change, alleles which have been neutral in their impact may become advantageous or disadvantageous – see the story of the oysters of Malpeque Bay, **Book 1 Section 3.2.2**.

Questions

1 In a population of plants, 95 out of 200 individuals express the homozygous recessive phenotype for hairy leaves. Calculate:
   (a) The percentage of the population that is homozygous for the recessive phenotype.
   (b) The percentage of the population that is heterozygous.
   (c) The frequency of homozygotes for the dominant phenotype in the population.

2 Grey fur in mice is recessive to brown fur. In a population of 150 mice, 126 of them have brown fur. Give the expected frequency of the alleles for homozygous recessive, homozygous dominant and heterozygotes in the population of mice. Explain your workings.

3 Explain why the Hardy-Weinberg equilibrium is rarely observed in living organisms across the generations.

Key definitions

A **population** is a breeding group of individuals of the same species occupying a particular habitat and a particular niche.

The **gene pool** is the sum total of all the genes in a population at a given time.

The **Hardy-Weinberg equilibrium** is the mathematical relationship between the frequencies of alleles and genotypes in a population. The equation used to describe this relationship can be used to work out the stable allele frequencies within a population.

Gene flow describes the migration of either whole organisms or genetic material into or out of a population and into another population, tending to make different populations more alike, but changing the allele frequencies within each individual population all the time.

By the end of this section, you should be able to...

● explain how allele frequencies are influenced by population bottlenecks and the founder effect

A large population is needed to maintain a large and diverse gene pool. Even within a large population, allele frequencies are affected by factors such as mutations and non-random mating. If the size of the population is severely reduced, so is the gene pool, and allele frequencies can change dramatically.

Population bottlenecks

The size of a population may be dramatically reduced by an environmental disaster, a new disease, hunting by humans or other very efficient predators, or habitat destruction. This is called a **population bottleneck** and it causes a severe decrease in the gene pool of the population. Many of the gene variants present in the original population are lost, so the gene pool shrinks and the allele frequency changes dramatically. In almost all cases, genetic diversity is greatly reduced.

After a catastrophic event the remaining small population is vulnerable to the complete loss of some alleles, and a single mutation or new individual can have a bigger effect than usual as a result. As the population recovers it may become so different from the original population genetically that it becomes a new species.

Cheetahs are the fastest land animals, capable of bursts of speed of up to 95 km h⁻¹. Unfortunately they have very little genetic diversity. At the end of the last Ice Age, many of the largest mammalian species, including woolly mammoths, cave bears and giant deer, became extinct and DNA evidence shows scientists that cheetahs may well have come very close to extinction at the same time. Although cheetah population numbers have recovered to some extent, their genetic diversity has not. Because they are all descended from this ancestral population bottleneck, the gene pool is very small. All cheetahs have about 99% of their alleles in common; there is little genetic diversity and the allele frequency is static. As a result fertility is low, and they are very vulnerable to any environmental change such as climate changes or a new disease.

Did you know?

Human-imposed bottlenecks

Heavy hunting of northern elephant seals meant that by the end of the nineteenth century there were only about 20 individuals left. Their population has recovered to over 30 000 but their allele frequency is still low. They have far less genetic diversity than southern elephant seals, which were much less intensively hunted and so have retained a much bigger gene pool.

fig A These northern elephant seals are all related to around 20 individuals who survived human hunters. As a result the gene pool is very small.

The founder effect

The **founder effect** is the loss of genetic variation that occurs when a small number of individuals leave the main population and set up a separate new population, producing a voluntary 'population bottleneck'. The alleles carried by the individuals who leave the main population are unlikely to include all the alleles, or at the same frequencies, as the original population. Any unusual genes in the founder members of the new population may become amplified as the population grows.

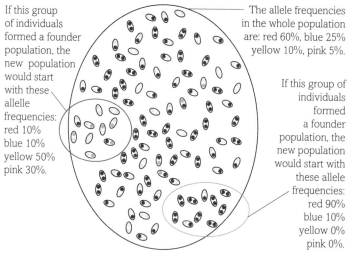

If this group of individuals formed a founder population, the new population would start with these allele frequencies: red 10% blue 10% yellow 50% pink 30%.

The allele frequencies in the whole population are: red 60%, blue 25% yellow 10%, pink 5%.

If this group of individuals formed a founder population, the new population would start with these allele frequencies: red 90% blue 10% yellow 0% pink 0%.

fig B A model of the founder effect.

The founder effect is demonstrated clearly amongst the Amish, an American religious sect that exists in three isolated communities. One of the groups has a high frequency of a very rare genetic disorder known as Ellis-van Creveld syndrome, as one of the founder members carried the gene. More cases of Ellis-van Creveld syndrome have been found in this one small population than in the whole of the rest of the world, providing a clear demonstration of the role of the founder effect in bringing about a dramatic change in allele frequencies in a population. The new population's gene pool is significantly different from that of the original population.

Did you know?

The founder effect

When Mr and Mrs Samuel King immigrated to Pennsylvania in 1744, as members of a small group of 200 people who founded an Amish community, they could have had no idea of the legacy they took with them. One of them was heterozygous for Ellis-van Creveld syndrome. This rare genetic disease results in a type of dwarfism where the limbs are shortened, there are often extra fingers and toes, and people often die young as a result of associated heart problems. The Kings produced many children, who in turn were also particularly prolific. This raised the frequency of the allele within the population way above its normal level. Purely by chance, at least one other member of the founding group also possessed this recessive gene in the heterozygous form. As a result of interbreeding in this isolated Amish community, 1 in 14 individuals in the population now carry the gene. This results in a distressingly high number of affected births. In 1964 there were 43 cases of Ellis-van Creveld syndrome in a population group of 8000. In the general population the incidence of this genetic condition can be as low as 1 in 200 000 births.

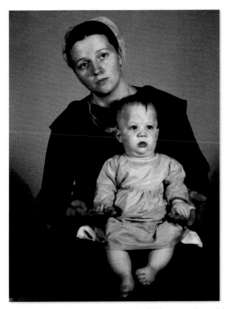

fig C This child shows the shortened limbs and extra fingers typical of Ellis-van Creveld syndrome. It is extremely rare in the population as a whole, but common in this isolated population of Amish people.

Questions

1 If 43 individuals with Ellis-van Creveld syndrome are found in a population of 8000, use the Hardy-Weinberg formula to work out the frequency of the Ellis-van Creveld allele in the population and the numbers with the homozygous and heterozygous forms of the dominant phenotype.

2 Explain the difference between a genetic bottleneck and the founder effect.

3 The lions of the Ngorongoro crater in Tanzania are isolated from other populations of lions. In 1962 the population was almost wiped out by disease and only 9 females and 1 male were left. The current population of crater lions is 75–125 animals. They have much lower genetic diversity than lions in the much bigger populations of the nearby Serengeti plains, and have problems of low fertility. What do these observations suggest about the gene pool and allele frequencies of these two populations of lions?

Key definitions

A **population bottleneck** is the effect of an event or series of events that dramatically reduces the size of a population and causes a severe decrease in the gene pool of the population, resulting in large changes in allele frequencies and a reduction in genetic diversity.

The **founder effect** is the loss of genetic variation that occurs when a small number of individuals become isolated, forming a new population with allele frequencies not representative of the original population.

By the end of this section, you should be able to...

● explain how selection pressures acting on the gene pool change allele frequencies in a population, including stabilising selection and disruptive selection

● explain how sometimes changes in allele frequencies can be the result of chance rather than selection, including genetic drift

The inheritance of individual genes may seem a long way from the complex evolutionary events that have led to the vast array of modern species. However, if we consider the balance of different variants of genes within a population, and how they change, the role of genes in evolution becomes easier to understand.

Selection for change or stability

Natural selection acts at the phenotype level of each individual organism. Some individuals carry alleles that result in phenotypes that give them some sort of advantage, so the alleles responsible for this phenotype become more common in a population.

Stabilising selection

Natural selection is often presented as the selection of new and advantageous features in a population. In fact, it often acts to conserve genotypes that are already proving to be successful. This reduces variation in a population, so that the frequency of a few alleles is very high but other alleles are greatly reduced. This **stabilising selection** has resulted in many organisms that have changed very little over time, including ferns, sharks and lungfish. Whenever the environment of an area has been stable for a long time, the effects of stabilising selection can be seen.

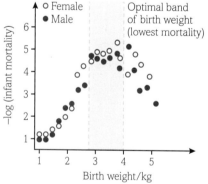

fig A Graph showing the effect of birth weight on infant mortality and the stabilising effect this has on birth weight.

For example, human babies with low birth weights, often born prematurely, face many problems including temperature regulation, breathing and infections. There are often problems delivering very large babies and the baby, the mother or both may die. Babies born within the 2.7–4.0 kg (6.0–9.0 lbs) range are most likely to survive and reproduce themselves. As a result

of stabilising selection, this is by far the most common size for newborn babies (see **fig A**).

Directional selection

Directional selection occurs whenever environmental pressure is applied to a population. For example, when populations of insects are sprayed with chemical insecticides, the chemicals may have a devastating effect initially, but directional selection ensures that within relatively few generations resistant individuals become common in the population. Directional selection results in a change in the allele frequency of a population as the majority of the population evolves an advantageous phenotype, e.g. warfarin resistance in rats (see **Section 8.2.1**).

In another example, the rabbit disease myxomatosis was introduced into Britain in 1953 and almost wiped out the rabbit population over the following 10 years. Fifty years later, rabbits are common once more. Many of them carry an allele that makes them immune to the ravages of myxomatosis as the frequency of that allele in the rabbit gene pool has increased enormously. The same process has been observed in Australia, New Zealand and all other countries where this rabbit disease has been introduced as a control measure.

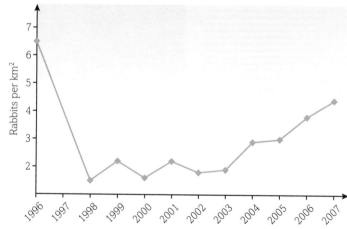

fig B Graph to show the effect of introducing myxomatosis into an area of New Zealand in 1998. The recovery of the rabbit population reflects directional selection as the frequency of the myxomatosis-resistant allele became more prevalent in the population.

Disruptive selection

Disruptive selection (or **diversifying selection**) is a variation of directional selection where the outcome is an increase in the diversity of the population rather than a trend in one particular direction (see **fig B**). It occurs when conditions are very diverse and small subpopulations evolve different phenotypes suited to their very particular surroundings. It often results in the evolution of new species. The evolution of Darwin's finches and the other unique organisms of the Galápagos Islands, and the cichlid fish of the African Great Lakes (see **Book 1 Section 3.2.4**), are good examples of disruptive selection in action.

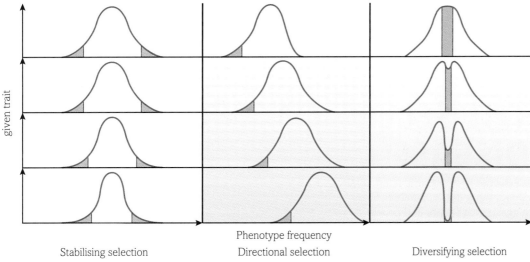

fig C The effects of stabilising, directional and diversifying selection on populations. Individuals that fall within the red region on each graph are being selected against and leave fewer offspring.

Genetic drift – change due to chance

Evolution or change in the gene pool does not always come about as a result of selection pressures. The alleles that are passed on to offspring through sexual reproduction are a random selection of the alleles available. As a result, gene frequencies will sometimes increase or decrease by chance, not because they confer any advantage or disadvantage on the offspring. These random changes in the gene pool of a population are known as **genetic drift**.

In a large population, random fluctuations in the gene pool make relatively little difference to the allele frequencies in the population. However, in small populations, genetic drift can have a major effect. If an allele only occurs a few times, and those individuals are lost, that variant is lost from the population for good. So genetic drift plays an important role when the gene pool is reduced, for example, after a population crash and genetic bottleneck, or in the founder effect.

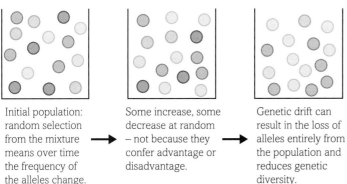

Initial population: random selection from the mixture means over time the frequency of the alleles change. → Some increase, some decrease at random – not because they confer advantage or disadvantage. → Genetic drift can result in the loss of alleles entirely from the population and reduces genetic diversity.

fig D Genetic drift – in a small population, random selection and random events can remove alleles from the population entirely.

The allele frequencies within any population are affected by both selection pressures and random factors. As a result, the gene pool changes and evolution takes place.

Questions

1 What are the main similarities and differences between the effects of stabilising and disruptive selection pressures on a population?

2 Evolution can be described as a change in the gene pool of an organism as a result of a change in selection pressures due to a change in the environment. What are the limitations of this model?

3 Why do small populations of organisms often have low genetic diversity, and what are the advantages of large genetic diversity?

Key definitions

Stabilising selection is the natural selection acting to conserve what is already present in a population, reducing variation in a population, so that the frequency of some alleles is very high but other alleles are greatly reduced.

Directional selection is the 'classic' natural selection that occurs whenever environmental pressure is applied to a population, showing a change from one phenotypic property to a new one more advantageous in the circumstances.

Disruptive selection/diversifying selection gives an increase in the diversity of the population rather than a trend in one particular direction. It is common when conditions are diverse and small subpopulations evolve different phenotypes suited to their very particular surroundings, and often results in the evolution of new species.

Genetic drift describes random changes in the gene pool of a population that occur by chance, not because they confer any advantage or disadvantage on the offspring.

THE SELECTION PRESSURE OF DISEASE

Are humans still evolving? A resource produced by Wellcome Genome Campus: Public Engagement suggests that we are and provides the evidence. Part of their material is reproduced here.

ARE HUMANS STILL EVOLVING?

For much of nature, natural selection and 'survival of the fittest' still play a dominant role; only the strongest can survive in the wild. As little as a few hundred years ago, the same was true for humans, but what about now?

In the 21st century, with the availability of better healthcare, food, heating and hygiene, the number of 'hazards' many of us experience in our lives has dramatically reduced. In scientific terms, these hazards are referred to as selection pressures…

The question is, now we have fewer selection pressures and more help in the form of medicine and science, will evolution stop altogether for humans? Has it stopped already?

Genetic studies have demonstrated that humans are still evolving. To investigate which genes are undergoing natural selection, researchers looked into the data produced by the International HapMap Project and the 1000 Genomes Project.

The strongest evolutionary pressure of all comes from infectious diseases. … People who are able to survive infections are more likely to pass on their genes to their offspring. However, genes that provide an advantage against one disease may not provide an advantage when faced with another.

The caspase-12 gene

When infectious diseases became more common in human populations … people with a genetic advantage were more likely to survive and reproduce. As a result, these genetic advantages were selected for, allowing more people to survive and fight disease. In some cases, a genetic advantage resulted from losing the full activity of a gene.

A good example of this is the caspase-12 gene. Caspase-12 works as a part of our immune system, responding specifically to bacterial infection.

In a study carried out by researchers at the Wellcome Genome Campus Institute in 2005, it was suggested that the caspase-12 gene was gradually inactivated in the human population, because the active gene can result in a poorer response to bacterial infection. People with fully functional caspase-12 were at a much higher risk of a fatal bacterial infection (sepsis) if bacteria entered the bloodstream, than people with the inactive version of the gene.

Before improved hygiene and antibiotics, survival of severe sepsis would have been a strong selective force for the inactive gene, which would have been greatly favoured. Today, people with two copies of the inactive gene are eight times more likely to escape severe sepsis if suffering with an infectious disease and three times more likely to survive.

But … if it is so good to have the inactive gene, why did our ancestors have an active form in the first place? It may be because in some areas of the world having the active gene carries an equal advantage to carrying the inactive gene… What is clear however, is that all organisms are dynamic and will continue to adapt to their unique environments to continue being successful. In short, we are still evolving.

fig A Human beings – evolution in action.

Where else will I encounter these themes?

Book 1   5.1   5.2   6.1   6.2   6.3   7.1   7.2

This is an extract from an extensive web resource produced by the Wellcome Trust Genome Campus: Public Engagement Team. The work carried out at the Wellcome Trust Genome Campus includes sequencing the genomes of human beings and many other organisms. Research is carried out in many areas, including evolution and both communicable diseases, such as sepsis, and non-communicable ones, such as cancer. This resource has been developed specifically for school students. Read through the text and answer the following questions:

1. Does this extract succeed in explaining how infectious diseases act as a selection factor in human evolution? Comment on aspects that are particularly helpful and suggest any improvements that you think would be useful.

2. You can find the entire article on human evolution at http://www.yourgenome.org/stories/are-humans-still-evolving, along with other examples of evidence of human evolution related to health and disease. Visit the website and read the whole article. Return to your answers to question 1 and reconsider them. Note down any changes in your opinions.

> Extracts, quotes and summaries are a useful way of presenting key information in a quick-to-read format. However, it is important to be aware that they may not always present the full picture.

Now let us examine the biology. You already know about the structure of DNA and protein synthesis, cell division, mutations, sexual reproduction, DNA sequencing, natural selection, bacteria as pathogens and the response of the body to infection. This will help you answer the questions below.

3. Using your knowledge of communicable diseases, suggest reasons why infections by different types of pathogens may have increased as the population increased.

4. The caspase-12 gene is involved in the immune response of the body. Research from mice models and human data suggests that caspase-12 codes for the production of enzymes that reduce the numbers of cytokines in the system. The inactive form of the gene enhances cytokine activity. The active form of the gene reduces the inflammatory response of affected individuals and the responses to endotoxins produced by bacteria. Using your knowledge of the working of the immune system, suggest an explanation for these observations.

> As you saw in **Chapter 7.3**, there are ethical considerations surrounding the use of animals for scientific research. However, it is important to also consider the many benefits that the use of animal models has in understanding diseases and developing new treatments.

Activity

Much of the work on communicable and non-communicable diseases and genomics has been done using animal models. Animal models are invaluable in research. Using this book as your starting point, plan a podcast, presentation or newspaper article explaining why animal models are so important for scientists when they try to understand the mechanisms of disease.

● Wellcome Genome Campus Public Engagement / www.yourgenome.org

Exam-style questions

1 In the Hardy-Weinberg equation, $p^2 + 2pq + q^2 =$
   A 0
   B 1
   C 0.5
   D 2   [1]
                                              [Total: 1]

2 p^2 represents the:
   A frequency of individuals in a population that have the dominant phenotype.
   B frequency of individuals in a population that have the homozygous dominant genotype.
   C frequency of individuals in a population that carry a dominant allele.
   D frequency of individuals in a population that have the homozygous recessive genotype.   [1]
                                              [Total: 1]

3 Population bottlenecks are a threat to populations because:
   A there are more individuals able to breed.
   B all allele frequencies are increased.
   C some alleles are likely to disappear from the population.
   D all allele frequencies are decreased.   [1]
                                              [Total: 1]

4 An increase in which of these would **not** cause a population bottleneck?
   A disease
   B poachers
   C predators
   D prey   [1]
                                              [Total: 1]

5 Maple Syrup Urine disease (MSUD) is caused when three amino acids (leucine, isoleucine and valine) cannot be metabolised. In a general human population 1 out of 200 individuals are born with the MSUD allele. Individuals who have this disease can have neurological damage and urine with the characteristic maple syrup smell.

   (a) Using the Hardy-Weinberg equation, where p represents 'normal' and q is the MSUD allele, calculate the value of p and q to three significant figures.   [2]

   (b) Using your values for p and q, calculate the number of individuals expected in a population of 100 000 who are:
       (i)   homozygous dominant   [2]
       (ii) heterozygous   [2]
       (iii) homozygous recessive   [2]

   (c) Some Amish populations in North America have an MSUD allele frequency (q) of 0.0754.

       Calculate the number of individuals in Amish population of 100 000 who might be expected to have MSUD.   [2]

   (d) Suggest reasons for the differences in the occurrence of MSUD in the general and Amish populations.   [5]
                                              [Total: 15]

6 The incidence of cystic fibrosis (CF) in several groups of humans is indicated in the table below.

Group of humans	Carrier frequency of cystic fibrosis allele
Africans	1 in 65
American Hispanics	1 in 46
Asians	1 in 90
Caucasian	1 in 26

   (a) Compare and contrast the incidence of the CF allele between the Caucasian and the Asian groups.   [2]

   (b) Cholera is a bacterial infection from contaminated water. It has been hypothesised that early Europeans (part of the Caucasian group) who carried the CF allele were less likely to die from cholera.

       State what effect this would have had on the allele frequency of CF in early Europeans.
       Explain your answer.   [2]

   (c) In 1959 the median age for survival of CF was six months. Infants born now in developed countries have a median life expectancy of 37 years. However, in developing countries, the median life expectancy is around 10 years.
       (i) State what is meant by the phrase 'median life expectancy'.   [1]
       (ii) Predict what will happen to the frequency of CF alleles in:
            developed countries
            developing countries.   [2]
                                              [Total: 7]

7   Sickle cell disease is a recessive genetic condition which results in the deformation of red blood cells. An amino acid substitution, from glutamic acid to valine, occurs in the β chain of haemoglobin, and at low oxygen levels the haemoglobin crystallises resulting in the red blood cell changing its shape.

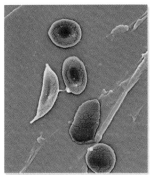

(a) Explain what caused the amino acid substitution in the β chain of haemoglobin.   [2]

(b) Sickle cell disease is a serious condition, causing anaemia, oxygen deprivation and severe pain. Untreated children die before they are five years old. The allele is carried at various frequencies within groups of humans as shown in the table.

Group of humans	Frequency of allele in population
African	1 in 3
African descent in America	1 in 13
Northern European	1 in 63

(i)   Using the Hardy-Weinberg equation, calculate the percentage of people who might be expected to have sickle cell disease in each group.   [6]

(ii) Two parents are heterozygous for the sickle cell allele. Draw a genetic diagram to show the possible genotypes and phenotypes of their children. What is the probability of these parents having a child with sickle cell anaemia?   [4]

(iii) Having at least one copy of the sickle cell allele confers some protection against a disease called malaria. Untreated malaria can be fatal, as can sickle cell disease. Using this information and your own knowledge, explain why individuals with a heterozygous genotype are more common in regions where malaria occurs.   [3]

[Total: 15]

8   The Isle of Man, circled in the map below, lies in the Irish Sea between Great Britain and Ireland. It has a famous population of Manx cats.

Manx cats have no tail. This is the result of being heterozygous for a gene, M.

Normal tails are the result of kittens being homozygous recessive.

Embryos that are homozygous dominant die in the uterus, resulting in miscarriages.

(a) Explain why two cats with normal tails are unlikely to produce a kitten with a Manx phenotype.   [1]

(b) Two tailless Manx cats are bred together and produce six live kittens. Using a genetic diagram, calculate the expected numbers of genotypes and phenotypes of the kittens.   [4]

(c) There is a much higher frequency of tailless cats on the Isle of Man than on the mainland. Suggest an explanation for this.   [2]

(d) Explain why the frequency of the Manx tail allele would be unsuitable for modelling by the Hardy-Weinberg equation.   [4]

[Total: 11]

Chemical control in mammals and plants

Introduction

Chemical control systems in animals and plants are often closely linked to environmental cues. In temperate countries, such as the UK, there is often a surge of sex hormones in animals as the days lengthen. This drives the frantic courtship, territorial singing, nest building and care of the chicks that we see in birds every spring, for example. Plants are similarly attuned to their environment. Complex chemical interactions in response to the long winter nights and short summer hours of darkness control when many plants flower. These responses are exploited commercially – huge greenhouses are filled with plants exposed to artificial days and nights, so they flower not in response to natural cues, but to the time when we are most likely to buy them!

In this chapter you will consider the importance of homeostasis in living organisms. You will look at the basis of hormonal control in mammals, including negative and positive feedback systems. You will study two main mechanisms by which hormones have their effect – through a second messenger or by a direct effect on transcription in the nucleus.

Hormones play an important role in the control of many aspects of plant biology and you will be learning about a number of these. Hormones control the growth of plants and you will be considering this at a cellular level. You will discover how two or more hormones often act antagonistically and how the balance between them determines the form of the plant. Finally, you will be looking at the role of phytochromes in plants and the control of flowering and photomorphogenesis.

All the maths you need

- Construct and interpret frequency tables and diagrams, bar charts and histograms (*e.g. interpret data on day length responses in plants*)
- Translate information between graphical, numerical and algebraic forms (*e.g. interpretation of graphs of light absorption of different phytochromes*)

What have I studied before?

- The principles of the nervous system
- Hormonal control of blood sugar levels
- The importance of hormones in reproduction
- Tropisms in plants
- The importance of mineral ions
- The cell membrane and receptors within the membrane
- Transport across membranes
- Transcription factors

What will I study later?

- The structure and function of the mammalian nervous system, which is composed of the central and peripheral nervous systems
- The structure of the spinal cord and the location and main functions of the medulla oblongata, the cerebellum, the cerebrum and the hypothalamus
- The division of the peripheral nervous system into autonomic and voluntary systems, and the further division of the autonomic nervous system into the sympathetic and parasympathetic systems that act antagonistically
- Nervous transmission and the way the resting potential is generated in an axon
- How an action potential is formed and propagated along an axon
- How the speed of transmission along myelinated nerves is greater than along non-myelinated nerves, including salutatory conduction
- The structure and function of the synapse, including neurotransmitters and the formation of excitatory and inhibitory post-synaptic potentials
- The effect of certain drugs on the nervous system by their influence on synaptic transmission
- The detection of light by mammals
- The control of the heart rate in mammals
- Osmoregulation and temperature regulation

What will I study in this chapter?

- That homeostasis is the maintenance of a state of dynamic equilibrium and the importance of negative and positive feedback control in maintaining that equilibrium
- The importance of maintaining pH, temperature and water potential in the body
- The principles of mammalian hormone production by the endocrine glands
- The mode of action of hormones on their target cells including receptors, the release of second messengers that activate enzymes and the binding of hormones to transcription factors
- Chemical control in plants brought about by plant hormones such as auxins, cytokinins and gibberellins acting in different ways
- How auxin acts on cell elongation, suppression of lateral buds and promotion of root growth
- How phytochrome controls flowering and photomorphogenesis

All living organisms need control systems. For example, the complex biochemistry of a cell needs to be controlled to make sure that the right products are available at the right time, and all living organisms need to grow, but only at the appropriate time and in the right places. Getting food and reproducing are also important biological driving forces. To control these processes successfully, a system of communication and regulation within the organism is needed.

The lifestyles of animals and plants are very different in some ways, although in others they are very similar indeed. Both animals and plants rely heavily on communication systems based on chemical signals and animals also have a rapid response system based on electrical signals. Here you will be considering the chemical coordination systems in both animals and plants, using mammals as a starting point.

Homeostasis

Mammalian cells are very sensitive to change, so whatever happens in the life of a mammal, for example during vigorous physical exercise or in extremes of external temperatures, the internal conditions of the body must be controlled within a narrow range. This is a dynamic equilibrium and involves matching the supply of oxygen and glucose to the continually changing demands of the body, while removing carbon dioxide and maintaining an even temperature and pH. The maintenance of a state of dynamic equilibrium through the responses of the body to external and internal stimuli is known as **homeostasis**. The main homeostatic mechanisms in mammals include systems that respond to changes in both external and internal conditions and control the pH, temperature and water potential of the body within narrow limits.

The pH levels of the body must be maintained so that the structures of protein molecules remain stable. This allows enzymes to function at their optimum activity and the structure of cell membranes to be maintained. You saw how the pH of the blood and body fluids is controlled by the transport of carbon dioxide away from the tissues of the body in **Book 1 Section 4.3.3**.

The core temperature of the body needs to be stable to maintain the optimum activity of the enzymes that control the rate of cellular reactions. A stable temperature also maintains the integrity of the membranes, so they can control the movement of substances into and out of the cells. In humans, this temperature is around 37 °C.

The water potential of the body fluids must remain within narrow limits to avoid osmotic effects that could damage or destroy the cells.

Homeostasis involves a high level of coordination and control. The nervous and chemical control systems interact to maintain a dynamic equilibrium in the body. Changes in the body are detected by **sensors**, also called **receptors**. They send messages to **effectors** that either work to reverse the change or to increase it in a number of different feedback systems. Effectors are usually muscles or glands.

Feedback systems

The communication in a feedback system may be by hormones (chemical messengers) or by nerve impulses (electrical messages). There is often a small overshoot or undershoot as a feedback system corrects, and that needs to be corrected as well, so most body system levels oscillate slightly around the ideal level in a dynamic equilibrium. In some cases there are separate mechanisms for controlling changes in different directions, for example in the control of blood sugar levels, and this gives a particularly sensitive response and a greater degree of control.

Negative feedback systems

Most of the feedback systems in mammals are **negative feedback systems**. They provide a way of maintaining a

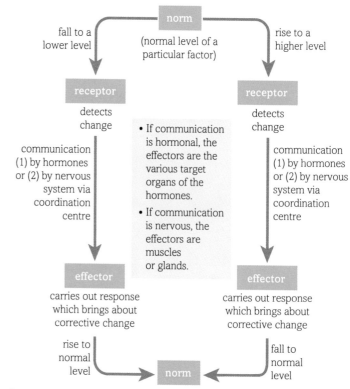

fig A This is a generalised diagram of the type of negative feedback system that plays a vital part in homeostasis.

condition, such as the concentration of a substance, within a narrow range. A change in conditions is registered by receptors and as a result effectors are stimulated to restore the equilibrium. If the concentration goes up, the effectors bring it down again, and vice versa. Examples of negative feedback loops are seen in the production of many hormones and in temperature regulation in mammals (see **Section 9.3.4**).

Positive feedback systems

In a **positive feedback system**, effectors work to increase the effect that has triggered the response. These are less common in biological systems but an example is the contractions of the uterus during labour. The pressure of the baby's head on the cervix causes the release of chemicals that increase the contraction of the uterus, so the head is then pushed down even harder.

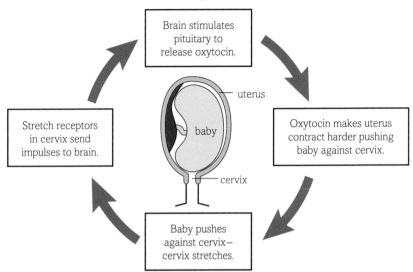

fig B The dilation of the cervix and delivery of a baby – one of a small number of positive feedback systems seen in mammals.

Questions

1 What is homeostasis and why is it so important in living organisms?

2 In the past, homeostasis was described as the maintenance of a steady internal state. Why do you think this definition has changed?

3 Suggest why there are many negative feedback systems in mammals and relatively few positive feedback systems.

Key definitions

Homeostasis is the maintenance of a state of dynamic equilibrium in the body, despite changes in the external or internal conditions.

Sensors/receptors are specialised cells that are sensitive to specific changes in the environment.

Effectors are systems (usually muscles or glands) that either work to reverse, increase or decrease changes in a biological system systems.

Negative feedback systems provide a way of maintaining a condition, such as the concentration of a substance, within a narrow range. A change in conditions is registered by receptors and as a result effectors are stimulated to restore the equilibrium.

Positive feedback systems are where effectors work to increase an effect that has triggered a response.

By the end of this section, you should be able to...

● explain the production of mammalian hormones by endocrine glands, and their mode of action involving receptors on target cells

Some of the body functions of both animals and plants require long-term stimulation of tissues, for example, for growth and sexual development. Sometimes it is necessary to send messages that have an effect on many different areas of the body simultaneously. Plants rarely need rapid responses, but they need coordination and control of their cells just as much as animals do (see **Section 9.1.4**).

For all of these situations, chemical messages are the answer. They can have an effect over a long period of time. They can reach the entire body as they are carried to their target organs in the transport system of an animal or plant. In plants, chemical control is the main system of coordination. In animals, it complements and interacts with the nervous system (see **Chapter 9.2**).

Chemical control in mammals

One of the main forms of chemical control in animals is brought about through **hormones**. These are organic chemicals produced in **endocrine glands** and released into the blood. They are carried through the transport system to parts of the body where they bring about changes, which may be widespread or highly targeted. Hormones are usually either proteins or peptides (e.g. insulin, antidiuretic hormone) or steroids (e.g. the sex hormones oestrogen and testosterone).

The changes your body underwent during puberty, the sensations you experience before an interview or an exam, and the way the amount of urine produced by your kidneys varies with the amount you drink and the weather conditions, are all changes brought about by the action of hormones, and there are many more.

Once a hormone enters the bloodstream, it is carried around in the blood until it reaches the target organ or organs. The cells of the target organs have specific receptor molecules on the surface of their membranes that bind to the hormone molecules. This brings about a change in the membrane and elicits a response.

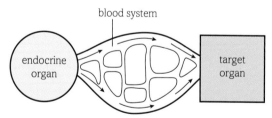

fig A The pathway followed by a hormone from the endocrine gland, where it is produced, to the cells of the target organ.

Where are the endocrine glands?

Endocrine glands are found around the body, often in association with other organ systems. Several of the glands have more than one function. For example, the ovaries produce ova as well as hormones, and the pancreas is both an **exocrine gland** producing digestive enzymes and an endocrine gland producing the hormone insulin. The glands all have a rich blood supply, with plenty of capillaries within the glandular tissue itself. The sites of the main endocrine glands in humans and the hormones they secrete are shown in **fig B**. Most mammals have very similar endocrine glands.

The exocrine glands that produce the secretions of the gut release their juices along small tubes or ducts. The endocrine glands, which produce hormones, do not have ducts – they release the hormones directly into the bloodstream.

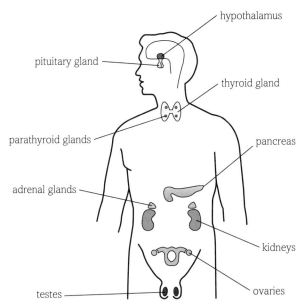

fig B The endocrine organs of the body. Although not large, these organs have a profound effect on the processes of life through their endocrine secretions.

Questions

1 What is a hormone?

2 (a) Give two features of a typical endocrine gland that are directly related to its function.
   (b) How do endocrine glands differ from exocrine glands?

Key definitions

Hormones are organic chemicals produced in endocrine glands and released into the blood and carried through the transport system to parts of the body where they bring about changes, which may be widespread or very targeted. They are usually either proteins, parts of proteins such as polypeptides, or steroids.

Endocrine glands produce hormones. They do not have ducts. They release the hormones directly into the bloodstream.

Exocrine glands produce chemicals (e.g. enzymes) and release them along small tubes or ducts.

By the end of this section, you should be able to...

● describe the two main modes of hormone action

Hormone release systems

The endocrine system interacts very closely with the nervous system (see **Chapter 9.2**). Some hormones are released as a result of direct stimulation of the endocrine gland by nerves. For example, the adrenal medulla of the adrenal glands releases adrenaline when it is stimulated by the sympathetic nervous system. The control of hormone release by the nervous system is relatively simple. If the gland is stimulated, hormone is released. If it is not stimulated, no hormone is released. The level of stimulation determines the level of response.

However, many hormones are released from endocrine glands in response to another hormone or chemical in the blood. For example, the pituitary gland in the brain secretes several hormones that directly stimulate other endocrine glands. Also, changes in chemicals in the blood, such as glucose and salt, can stimulate the release of hormones, which in turn act to regulate the levels of those chemicals.

When hormones are released in response to a chemical stimulus such as another hormone or glucose, secretion is controlled by a negative feedback loop (see **fig A**). The presence of the appropriate chemical in the blood stimulates the release of the hormone. As a result of the rise in the hormone levels, the amount of stimulating chemical in the blood drops. Therefore, the endocrine gland receives less stimulation and so the hormone levels drop. This gives a very sensitive level of control that can be adjusted constantly to the needs of the body.

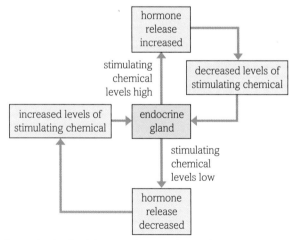

fig A A negative feedback loop in a hormonal control system.

The pituitary gland – controlling hormone release

The **pituitary gland** in the brain produces and releases secretions that affect most of the other endocrine glands in the body. It has an anterior lobe and a posterior lobe. Control of the pituitary itself falls largely to the **hypothalamus**.

The hypothalamus is the small area of brain directly above the pituitary gland. Although it is a relatively small area, it carries out a variety of functions. One of these is to monitor the blood levels of a number of metabolites and hormones. In response to the levels of these chemicals, the hypothalamus controls the activity of the pituitary gland.

How is this control brought about? A look at **fig B** shows the anatomical relationship between the hypothalamus and the pituitary. As an embryo forms, the posterior pituitary lobe develops as an outgrowth of the hypothalamus itself, whilst the anterior lobe grows out from the roof of the mouth. The two parts then fuse and the connection with the roof of the mouth is lost. The two different origins of the parts of the gland are reflected in their rather different functions.

The hypothalamus contains **neurosecretory cells**. These are nerve cells that produce secretions from the ends of the axons. One group of these cells (neurosecretory cells 1) produce substances that stimulate or inhibit the release of hormones from the anterior pituitary. They are known as releasing factors or release-inhibiting factors, depending on what they do. The other group of cells (neurosecretory cells 2) produce secretions that are stored in the posterior pituitary and released later as hormones.

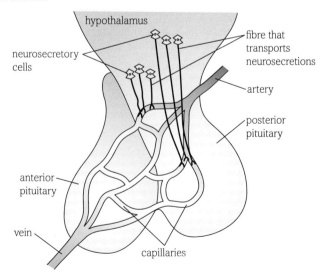

fig B The close structural relationship between the hypothalamus and the pituitary gland is reflected in their function, with the hypothalamus producing neurosecretions that control both lobes of the pituitary gland.

The pituitary gland, under the control of the hypothalamus, produces six hormones from the anterior lobe and two from the posterior lobe. These range in function from controlling the secretions of the thyroid gland to the control of growth, and from sexual development to the control of urine volume. The hormones produced and the role they play are described in **fig C**.

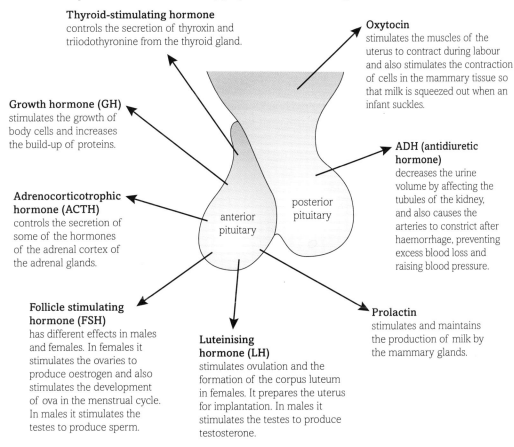

Thyroid-stimulating hormone controls the secretion of thyroxin and triiodothyronine from the thyroid gland.

Oxytocin stimulates the muscles of the uterus to contract during labour and also stimulates the contraction of cells in the mammary tissue so that milk is squeezed out when an infant suckles.

Growth hormone (GH) stimulates the growth of body cells and increases the build-up of proteins.

ADH (antidiuretic hormone) decreases the urine volume by affecting the tubules of the kidney, and also causes the arteries to constrict after haemorrhage, preventing excess blood loss and raising blood pressure.

Adrenocorticotrophic hormone (ACTH) controls the secretion of some of the hormones of the adrenal cortex of the adrenal glands.

posterior pituitary

anterior pituitary

Follicle stimulating hormone (FSH) has different effects in males and females. In females it stimulates the ovaries to produce oestrogen and also stimulates the development of ova in the menstrual cycle. In males it stimulates the testes to produce sperm.

Luteinising hormone (LH) stimulates ovulation and the formation of the corpus luteum in females. It prepares the uterus for implantation. In males it stimulates the testes to produce testosterone.

Prolactin stimulates and maintains the production of milk by the mammary glands.

fig C The hormones of the pituitary gland, particularly those from the anterior lobe, largely have their effect by stimulating other endocrine organs elsewhere in the body.

How do hormones have their effects?

Some hormones act by binding to specific receptor sites on the membrane of their target cells. At this point the hormone needs to affect the target cell in some way to bring about the desired change in activity. The two main modes of action by which hormones have their effects are:

Release of a second messenger

Some hormones are not lipid soluble and cannot cross the cell membrane. Examples include adrenaline, glucagon and follicle stimulating hormone. The hormone molecule binds to a receptor in the cell membrane. This triggers a series of membrane-bound reactions that result in the formation of a second chemical messenger inside the cell. This second messenger then activates a number of different enzymes within the cell, altering the metabolism. The most common second messenger is a substance called **cyclic AMP (cAMP)**, which is formed from ATP (see **fig D**). Cyclic AMP triggers a number of responses in the cell, including increased cellular respiration, increased contraction of muscle cells, relaxation of smooth muscle in blood vessels and so on. This is how adrenaline, the hormone associated with the 'fight or flight' response, has its effects.

adrenaline

adrenaline (first messenger) binds to receptor

cell surface membrane

adrenaline receptor in cell membrane

activates membrane-bound proteins

membrane-bound proteins

activates enzyme

adenylate cyclase

ATP

cAMP

second messenger

triggers responses in cell e.g. glycogen breakdown, increased respiration etc.

fig D Hormones such as adrenaline have their effect on cell metabolism via a second messenger system.

The hormone enters the cell

A lipid soluble hormone passes through the membrane and acts as the internal messenger itself. Inside the cell the hormone binds to a receptor and the hormone–receptor complex passes through the pores of the nuclear membrane into the nucleus. The hormone bound to the receptor acts as a transcription factor (see **Section 7.2.1**), regulating gene expression and switching sections of the DNA on or off. This is the mode of action of the lipid-soluble steroid hormones such as oestrogen and testosterone (see **fig E**).

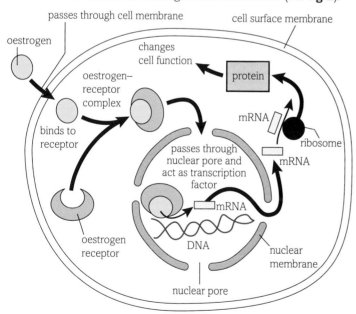

passes through cell membrane

cell surface membrane

oestrogen

changes cell function

oestrogen–receptor complex

protein

mRNA

binds to receptor

ribosome

passes through nuclear pore and act as transcription factor

mRNA

mRNA

oestrogen receptor

DNA

nuclear membrane

nuclear pore

fig E Steroid hormones, such as oestrogen, act as transcription factors to have their effect.

Questions

1 Compare the control of hormone release by nervous stimulation and by chemical stimulation.

2 Protein and peptide hormones are not lipid soluble, so they cannot enter a cell directly. Explain how hormones such as adrenaline still manage to have a wide effect on the biochemistry of the cell.

3 Steroid hormones act as transcription factors in the nucleus of a cell. Discuss this statement and point out any inaccuracies.

Key definitions

The **pituitary gland** is a small gland in the brain that has an anterior lobe and a posterior lobe and produces and releases secretions that affect the activity of most of the other endocrine glands in the body.

The **hypothalamus** is a small area of brain directly above the pituitary gland that controls the activities of the pituitary gland and coordinates the autonomic (unconscious) nervous system.

Neurosecretory cells are nerve cells that produce secretions from the ends of their axons. These secretions either stimulate or inhibit the release of hormones from the anterior pituitary, or are stored in the posterior pituitary and then later released as hormones.

Cyclic AMP (cAMP) is a compound formed from ATP that is produced when protein hormones such as adrenaline bind to membrane receptors and acts as a second messenger in cells.

By the end of this section, you should be able to...

● explain how chemical control in plants is brought about by plant hormones (plant growth substances), such as auxins, cytokinins and gibberellins

● describe some of the effects of auxin including cell elongation, suppression of lateral buds (apical dominance) and promotion of root growth

● explain how plant hormones often interact with each other, as shown by the antagonistic actions of cytokinin and auxin on apical dominance

As you have seen, specific chemicals released by cells in response to a stimulus can act as messages. Plants rely on chemical messages for communication between different parts, allowing them to respond to factors such as light and gravity. The chemicals move from cell to cell, and also through the plant transport system. They act as plant hormones.

Which stimuli affect plants?

Although most plant movements are very slow, plants respond to a variety of stimuli, most of which are external environmental cues that have a direct impact on the well-being of the plant.

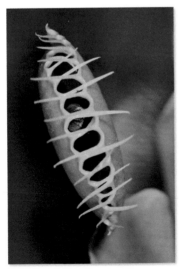

fig A Chemical control systems allow plants to respond to light and gravity. Some plants also respond to touch. The slow responses of the bean plant and the fast responses of a Venus fly trap are both examples of sensitivity to touch.

Plants are sensitive to light, and not simply the presence or absence of it. They respond to the direction from which light comes, the intensity of that light and the length of daily exposure to it. Light affects how much they grow, the direction of growth and when they reproduce. Plants are also sensitive to gravity, water and temperature, and in some cases to touch and chemicals. Different parts of the same plant may react differently to the same stimulus. For example, shoots grow towards light but roots grow away from it.

As well as responding to external stimuli, plants also respond to internal chemical signals. Most of the responses of plants are concerned either directly or indirectly with maximising the opportunities for photosynthesis and reproduction.

Plant responses

Chemical control in plants is brought about by a number of different chemicals produced in response to specific stimuli. Many of these chemicals act as plant growth regulators, but they do more than simply control the growth of plants.

At one time these chemicals were called plant growth substances, but plant scientists now recognise them as plant hormones, analogous to animal hormones. They are produced in one area of the plant, are transported around the body of the plant and have their effect on cells elsewhere. Some of their effects involve growth, but many do not. Animals respond to nervous and chemical messages in a variety of ways that include the release of further chemicals, the contraction of muscle cells, and growth. Plants also respond to chemical messages in a number of different ways. In some cases, growth is stimulated, while in others it is inhibited. For example, sometimes one side of a plant grows more than the other in response to a particular stimulus, resulting in the bending of the shoots or roots.

These directional growth responses to specific environmental cues are known as **tropisms**. Other plant hormones affect the differentiation and development of plant cells and processes, such as fruit ripening and leaf fall.

How plants grow

Growth is a permanent increase in the size of an organism or of some part of it. It is brought about by cell division and the assimilation of new material within the cells that result from the division, followed by cell expansion (see **fig B**). The main areas of cell division in plants are known as the meristems and they occur just behind the tip of a root or shoot. The regions of cell division and cell elongation are particularly sensitive to plant growth substances. These chemical messages act in a number of ways – for example, they can make it easier for the cellulose walls to be stretched, in turn making it easier for the cells to expand and grow.

meristematic cell

cytoplasm

cell wall

division

nucleus

assimilation

vacuoles develop

elongation

cellulose wall stretches

large central vacuole

fig B Plants respond to a variety of stimuli by differential growth. The growth-regulating chemicals generally seem to affect cell elongation, making it easier for the cellulose cell wall to be stretched, although some hormones affect the cell division stage, increasing the number of divisions that occur.

Auxins

Auxins, for example indoleacetic acid (IAA), are powerful growth stimulants that are effective in very low concentrations. Auxins are produced in young shoots and always move down the plant from the shoots to the roots. This movement involves some active transport and calcium ions. Auxins are involved in apical dominance, where they suppress the growth of lateral shoots so that one main stem grows fastest. In low concentrations, they promote root growth. The more auxin that is transported down the stem, the more root growth occurs. If the tips of the stems are removed, removing the source of auxins, the stimulation of root growth is removed and root growth slows and stops. Auxins are also involved in the tropic responses of plant shoots to unilateral light. The response of a plant to auxins often depends on both the concentration of the hormone and the region of the plant.

Auxins in the garden

Gardeners and horticulturists make use of auxins in a number of ways.

Taking cuttings: applying rooting powder containing different auxins to the cut end of stems when a gardener takes cuttings encourages the development of new adventitious roots and so helps the cutting to form a new independent plant.

Weedkillers: synthetic auxins mimic the effects of natural plant hormones, but cannot be synthesised and regulated by the plants. They are used as weedkillers. They are taken up much more effectively by broadleaved dicot plants than by narrow-leaved monocots, meaning that they are selectively taken up by the broadleaved weeds in lawns and cereal crops. The hormone analogues are also taken down into the roots, so they affect the whole plant. They interfere with the growth of the plant, so that different parts grow at different rates and the internal metabolism is so distorted that the plant dies. These hormone weedkillers are selective to dicot plants, and are also relatively harmless to pets and garden birds.

How do auxins work in a plant?

Auxins seem to affect the ability of the plant cell walls to stretch. For example, IAA is made in the tip of the shoot and diffuses back towards the zone of elongation. The molecules of IAA bind to specific receptor sites on the cell surface membranes, activating the active pumping of hydrogen ions into the cell wall spaces. This changes the hydrogen ion concentration, providing the optimum pH of around 5 for enzymes that break bonds between adjacent cellulose microfibrils (see **Book 1 Section 1.2.2**). This allows the microfibrils to slide past each other very easily, keeping the walls very plastic and flexible. The cells absorb water by osmosis and, as a result of turgor pressure, the very flexible cell walls stretch allowing the cells to elongate and expand. Eventually, as the cells mature, the IAA is destroyed by enzymes, the pH of the cell walls rises, the enzyme is inhibited and bonds form between the cellulose microfibrils. As a result, the cell wall becomes more rigid and the cell can no longer expand (see **fig C**).

This basic model of the way plants grow was based on shoots kept entirely in the dark or in full illumination. However, in real life plants are usually in a situation where the light on one side is stronger than the other. Research shows that the side of a shoot exposed to light contains less auxin than the side that is not illuminated. Light seems to cause the auxin to move laterally across the shoot, producing a greater concentration on the unilluminated side. This movement means the shoot tip acts as a photoreceptor. More of the hormone diffuses down to the region of cell elongation on the dark side. This stimulates cell elongation and therefore growth on the dark side, resulting in the shoot bending towards the light. Once the shoot is growing directly towards the light, the unilateral stimulus is removed. The asymmetric transport of auxin ends and the shoot grows straight towards the light. The original theory was that light destroyed the auxin. However, more recent evidence suggests that the levels of auxin in shoots are much the same, regardless of whether they have been kept in the dark or under unilateral illumination. Although scientific work on tropisms began over a century ago, with Charles Darwin and his son Francis, there is still a great deal to learn.

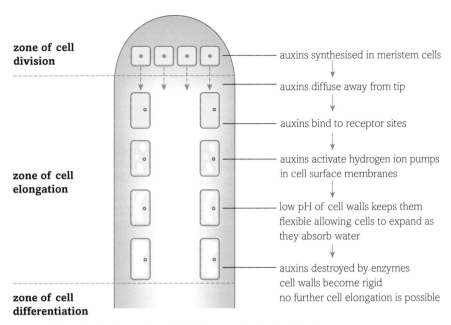

fig C A summary of the role of auxins, such as IAA, in the growth of a plant shoot.

More plant hormones

Auxins do not act alone to control and coordinate plant responses. There are a number of other classes of plant hormones that have different effects on plant cells. They work individually and also interact to give finely-tuned responses to an ever-changing environment:

- **Gibberellins**: these compounds act in several ways, including as growth regulators. They affect the internodes of stems, stimulating elongation of the growing cells. They also promote the growth of fruit. They are involved in breaking dormancy in seeds and in germination, as they stimulate the formation of enzymes in seeds. For example, they stimulate the production of amylase, which breaks down starch stores in cereal plants. This makes glucose available for respiration in the embryo plant as it develops as the seed germinates. Gibberellins also stimulate bolting, a period of sudden rapid growth and flowering, in biennial plants.

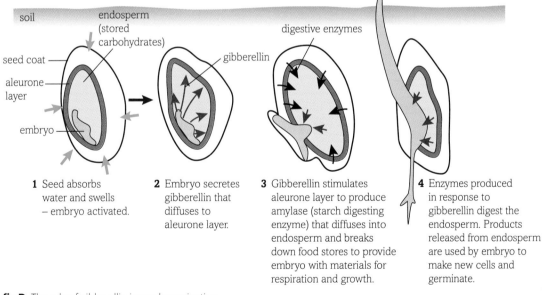

1 Seed absorbs water and swells – embryo activated.

2 Embryo secretes gibberellin that diffuses to aleurone layer.

3 Gibberellin stimulates aleurone layer to produce amylase (starch digesting enzyme) that diffuses into endosperm and breaks down food stores to provide embryo with materials for respiration and growth.

4 Enzymes produced in response to gibberellin digest the endosperm. Products released from endosperm are used by embryo to make new cells and germinate.

fig D The role of gibberellin in seed germination.

- **Cytokinins**: these are growth regulators that promote cell division in the apical meristems and the cambium through interactions with auxins. They promote lateral bud development, which can overcome apical dominance if the leading shoot is removed or damaged. Cytokinins work synergistically with ethene in the **abscission** of leaves, flowers and fruits.

Learning tip

As part of your study of this topic, you will carry out **Core Practical 14: Investigate the effect of gibberellin on the production of amylase in germinating cereals using a starch agar assay**. Make sure you have a good understanding of this practical as your understanding of the experimental method may be assessed in your examination.

Did you know?

Ethene is the only gaseous plant hormone. It promotes fruit ripening, and the fall of ripe fruits and leaves from plants. It is also involved in the development of aerenchyma in water plants. If you put a ripe banana in a bag of green bananas, the unripe bananas will ripen fast in response to the ethene produced by the ripe fruit.

Synergy and antagonism

Most plant hormones do not work in isolation, but in an interaction with other substances. This adaptation means that very fine control over the responses of the plant can be achieved. The growth regulators interact in one of two ways. If they work together, complementing each other and giving a greater response than would otherwise be the case, the interaction is known as synergy. Auxins and gibberellins work synergistically in the growth of stems. If the substances have opposite effects, for example one promoting growth and one inhibiting it, the balance between them will determine the response of the plant. This is known as antagonism. Auxins and cytokinins work antagonistically on the maintenance of apical dominance.

Apical dominance

The balance between auxins and cytokinins is very important in the overall development of the plant. Their most important interaction is in the maintenance of apical dominance. In apical dominance, one lead shoot in a young plant grows bigger and faster than all the others and the growth of all the other lateral buds is inhibited, as a result of the high auxin levels produced by the first shoot to emerge. The auxin acts antagonistically with cytokinin, which stimulates the development of the lateral buds.

In the natural growth of the plant, as the first shoot grows away, the inhibition of the auxin is reduced and cytokinin becomes dominant, so the lateral buds lower down the plant begin to grow. If the apical bud is removed, the auxin inhibition on the lateral buds is removed and the cytokinin can have its full effect and many lateral buds grow rapidly. If auxin is artificially added to the cut apical stem, the antagonistic effect returns and lateral shoot growth slows again.

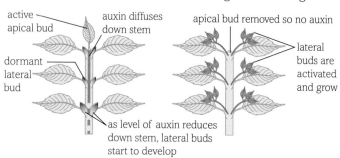

fig E Apical dominance – the result of interactions between auxins and cytokinins in plant growth.

Questions

1 List as many of the different environmental stimuli that elicit a response in plants as you can. For each, try to explain why it is important for the plant to respond to that stimulus.

2 Growth in animals usually stops at a certain stage. The meristems of plants, where growth occurs, remain active throughout the life of the plant. Explain why this difference is important in the way the organisms respond to stimuli.

3 Make a table to show five different types of plant hormones. Summarise the way they work and indicate synergistic or antagonistic interactions.

Key definitions

Tropisms are plant growth responses to environmental cues.

Auxins are plant hormones that act as powerful growth stimulants (e.g. indoleacetic acid) and are involved in apical dominance, stem and root growth, and tropic responses to unilateral light.

Gibberellins are plant hormones that act as growth regulators, particularly in the internodes of stems by stimulating elongation of the growing cells; they also promote the growth of fruit and are involved in breaking dormancy in seeds and in germination.

Cytokinins are plant hormones that promote cell division in the apical meristems and the cambium through interactions with auxins. They promote lateral bud development, which can overcome apical dominance, and work synergistically with ethene in the abscission of leaves, flowers and fruits.

Abscission is the shedding of leaves, flower parts or fruits from a plant after the formation of an abscission zone across the stem attaching the organ to the plant.

Phytochrome and flowering

By the end of this section, you should be able to...

● explain how phytochrome controls flowering

Light is vital for plants. Without light, the metabolism of a plant is severely disrupted and no chlorophyll is formed. Without chlorophyll, no photosynthesis takes place and with prolonged light deprivation, the plant dies. Day length (or night length) is the environmental cue that determines changes such as bud development, flowering, fruit ripening and leaf fall. Plants have evolved very sensitive mechanisms for detecting light, and the process by which plant development is controlled by the levels and type of light is known as **photomorphogenesis**.

Sensory systems in plants

For some time it has been known that the seeds of many plants will germinate only if they are exposed, even if very briefly, to light. Researchers in the US Department of Agriculture showed that **red light** (wavelength 620–700 nm) is the most effective at stimulating germination in lettuce seeds, while **far red light** (wavelength 700–800 nm) actually inhibits germination.

If the seeds are exposed to a flash of red light they will germinate. If they are exposed to a flash of red light followed by a flash of far red light, they will not germinate. With a series of flashes of light, it is the colour of the final flash that determines whether or not the seeds will germinate (see **fig A**). Scientists hypothesised the existence of a plant pigment that reacts with different types of light, and in turn affects the responses of the plant, acting as part of the system that controls photomorphogenesis. In 1960, this pigment was isolated from plants and called **phytochrome**.

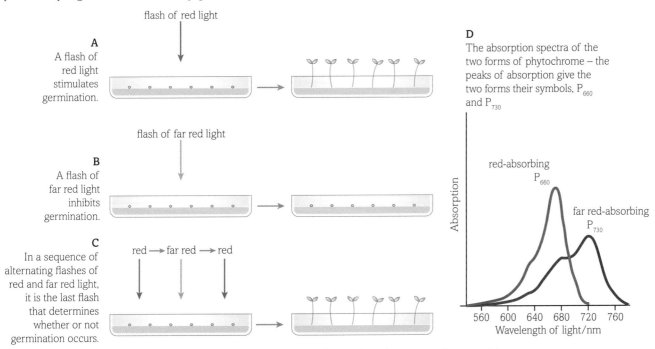

A A flash of red light stimulates germination.

flash of red light

B A flash of far red light inhibits germination.

flash of far red light

C In a sequence of alternating flashes of red and far red light, it is the last flash that determines whether or not germination occurs.

red → far red → red

D The absorption spectra of the two forms of phytochrome – the peaks of absorption give the two forms their symbols, P_{660} and P_{730}

red-absorbing P_{660}

far red-absorbing P_{730}

Absorption

560   600   640   680   720   760
Wavelength of light/nm

fig A The idea of a phytochrome light receptor system in plants moved from an hypothesis to a well-accepted theory, once the predicted light-sensitive chemical was extracted and analysed. Graph courtesy of Science and Plants for Schools.

Phytochromes

Plants make a number of different phytochromes, but they all respond to light in the same way. A phytochrome is a blue-green pigment that exists in two interconvertible forms: P_r (or P_{660}) absorbs red light; P_{fr} (or P_{730}) absorbs far red light (see **fig B**). When one form of the pigment absorbs light, it is converted reversibly into the other form. As a seedling germinates, it makes P_r. As soon as it breaks through the surface of the soil and is exposed to red light, some of the new pigment is converted into P_{fr} and from then on the two interconvertible forms exist in the plant.

The length of time it takes for one form of the pigment to be converted into the other depends on the light intensity. In low light intensity it takes minutes, but in high light intensity it takes seconds. In the dark, P_{fr} is converted to P_r very slowly but no P_r is converted back. P_r is the more stable form of the pigment, but it is the P_{fr} that is biologically active.

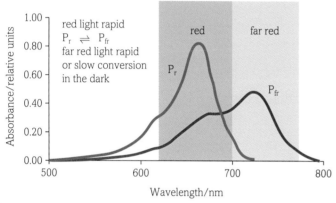

fig B The photoconversion of phytochrome, which plays a vital role in the coordination of plant growth and development. Courtesy of Science and Plants for Schools.

The balance between the two forms of phytochrome is affected by varying periods of light and dark, and that in turn affects the plant metabolism, including flowering patterns. As normal sunlight contains more red light than far red light, the usual situation in a plant during daylight hours is for most of the phytochrome to be in the far red form, P_{fr}. If the night period is long enough, it is all converted back into the red form, P_r. This can be used to explain the control of the germination of the lettuce seedlings seen in **fig A**, where a flash of red light produces biologically active P_{fr}, which triggers germination. A flash of far red light converts P_{fr} back to the inactive P_r before it has an effect.

Phytochromes enable plants to respond to environmental cues such as change in day length. In some cases phytochromes have a stimulating effect on growth in plants, in others an inhibitory effect. Exactly how phytochromes influence the responses of the plant is still not fully understood, but as you will see in **Section 9.1.6**, the evidence increasingly suggests that the presence of phytochromes stimulates the production of other growth regulators and plant hormones, bringing about the response to light.

Developing ideas about photoperiodism

In temperate regions such as the UK, the period of daylight can vary from about 9 to 15 hours throughout the year. The lengths of the days and nights give important environmental cues to living organisms, directing their growth, development and behaviour. The amount of time that an organism is exposed to light during a 24-hour period is known as the photoperiod. In plants, one of the most clearly affected activities is flowering, and scientists have developed models of how plants sense and respond to day length cues.

Scientists found that day length appeared to be the environmental cue affecting flowering in many plants. Plants flowering when days are short and nights are long became known as **short-day plants** (SDPs), and include, for example, strawberries, chrysanthemums and the tobacco plant. Plants flowering in relatively long days and short nights are known as **long-day plants** (LDPs), and include, for example, snapdragons, cabbages and henbane. It can be very difficult to decide whether a plant is a short- or long-day plant as the two groups merge. Some plants, such as cucumbers, tomatoes and pea plants, are unaffected by the length of the day and are known as **day-neutral plants** (DNPs). These are usually plants that have evolved in tropical regions where the day length is the same all year round. As a result, they are adapted to use different cues, such as the amount of available water, as the triggers for flowering.

Different flowering patterns allow plants to take advantage of different circumstances. In temperate regions, short-day plants tend to flower in spring and autumn, when the light-shading canopy of leaves either has not developed or has fallen off. They also flourish nearer the equator, where the days are never longer than about 12 hours.

Long-day plants flower in the summer in temperate regions, and are found further from the equator where in some seasons there are very long days.

Scientists subsequently discovered that the length of the period of *darkness* is actually the environmental cue affecting flowering, not day length! It was demonstrated that if a short-day plant such as a cocklebur has the long night (period of darkness) interrupted by flashes of light, they do not flower.

How is the signal received?

All the research on photoperiodism points to the involvement of the phytochromes in the sensitivity of the flowering pattern of plants to the photoperiod. The changes in flowering patterns that can be brought about by disturbing the dark periods can also be affected by red or far red light alone. Red light inhibits the flowering of short-day plants, but if the red light is followed by far red light, the inhibition is lifted. It is the balance of the P_r and P_{fr} that is key.

The current hypothesis is that, in short-day plants, the biologically active molecule P_{fr} inhibits flowering, and a lack of P_{fr} allows flowering to occur. During long periods of darkness, the levels of P_{fr} fall, as it is almost all converted to P_r. This allows flowering to take place.

In long-day plants, the situation is reversed, and it appears that high levels of P_{fr} stimulate flowering. The nights are short so relatively little P_{fr} is converted back to P_r. As a result, relatively high P_{fr} levels are maintained all the time, stimulating flowering.

Experiment A

Cocklebur, a short-day plant, will not flower if kept under long days and short nights.

If even one leaf is masked for part of the day – thus shifting that leaf to short days and long nights – the plant will flower; note the burrs.

Experiment B

1 Graft five cocklebur plants together and keep under long days and short nights, with most leaves removed.

2 Induce a leaf by long nights/short days.

3 If a leaf on a plant at one end of the chain is subjected to long nights, all of the plants will flower.

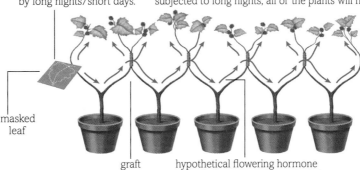

burrs (fruit)

masked leaf

masked leaf

graft

hypothetical flowering hormone spreads through all the grafted plants

fig C Evidence like this is building a clear picture of how the photoperiodic response in the leaves is transferred to the flowering regions of the plant.

Day-neutral plants evolved in tropical conditions where the levels of P_r and P_{fr} are similar all year round, so even in temperate regions they do not respond to changes in day length by flowering. They have different flowering triggers.

Scientists know that phytochromes are only part of the story and control of flowering is very complex, but this provides a useful working model.

The detection of the photoperiod seems to take place in the leaves of the plant. The presence of a plant hormone known as **florigen** was first hypothesised by the Russian plant physiologist Mikhail Chailakhyan in the 1930s. It was thought that florigen was made in response to the changing levels of phytochromes and carried in the plant transport system to the flower buds. The evidence included the following findings:

*   If the whole plant is kept in the dark, apart from one leaf which is exposed to the appropriate periods of light and dark, flowering occurs as normal. A plant kept in total darkness does not flower (see **fig C**, experiment A).

*   Using the same experimental set-up, if the photoperiodically exposed leaf is removed immediately after the stimulus, the plant does not flower. If the leaf is left in place for a few hours, it does flower.

*   If two or more plants are grafted together and only one exposed to appropriate light patterns, all the plants will flower (see **fig C**, experiment B).

*   In some species, if a light-stimulated leaf from one plant is grafted onto another plant, the new plant will flower.

For years no one could isolate the theoretical hormone and the florigen theory fell into disrepute. However, recently scientists have shown that when a leaf is exposed to a given amount of light and dark, a particular form of mRNA is produced in the leaf, linked with a gene associated with flowering (the FT gene or Flowering Locus T). It is known as FTmRNA. There were concerns that a large molecule like FTmRNA could not be florigen, as it would not be able to leave the cell. Now scientists have shown that FTmRNA can move from cell to cell to the transport tissues through the plasmodesmata. They have also shown that FTmRNA travels from the leaves, where it is formed, to the apex of the shoot, where other genes associated with flowering are activated. So at the moment FTmRNA looks as if it may well be florigen!

Questions

1   (a) What is phytochrome?
    (b) Summarise the role of phytochrome in the response of plants to differences in day length.

2   How does the evidence presented support the idea of the plant hormone florigen? Take each bullet point in turn and explain its relevance.

3   Produce a flowchart to show how phytochrome and florigen interact to bring about flowering in a plant.

Key definitions

Photomorphogenesis is the process by which the form and development of a plant is controlled by the levels of and type of light.

Red light has a wavelength of 580–660′ nm, which is detected by plants using phytochromes.

Far red light has a wavelength of 700–730 nm, which is detected by plants using phytochromes.

Phytochrome is a plant pigment that reacts with different types of light, and in turn affects the responses of the plant.

Short-day plants (SDPs) are plants flowering when days are short and nights are long.

Long-day plants (LDPs) are plants flowering when days are long and nights are short.

Day-neutral plants (DNPs) are plants where flowering is not affected by the length of time they are exposed to light or dark.

Florigen is a hypothetical plant hormone which is involved in the photoperiodic response. It may be FTmRNA.

Phytochrome and photomorphogenesis

By the end of this section, you should be able to ...

● explain how phytochrome controls photomorphogenesis

The whole shape and form of a plant is dependent on its need for light. What part do phytochromes play in this process?

Photomorphogenesis

Any plants that are grown in the dark, or become heavily shaded by other plants, become **etiolated** – they grow rapidly, using up food reserves in an attempt to reach the light. As a result, the plants end up tall and thin, with fragile stems, long internodes and small, pale, yellowish leaves as no chlorophyll is formed. Etiolation seems to be a survival mechanism – all of the resources of the plant go into growing up towards the light needed for photosynthesis. Once the plant reaches the light, growth slows and the leaves turn green as chlorophyll forms.

This is similar to the changes that take place as a seed germinates and grows. Almost all seeds germinate under the ground, so the early stages of growth take place in the dark and are etiolated. The changes that take place when a plant becomes etiolated, and the reverse of etiolation when germinating seedlings break through the soil, appear to be controlled by phytochrome.

Germination and photomorphogenesis

Phytochrome is synthesised as P_r. When a seedling emerges from a seed underground it only contains P_r as it has not been exposed to light. The early seedling emerging from the seed has a cotyledon (monocots) or a hooked apical shoot (dicots) and shows typical characteristics of etiolation. These include:

* rapid stem lengthening but very little thickening – the seedling grows as tall as possible as fast as possible to reach the light

* relatively little root growth – just enough to act as an anchor and obtain water

* no leaf growth – the leaves remain furled so no energy is wasted producing leaf tissue that is useless underground

* no chlorophyll – the seedling is white or pale yellow, so no energy is wasted producing chlorophyll that is useless in the dark.

Once the tip of the new shoot breaks through the soil surface into the light, a series of changes take place:

* the elongation of the stem slows down

* the stem straightens

* the cotyledons and/or first leaves open

* chlorophyll forms and the seedling begins to photosynthesise.

The changes of photomorphogenesis are controlled by phytochrome interconversion. In the seed there is plenty of P_r but no P_{fr}. Without P_{fr}, the internodes grow but the leaves do not and no chlorophyll forms. Once the plant is exposed to light, P_r is rapidly converted to P_{fr} and levels of P_{fr} build up quickly. P_{fr} inhibits the lengthening of the internodes (stem between the leaf nodes), so internode growth slows. It stimulates leaf development and the production of chlorophyll. The seedling turns green and begins to photosynthesise. These changes begin even before the seedling breaks through the surface of the soil, as a little light penetrates through the surface of the soil and begins the transformation. As a result, the chloroplasts are maturing and the seedling is often green and ready to photosynthesise as it breaks through the soil.

fig A The difference between the normal seedlings on the left, which have emerged into the light, and the etiolated seedlings on the right, grown in the dark, shows the importance of phytochromes on the photomorphogenesis of these young plants.

Phytochrome, photomorphogenesis and tropisms

Until recently, the changes in plants controlled by phytochrome and the responses of plants to light controlled by auxins and other hormones were thought to be quite separate processes. However, research is increasingly showing a link between phytochrome and both phototropisms and geotropisms. For example, it appears that phototropisms cannot take place in very young shoots until phytochrome has been activated, and that geotropisms are similarly dependent on phytochrome actions.

Phytochrome as a transcription factor

Scientists are only just beginning to understand how phytochrome can change so many things, from triggering flowering to the production of chlorophyll and the growth of the stem. Evidence is beginning to build up suggesting that P_{fr} acts as a transcription factor, involved in switching genes on and off in the nuclei of plant cells.

Researchers have produced recombinant DNA (see **Section 7.3.1**) linking the genes for the production of phytochrome to a gene for the production of **green fluorescent protein (GFP)**, originally from jellyfish. By inserting these hybrid genes into plant cells, the scientists produced plants with fluorescent phytochrome.

If seedlings were kept in the dark, the fluorescence linked to the inactive P_r was spread evenly through the cytoplasm. When the seedlings were exposed to red light and therefore converting labelled P_r to labelled P_{fr}, scientists observed that the fluorescence moved into the nucleus. What is more, in the nucleus the fluorescent P_{fr} did not form a single blob – it appeared as specks linked to the chromosomes.

The current model is as follows:

- When P_r is converted into P_{fr} in the presence of light, it moves into the nucleus through the pores in the nuclear membrane.
- In the nucleus, it binds to a nuclear protein known as the phytochrome-interacting factor 3 (PIF3).
- PIF3 is a known transcription factor.
- PIF3 only binds to P_{fr}. It does not bind to P_r.
- PIF3 only activates gene transcription and the formation of mRNA when it is bound to P_{fr}.

The hypothesis is that by binding to PIF3, P_{fr} activates different genes and so controls different aspects of growth and development in plants. There is still a lot of work to be done to understand how phytochrome has its effect, but our models for the control and coordination are becoming more sophisticated and more integrated all the time.

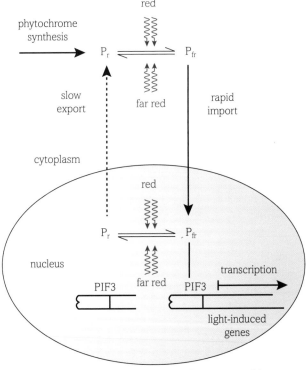

fig B Model of how P_{fr} might act as a transcription factor in plant cells. Courtesy of Science and Plants for Schools.

Questions

1 Produce a diagram to illustrate our current model of how the interconversion of phytochome affects the photomorphogenesis of a germinating seedling.

2 What are the effects of red light (P_{fr} build-up) and far red light (P_r build-up) on different parts of a plant? You may find it useful to produce a table summary.

Key definitions

Etiolated describes the form of plants grown in the dark, with long internodes, thin stems, small or unformed leaves and white or pale yellow in colour.

Green fluorescent protein (GFP) is the product of a gene often used as a marker in the production of recombinant DNA.

1 What name is given to hormone-secreting glands in mammals?
  A endocrine glands
  B exocrine glands
  C hypocrine glands
  D endothalamus glands [1]
  [Total: 1]

2 How are hormones transported to their target cells in mammals?
  A through plasmodesmata
  B in the red blood cells
  C in the plasma
  D through neurones [1]
  [Total: 1]

3 How do plant hormones travel through to their targets?
  A in the phloem
  B in the xylem
  C in the sclerenchyma
  D through plasmodesmata [1]
  [Total: 1]

4 The statements below refer to some effects of two groups of plant growth substances, auxins and gibberellins.

  Copy the table below. If the statement is correct place a tick (✔) in the appropriate box and if the statement is incorrect place a cross (✗) in the appropriate box. [5]

Effect	Auxins	Gibberellins
Promote cell elongation		
Promote root formation in cuttings and calluses		
Promote fruit growth		
Inhibit lateral bud development		
Promote the breaking of dormancy in seed		

  [Total: 5]

5 Peptide hormones, such as insulin and adrenaline, affect the activity of their target cells by binding to receptors in the cell surface membrane. This sets off cascade effects in the cells, resulting in the activation of enzymes.

  Steroid hormones, however, diffuse through the cell surface membrane. The diagram below shows the mechanism of action of a steroid hormone.

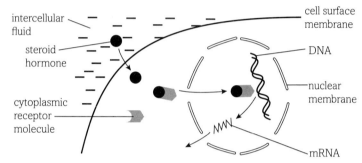

  (a) Name **one** steroid hormone that could act in this way. [1]

  (b) Explain why steroid hormones can diffuse through the cell surface membrane while insulin and adrenaline do not. [2]

  (c) Using the information in the diagram, and your own knowledge, explain how the presence of only a very small amount of steroid hormone in the intercellular fluid can result in a large change in the activity of the cell. [3]

  (d) Adrenaline has an effect on its target cells within seconds of its release from the adrenal glands, and its effects last only for a short time. Steroid hormones take much longer to act on their target cells, and their effects last longer. Explain why this is so. [2]

  [Total: 8]

6 Plants can detect and respond to environmental cues.

Cocklebur is a plant that flowers after it has been exposed to a sufficiently long period of darkness.

The minimum length of time in darkness needed to stimulate flowering is called the critical period.

An investigation was carried out into the effect of light and dark periods on cocklebur flowering.

Four plants, **A**, **B**, **C** and **D**, were exposed to light and dark periods of different length.

The presence or absence of flowers was recorded after several weeks.

The diagram below shows the pattern of light and dark periods for these plants and the effect on flowering.

Plant	Time/hours						Flowers present
	0	4	8	12	16	20	
A							Yes
B							No
C							No
D							No

Key

☐ Light

▨ Dark

(a) (i) Using the information in the diagram, give the critical period for flowering of cocklebur plants. [1]

(ii) Using the information in the diagram and your own knowledge of photoreceptors, explain why plant **B** has not flowered. [2]

(b) In a further investigation, plants **E** and **F** were exposed to six hours of darkness each day.

Part of a leaf on plant **F** was covered so that the leaf experienced eight hours of darkness each day.

The diagram below summarises the results of this investigation.

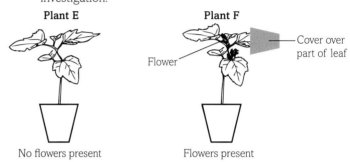

Plant E   Plant F

Flower   Cover over part of leaf

No flowers present   Flowers present

Explain the purpose of plant **E** in this investigation. [2]

(c) Using your own knowledge of photoreceptors, explain the results of these investigations. What do they suggest about the control of flowering in cocklebur plants? [4]

(d) Discuss benefits to plants of being able to respond to changes in day length. [3]

[Total: 12]

TOPIC 9
Control systems

9.2 The mammalian nervous system

Introduction

If someone loses a limb as a result of cancer, an accident, on active service, or is born without a limb, they may be provided with a prosthetic replacement. Until very recently, prosthetic limbs could look like the real thing, but their wearer had no control over them. This is all beginning to change. New prosthetic limbs are being developed that can be controlled by electrical signals in the nerves or muscles remaining in the stump. It may even be possible to control a prosthetic limb by thought. Already work is progressing using computer technology with amputees learning to control a limb on a computer avatar, before progressing to using a similar system integral to their prosthetic limb. Scientists are now moving towards linking sensors in the limbs to the remaining sensory nerves. Eventually, people may be able to both control and sense their prosthesis almost like their own limb.

In this chapter you will learn about the mammalian nervous system. You will consider the structure of individual neurones and consider how this structure is related to their functions. You will study the movements of ions into and out of axons that result in the passage of an action potential. You will learn about synapses, the junctions between neurones that have a complex structure that is tightly and clearly related to their functions. You will consider how scientists have found out how this amazing system works and look at the effect of various drugs on transmission across the synapse.

The sensory organs are vital in the functioning of the nervous system and you will consider the way the eye works in some detail. Finally, you will move on to the bigger picture – the structure and function of the central and peripheral nervous systems, including the basic anatomy of the spinal cord and the brain and how this is related to their functions. You will consider the importance within the body of the voluntary and autonomic nervous systems and the way they interact to bring about nervous coordination.

All the maths you need

- Construct and interpret frequency tables and diagrams, bar charts and histograms (*e.g. interpreting bar charts on efficacy of lidocaine*)
- Translate information between graphical, numerical and algebraic forms (*e.g. interpretation of action potentials*)

What will I study later?

- How osmoregulation is brought about in the mammalian body
- The gross and microscopic structure of the kidney
- How urea is produced in the liver and removed from the blood in the kidney by ultrafiltration
- That solutes are selectively reabsorbed in the proximal tubule of the kidney and how the loop of Henle acts as a countercurrent multiplier to increase the absorption of water
- How the pituitary gland and osmoreceptors in the hypothalamus, combined with the action of antidiuretic hormone, bring about negative feedback control of mammalian plasma concentration
- How the kidneys of some desert animals are adapted for life in a very dry environment
- That endotherms maintain their body temperature through metabolic processes, but ectotherms rely on the external environment
- How endotherms use both behavioural and physiological mechanisms to maintain their body temperature

What have I studied before?

- The principles of the nervous system
- The importance of mineral ions
- The cell membrane and receptors within the membrane
- Transport across membranes
- The production of ATP in cellular respiration

What will I study in this chapter?

- The structure and function of the mammalian nervous system, which is composed of the central and peripheral nervous systems
- The structure of the spinal cord and brain, including the location and main functions of the medulla oblongata, the cerebellum, the cerebrum and the hypothalamus
- The division of the peripheral nervous system into autonomic and voluntary systems, and the further division of the autonomic nervous system into the sympathetic and parasympathetic systems that act antagonistically
- Nervous transmission and the way the resting potential is generated in an axon
- How an action potential is formed and propagated along an axon
- How the speed of transmission along myelinated nerves is greater than along non-myelinated nerves, including saltatory conduction
- The structure and function of the synapse, including neurotransmitters and the formation of excitatory and inhibitory post-synaptic potentials
- The effect of certain drugs on the nervous system by their influence on synaptic transmission
- The detection of light by mammals, including the structure and function of the retina
- How vision in different light intensities results from the distribution of the rods and cones
- The control of the heart rate in mammals by the autonomic nervous system, including the role of baroreceptors and chemoreceptors, the cardiac centre in the medulla oblongata, and the sympathetic and parasympathetic nervous system

In **Chapter 9.1** you looked at the need for living organisms to have a system of control, and at the way coordination is achieved in plants using chemical communication systems. Many animals, including humans, need a more rapid and specifically-targeted system of communication, and this is where a nervous system comes into its own.

The basic structure of the nervous system

A nervous system is made up of interconnected **neurones** (nerve cells) specialised for the rapid transmission of impulses throughout the organism. Neurones carry impulses from special **receptor cells**, giving information about both the internal and the external environment. They also carry impulses to specialised **effector cells**, which are muscles or glands, which then bring about the appropriate response.

The simplest nervous systems are made up of receptor cells, neurones and the nerve endings associated with the effectors (see **fig A**). However, many nervous systems are much more complex. As well as single receptor cells, groups of receptors have evolved to work together in **sense organs** such as the eye and the ear, and simple nerve nets have been replaced by complex nerve pathways.

Neurones that only carry information from the internal or external environment into the central processing areas of the nervous system are known as sensory neurones. As animals increase in size and complexity, they develop more specialised concentrations of nerve cells, which form a **central nervous system (CNS)**. This is an area where incoming information from sensory neurones is processed, and from where impulses are sent out through motor neurones, which carry impulses to the effector organs. In vertebrates, the central nervous system consists of the brain and spinal cord.

Neurones are individual cells and each one has a long nerve fibre that carries the nerve impulse. Nerves are bundles of nerve fibres called **axons** or **dendrons** (see **fig A**). Some nerves carry only motor fibres and are known as motor nerves, some carry only sensory fibres and are known as sensory nerves, while others carry a mixture of motor and sensory fibres.

The parts of the nervous system that are not within the central nervous system are known as the **peripheral nervous system**. You will be learning more about this in **Section 9.2.7**.

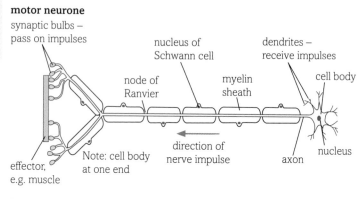

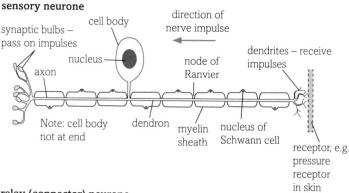

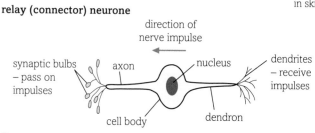

fig A All neurones have the same basic structure of a cell body, dendrites and axons or dendrons, but the detailed arrangements vary depending on their function.

The structure and function of neurones

Neurones are the basic unit of a nervous system and millions of neurones work together as an integrated whole in mammals such as ourselves. Neurones are cells specialised in the transmission of electrical signals called **nerve impulses**. They have a cell body that contains the cell nucleus, mitochondria, other organelles, and the rough endoplasmic reticulum (ER) and ribosomes, which are needed for the synthesis of the neurotransmitter molecules. The cell body has slender finger-like processes called **dendrites** that connect to neighbouring nerve cells. The most distinctive feature of all nerve cells is the nerve fibre, which is extremely long and thin and carries the nerve impulse (see **fig B**). Fibres that carry impulses away from the nerve cell body (in motor neurones) are called axons. Fibres that transmit impulses towards the cell body (in sensory neurones) are known as dendrons. Short relay or connector neurones are found in the CNS and they connect motor and sensory neurones. They are also known as bipolar neurones, because two fibres leave the same cell body.

Defining a nerve impulse

The current model of a nerve impulse is of a minute electrical event due to charge differences between the outside and inside of the neurone membrane. It is based on ion movements through specialised protein pores and by an active pumping mechanism.

Myelinated nerve fibres

Most vertebrate neurones are associated with another very specialised type of cell, the **Schwann cell** (see **fig B**). The Schwann cell membrane wraps itself repeatedly around the nerve fibre, forming a fatty layer known as the **myelin sheath**. There are gaps between the Schwann cells, known as the **nodes of Ranvier**. The myelin sheath is important for two reasons – it protects the nerves from damage and speeds up the transmission of the nerve impulse (see **Section 9.2.3**).

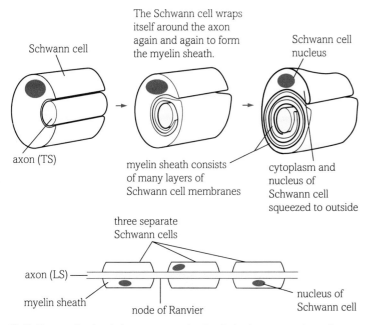

fig B The myelin sheath forms a protective, insulating layer around vertebrate nerve fibres. TS: transverse section; LS: longitudinal section.

Speedy nerve impulses

The role of the nerve cells is to quickly carry electrical impulses from one area of the body to another. The speed at which the impulses can be carried depends largely on two things. The first is the diameter of the nerve fibre – generally the thicker the fibre, the more rapidly impulses travel along it. The second is the presence or absence of a myelin sheath – myelinated nerve fibres can carry impulses much faster than unmyelinated ones.

Invertebrates do not have myelin sheaths on their nerve fibres, and many of their axons and dendrons are less than 0.1 mm in diameter. As a result, many invertebrate nerve impulses travel quite slowly at around 0.5 m s^{-1}. But there are times when even a relatively slow-moving invertebrate needs to react quickly to avoid danger. Many invertebrate groups have evolved a number of giant axons, which are nerve fibres with diameters of around 1 mm. These allow impulses to travel at approximately 100 m s^{-1}, fast enough for most escape strategies to have a chance of success. Much of the research into how axons work has been done on these giant axons, because they are relatively easy to work with and there are fewer ethical issues when working with invertebrates.

Vertebrates have both unmyelinated and myelinated nerves. The voluntary motor neurones that transmit impulses to voluntary muscles, for example to control movement, are all myelinated. However, the autonomic neurones that control involuntary muscles, like those of the digestive system, have some unmyelinated fibres. The effect of the myelin sheath is to speed up the transmission of a nerve impulse without the need for giant axons which take up quite a lot of space. A more versatile network of relatively small nerve fibres can carry impulses extremely rapidly, at speeds of up to 120 m s^{-1}.

Investigating nerve impulses

To look at the events of a nerve impulse it is easiest to consider a 'typical' nerve fibre – ignoring for the moment size, myelination or type. Since nerve impulses are electrical events, albeit very small ones, one of the most effective ways of investigating them has been to record and measure the tiny electrical changes. Early work was done using a pair of recording electrodes placed on a nerve, which was then given a controlled stimulus. The impulses that resulted were recorded by the electrodes and displayed on a screen.

External electrodes record the responses of entire nerves, made up of large numbers of different nerve fibres. The fibres are of varying diameter and sensitivity, and so the results of the recordings can be difficult to interpret correctly and the technique has been of limited value.

To make real progress, recordings had to be made from within individual nerve fibres. Much of the work on nerve fibres has been done using axons, because they are relatively easy to use. Sensory nerve fibres often run from a sense organ in the head directly to the brain or from individual sensory receptors in the skin to the spinal cord, making them difficult to access. Motor axons, on the other hand, run directly to muscles, often in large motor nerves. This makes them relatively easy to get at, and the effect of stimulating them can be seen immediately with the twitch of a muscle. Much of your work on nerve impulses refers to axons – but the same events are seen in all nerve fibres.

Most axons are extremely small – around 20 μm in diameter – so making a recording from inside is not an easy procedure. Over 70 years ago Alan Hodgkin and Andrew Huxley began work on the giant unmyelinated axons of the squid, which are around 0.5 mm in diameter. These axons allow for very rapid nerve transmission to particular muscles, in situations when the squid needs to move in a hurry.

Hodgkin and Huxley used very fine glass microelectrodes inserted into the giant axon. Another electrode recorded the electrical potential from the outside. This allowed the changes that occur during the passage of an individual nerve impulse to be recorded accurately for the first time (see **fig C**). This technique has been refined so that now internal electrodes can be used with almost any nerve fibre.

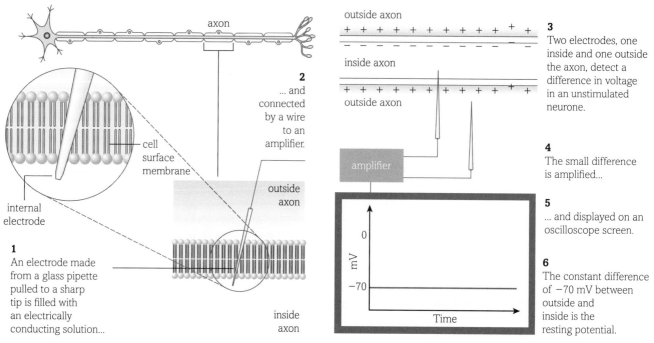

fig C This apparatus, with an internal and external electrode, has been used to discover many of the secrets of how neurones work. Here you can see the resting potential of a neurone being measured – it is the potential difference across the membrane in millivolts.

Questions

1. Outline the differences between a nerve and a nerve fibre.

2. Summarise how the structure of a motor neurone is related to its function.

3. The speed of transmission of a nerve impulse is in part related to the diameter of the nerve fibre. Explain why vertebrate nerve fibres always have relatively small diameters, yet the speed of the transmission of the nerve impulses is much faster than that in invertebrates.

4. Squid giant axons are widely used in neurophysiology. Explain why.

Key definitions

Neurones are cells specialised for the rapid transmission of impulses throughout an organism.

Receptor cells are specialised neurones that respond to changes in the environment.

Effector cells are specialised cells that bring about a response when stimulated by a neurone.

Sense organs are groups of receptors working together to detect changes in the environment.

The **central nervous system (CNS)** is a specialised concentration of nerve cells where incoming information is processed and from where impulses are sent out through motor neurones, which carry impulses to the effector organs.

An **axon** is the long nerve fibre of a motor neurone, which carries the nerve impulse.

A **dendron** is the long nerve fibre of a sensory neurone, which carries the nerve impulse.

The **peripheral nervous system** includes the parts of the nervous system that spread through the body and are not involved in the central nervous system.

Nerve impulses are the electrical signals transmitted through the neurones of the nervous system.

Dendrites are the slender, finger-like processes that extend from the cell body of a neurone and connect with neighbouring neurones.

A **Schwann cell** is a specialised type of cell associated with myelinated neurones. It forms the myelin sheath.

The **myelin sheath** is a fatty insulating layer around some neurones produced by the Schwann cell.

The **nodes of Ranvier** are gaps between the Schwann cells that enable saltatory conduction.

By the end of this section, you should be able to...

- explain how the properties of the axon membrane and the transport of Na⁺ ions and K⁺ ions result in the resting potential

- explain how an action potential is formed and how it is propagated along an axon

The nervous system is based on the passage of nerve impulses, which are minute electrochemical events that depend on the nature of the axon membrane and the maintenance of sodium ion (Na^+) and potassium ion (K^+) gradients across that membrane.

Nerve impulses

The concentration of sodium ions, potassium ions and other charged particles outside the axon is different from that inside the axon and this is the basis of the nerve impulse.

The membrane of an axon, like any other cell surface membrane, is partially permeable. It is the difference in permeability of this membrane to positively charged sodium and potassium ions that gives it special conducting properties. The axon membrane is relatively impermeable to sodium ions, but quite freely permeable to potassium ions.

The resting potential

An axon is said to be 'at rest' when it is not conducting a nerve impulse. The extracellular concentration of ions is greater than the concentration in the axon's cytoplasm. This gradient is created by a very active sodium/potassium ion pump, often referred to as the sodium pump or sodium/potassium pump. This pump has an enzyme called Na^+/K^+ ATPase that uses ATP to move sodium ions out of the axon and potassium ions into the axon (see **Book 1 Section 4.1.4** on active transport). It pumps sodium ions out of the axon, lowering the concentration of sodium ions inside the axon and, due to the relative impermeability of the membrane to sodium ions, they cannot diffuse back in again. At the same time, potassium ions are actively moved into the axon by the pump, but then passively diffuse out again along the concentration gradient through open potassium ion channels. As a result, the inside of the cell is left slightly negative in charge relative to the outside – the membrane is **polarised**.

When you look at the events in neurones, the figures are always relative, comparing the inside of the nerve fibre to the outside solution. There is a potential difference across the membrane of around −70 mV, which is known as the **resting potential** (see **fig A**).

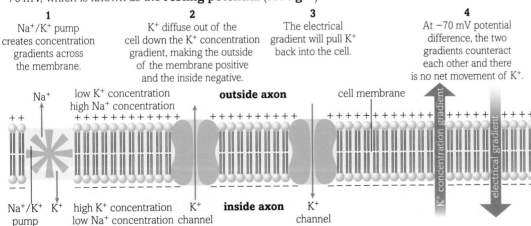

1
Na⁺/K⁺ pump creates concentration gradients across the membrane.

2
K⁺ diffuse out of the cell down the K⁺ concentration gradient, making the outside of the membrane positive and the inside negative.

3
The electrical gradient will pull K⁺ back into the cell.

4
At −70 mV potential difference, the two gradients counteract each other and there is no net movement of K⁺.

Na⁺
low K⁺ concentration
high Na⁺ concentration
outside axon
cell membrane

++

+ + + + + + + + + +   + + + + + + + + + +   + + + + + + + + + + + + + + + +   + + + +

Na⁺/K⁺   K⁺   high K⁺ concentration   K⁺   **inside axon**   K⁺
pump   low Na⁺ concentration   channel   channel

K⁺ concentration gradient
electrical gradient

fig A The resting potential of the axon is maintained by the sodium pump, the relative permeability of the membrane and the movement of potassium ions along concentration and electrochemical gradients.

The action potential

When an impulse travels along an axon, the key event is a change in the permeability of the cell membrane to sodium ions. This change occurs in response to a stimulus – for example light, sound, touch, taste or smell in a sensory neurone, or the arrival of a **neurotransmitter** in a motor neurone. In the experimental situation the stimulus is usually a minute and precisely controlled electrical impulse.

When a neurone is stimulated, the axon membrane shows a sudden and dramatic increase in its permeability to sodium ions. Specific sodium ion channels or **sodium gates** open up, allowing sodium ions to diffuse rapidly down their concentration and electrochemical gradients. As a result the potential difference across the membrane is briefly reversed, with the cell becoming positive on the inside with respect to the outside. This **depolarisation** lasts about 1 millisecond. The potential difference across the membrane at this point is about +40 mV. This is known as the **action potential**. Remember that these events happen in any nerve fibre, not just axons.

At the end of this brief depolarisation, the sodium ion channels close again and the excess sodium ions are rapidly pumped out by the active sodium pump. Also, the permeability of the membrane to potassium ions is temporarily increased as voltage-dependent potassium ion channels open as a result of the repolarisation. As a result, potassium ions (potassium channels) diffuse out of the axon down their concentration gradient and an electrochemical gradient, attracted by the negative charge on the outside of the membrane. The inside of the axon becomes negative relative to the outside once again. It takes a few milliseconds before the resting potential is restored and the axon is ready to carry another impulse (see **fig B**).

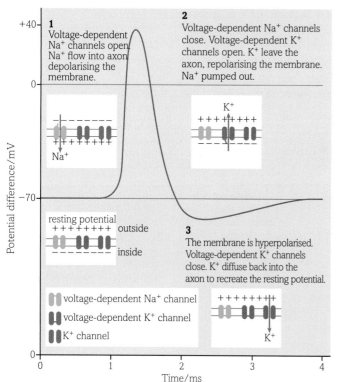

fig B The ionic changes during excitation of an axon result in an action potential.

The events of the action potential can be recorded clearly using the internal/external electrode combination you have already seen. The oscilloscope trace is often referred to as the 'spike', because of its shape. It clearly shows the change in the potential difference across the membrane with the inrush of sodium ions followed by a return to the resting potential as the permeability of the membrane changes again. The **threshold** for any nerve fibre is the point when sufficient sodium ion channels open for the rush of sodium ions into the axon to be greater than the outflow of potassium ions. Once the threshold has been reached, the action potential occurs. The size of this action potential is always the same – it is an all-or-nothing response.

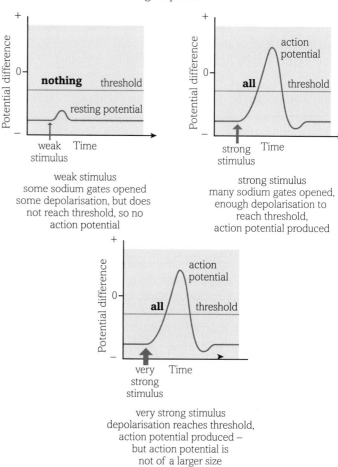

weak stimulus
some sodium gates opened
some depolarisation, but does not reach threshold, so no action potential

strong stimulus
many sodium gates opened, enough depolarisation to reach threshold, action potential produced

very strong stimulus
depolarisation reaches threshold, action potential produced – but action potential is not of a larger size

fig C There are two possibilities when a neurone is stimulated: if the stimulus is *too small*, the action potential does not happen; if the stimulus is *large* enough, the action potential will happen. If the stimulus increases, the action potential stays the same.

The recovery time of an axon (known as the **refractory period**) is the time it takes for an area of the axon membrane to recover after an action potential, that is the time it takes for ionic movements to repolarise the membrane and restore the resting potential. This depends both on the sodium/potassium pump and on the membrane permeability to potassium ions. For the first millisecond or so after the action potential, it is impossible to restimulate the fibre – the sodium ion channels are completely blocked and the resting potential has not been restored. This is known as the **absolute refractory period**. After this there is a period of several milliseconds when the axon may be restimulated, but it will only respond to a much stronger stimulus than before – the threshold has effectively been raised. This is known as the **relative refractory period**. During this time the voltage-dependent potassium ion channels are still open. It is not until they are closed that the normal resting potential can be fully restored.

The refractory period is important in the functioning of the nervous system as a whole. It limits the rate at which impulses may flow along a fibre to 500–1000 each second. It also ensures that impulses flow in only one direction along nerves. Until the resting potential is restored, the part of the nerve fibre that the impulse has just left cannot conduct another impulse. This means the impulse can only continue travelling in the same direction.

Did you know?

Building up the evidence

Some of the most convincing evidence for this model of the nerve impulse comes from work done using poisons. A metabolic poison such as dinitrophenol (DNP) prevents the production of ATP. It also prevents the axon from functioning properly.

When an axon is treated with a metabolic poison, the sodium pump stops working as ATP is used up and the resting potential is lost at the same rate as the decrease in concentration of ATP. This implies that the ATP is being used to power the sodium pump – when it runs out, the pump no longer works.

If the poison is washed away, the metabolism returns to normal and ATP production begins again. The resting potential is restored, meaning that the sodium pump has started up again with the return of ATP (see **fig D**). If a poisoned axon is supplied with ATP by experimenters, the resting potential will be at least partly restored. This again confirms our model, implying that the poison is acting by depriving the sodium pump of ATP rather than by interfering with the membrane structure and its permeability. If the latter were the case, even supplying ATP directly to the pump would have no effect, because ions would move freely across the membrane and a potential difference could not be maintained.

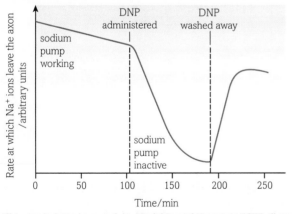

fig D This graph, based on work by Hodgkin and Keynes in 1955, illustrates clearly the effect of dinitrophenol (DNP) on the removal of sodium ions from the axon of a cuttlefish.

Questions

1 Describe the resting potential of a neurone and explain how it is maintained.

2 (a) Describe an action potential and explain the importance of the refractory period.
  (b) Explain how an action potential can most accurately be measured.

3 Explain how the graph in **fig D** provides evidence for the role of the sodium/potassium ATPase pump in maintaining the resting potential.

Key definitions

Polarised describes the condition of a neurone when the movement of positively charged potassium ions out of the cell down the concentration gradient is opposed by the actively produced electrochemical gradient, leaving the inside of the cell slightly negative relative to the outside.

The **resting potential** is the potential difference across the membrane of around –70 mV when the neurone is not transmitting an impulse.

A **neurotransmitter** is a chemical which transmits an impulse across a synapse.

Sodium gates are specific sodium ion channels in the nerve fibre membrane that open up, allowing sodium ions to diffuse rapidly down their concentration and electrochemical gradients.

Depolarisation is the condition of the neurone when the potential difference across the membrane is briefly reversed during an action potential, with the cell becoming positive on the inside with respect to the outside for about 1 millisecond.

The **action potential** is when the potential difference across the membrane is briefly reversed to about +40 mV on the inside with respect to the outside for about 1 millisecond.

The **threshold** is the point when sufficient sodium ion channels open for the rush of sodium ions into the axon to be greater than the outflow of potassium ions, resulting in an action potential.

The **refractory period** is the time it takes for ionic movements to repolarise an area of the membrane and restore the resting potential after an action potential.

The **absolute refractory period** is the first millisecond or so after the action potential when it is impossible to re-stimulate the fibre – the sodium ion channels are completely blocked and the resting potential has not been restored.

The **relative refractory period** is a period of several milliseconds after an action potential and the absolute refractory period when an axon may be re-stimulated, but only by a much stronger stimulus than before.

By the end of this section, you should be able to...

● explain why the speed of transmission along myelinated axons is greater than along non-myelinated axons, including the role of saltatory conduction

● explain the structure and function of a synapse, including the role of transmitter substances acetylcholine and noradrenaline

● explain the formation and effects of excitatory and inhibitory post-synaptic potentials

Once an action potential is set up in response to a stimulus, it will travel the entire length of that nerve fibre, which may be many centimetres or even metres long. The movement of the nerve impulse along the fibre is the result of local currents set up by the ion movements at the action potential itself. These ion movements occur both in front of and behind the action potential. They depolarise the membrane in front of the action potential sufficiently to cause the sodium ion channels to open. The sodium ion channels behind the action potential cannot open due to the refractory period of the membrane behind the spike. In this way the impulse is continually conducted in the required direction (see **fig A**).

Saltatory conduction

In myelinated neurones (see **Section 9.2.1**), the situation is more complex. Ions can only pass in and out of the axon freely at the nodes of Ranvier, which are about 1 mm apart. This means that action potentials can only occur at the nodes and so they appear to jump from one to the next (see **fig B**). The effect of this is to speed up transmission as the ionic movements associated with the action potential occur much less frequently, taking less time. It is known as **saltatory conduction**, from the Latin verb *saltare*, which means 'to jump'.

Synapses

Neurones must be able to intercommunicate. Receptors must pass their information into the sensory nerves, which in turn must relay the information to the central nervous system. Information needs to pass freely around the central nervous system and the impulses sent along the motor nerves must be communicated to the effector organs so that action can be taken. Wherever two neurones meet they are linked by a **synapse** (see **fig C**). Every cell in the central nervous system is covered with **synaptic knobs** from other cells – several hundred in some cases. Neurones never actually touch their target cell, so a synapse involves a gap between two cells – a gap which the nerve impulses must somehow cross.

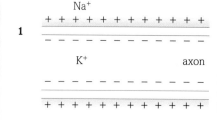

1 At resting potential there is positive charge on the outside of the membrane and negative charge on the inside, with high sodium ion concentration outside and high potasssium ion concentration inside.

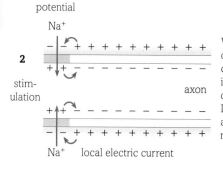

2 When stimulated, voltage-dependent sodium ion channels open, and sodium ions flow into the axon, depolarising the membrane. Localised electric currents are generated in the membrane.

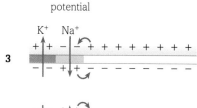

3 The potential difference in the membrane adjacent to the first action potential changes. A second action potential is initiated. At the site of the first action potential the voltage-dependent sodium ion channels close and voltage-dependent potassium ion channels open. Potassium ions leave the axon, repolarising the membrane. The membrane becomes hyperpolarised.

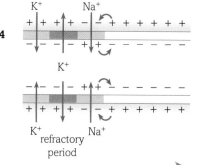

4 A third action potential is initiated by the second. In this way local electric currents cause the nerve impulse to move along the axon. At the site of the first action potential, potassium ions diffuse back into the axon, restoring the resting potential.

progress of the impulse

fig A By a combination of tiny local currents and the inhibiting effect of the refractory period, the action potential is conducted along a nerve fibre only in the direction in which it is needed to go.

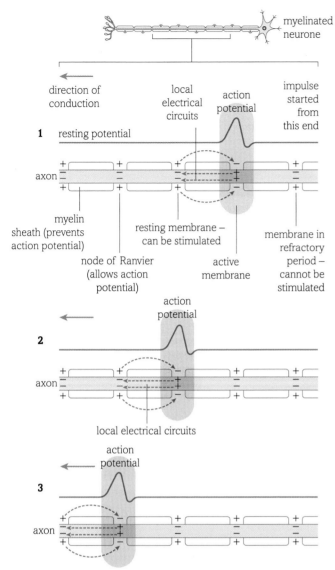

fig B Saltatory conduction. By 'jumping' from node to node along a myelinated nerve fibre, the nerve impulses in vertebrate neurones can travel very rapidly along very narrow nerve fibres. This allows for the development of complex but compact nervous systems.

The functioning of synapses depends on the movement of calcium ions. The arrival of an impulse at the synaptic knob increases the permeability of the **presynaptic membrane** to calcium ions as calcium ion channels open up. Calcium ions then move into the synaptic knob down their concentration gradient. The effect of the influx of calcium ions is to cause the **synaptic vesicles**, which contain a transmitter substance or neurotransmitter, to move to the presynaptic membrane. Each vesicle contains about 3000 molecules of transmitter. Some of the vesicles fuse with the presynaptic membrane and release the transmitter substance into the **synaptic cleft**. These molecules diffuse across the gap and become attached to specific protein receptor sites on the sodium channels of the post-synaptic membrane. This opens sodium ion channels in the membrane, resulting in an influx of sodium ions into the nerve fibre, causing a change in the potential difference across the membrane and an **excitatory post-synaptic potential (EPSP)** to be set up.

If there are sufficient of these EPSPs, the positive charge in the post-synaptic cell exceeds the threshold level, and an action potential is set up which then travels on along the post-synaptic neurone.

In some cases the neurotransmitter has the opposite effect. Different ion channels open in the membrane, allowing the inwards movement of *negative* ions, which makes the inside more negative than the normal resting potential. An **inhibitory post-synaptic potential (IPSP)** results, which makes it less likely that an action potential will occur in the post-synaptic fibre. These IPSPs play a part in the way we hear patterns of sound for example, and they are thought to be important in the way birds learn songs.

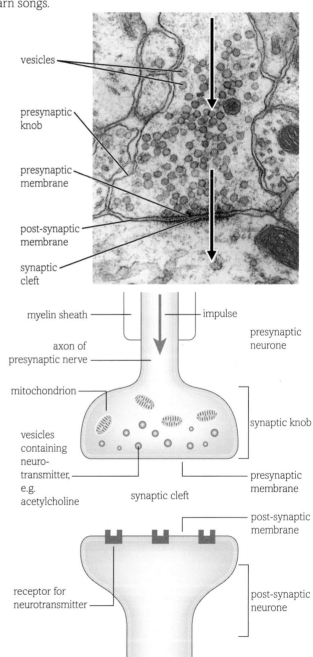

fig C The structure of the synapse, based on information revealed by the electron microscope.

Once the transmitter has had an effect, it is destroyed by enzymes in the synaptic cleft so that the receptors on the post-synaptic membrane are emptied and can react to a subsequent impulse. Something similar to a synapse occurs where a motor neurone meets an effector. For example, neuromuscular junctions are found between motor neurones and muscle cells. When the transmitter is released, it stimulates a contraction in the muscle cell.

What are the transmitter substances?

One common neurotransmitter, found at the majority of synapses in humans, is **acetylcholine (ACh)**. It is synthesised in the synaptic knob using ATP produced in the many mitochondria present. Nerves using acetylcholine as their transmitter are known as **cholinergic nerves**. Once the acetylcholine has done its job, it is very rapidly hydrolysed by the enzyme **acetylcholinesterase**. The enzyme is embedded in the post-synaptic membrane next to the acetylcholine receptors. Once the neurotransmitter has bound to the receptor and initiated a response, it is rapidly hydrolysed into acetate and choline. This ensures that it no longer affects the post-synaptic membrane and also releases the components to be recycled. The components rapidly diffuse across the synaptic cleft down a concentration gradient and are taken back into the synaptic knob through the presynaptic membrane. In the presynaptic knob they are re-synthesised into more acetylcholine. Acetylcholine is the neurotransmitter in all motor neurones, the parasympathetic nervous system (part of the autonomic nervous system, see **Section 9.2.7**), and cholinergic synapses in the CNS. It usually results in excitation at the post-synaptic membrane.

Not all vertebrate nerves use acetylcholine as their synaptic transmitter substance. Some, particularly those of the sympathetic nervous system (part of the autonomic nervous system, see **Section 9.2.7**), produce **noradrenaline** in their synaptic vesicles and are known as **adrenergic nerves**. The binding of noradrenaline to the receptors in the post-synaptic membrane depends on the concentration of the neurotransmitter in the synaptic cleft. As the release of noradrenaline from the presynaptic knob stops, levels in the synaptic cleft fall. Noradrenaline is then released from the post-synaptic receptors back into the synaptic cleft. Up to 90% of the noradrenaline is then taken up by the presynaptic knob, where much of it is repacked and reused when another action potential comes along. Noradrenaline is the neurotransmitter in the sympathetic nervous system and in adrenergic synapses in the brain.

Questions

1   Nerve transmission in mammals is usually faster than nerve transmission in invertebrates, although the nerve fibres usually have a much smaller diameter. How would you explain these observations?

2   Enzymes play several important roles in a synapse – what are they?

3   Produce flow diagrams to summarise the sequence of events at (a) a cholinergic synapse and (b) an adrenergic synapse.

4   (a) What is the difference between an EPSP and an IPSP?

    (b) Why are they important in the functioning of the nervous system?

Key definitions

Saltatory conduction is the process by which action potentials are transmitted from one node of Ranvier to the next in a myelinated nerve.

A **synapse** is the junction between two neurones that nerve impulses cross via neurotransmitters.

Synaptic knobs are the bulges at the end of the presynaptic neurones where neurotransmitters are made.

The **presynaptic membrane** is the membrane on the side of the synapse where the first impulse arrives and from which neurotransmitters are released.

Synaptic vesicles are membrane-bound sacs in the presynaptic knob which contain about 3000 molecules of neurotransmitter and move to fuse with the presynaptic membrane when an impulse arrives in the presynaptic knob.

The **synaptic cleft** is the gap between the pre and post synaptic membranes in a synapse.

The **excitatory post-synaptic potential (EPSP)** is the potential difference across the post-synaptic membrane caused by an influx of sodium ions into the nerve fibre, as the result of the arrival of a molecule of neurotransmitter on the receptors of the post-synaptic membrane that makes the inside more positive than the normal resting potential, increasing the chance of a new action potential.

The **inhibitory post-synaptic potential (IPSP)** is the potential difference across the post-synaptic membrane caused by an influx of negative ions as the result of the arrival of a molecule of neurotransmitter on the receptors of the post-synaptic membrane, which makes the inside more negative than the normal resting potential, decreasing the chance of a new action potential.

Acetylcholine (ACh) is a neurotransmitter found in the synapses of motor neurones, the parasympathetic nervous system and cholinergic synapses in the brain.

Cholinergic nerves use acetylocholine as the neurotransmitter in their synapses.

Acetylcholinesterase is an enzyme found embedded in the post-synaptic membrane of cholinergic nerves that breaks down acetylcholine in the synapes after it has triggered a post-synaptic potential.

Noradrenaline is a neurotransmitter found in the synapses of the sympathetic nervous system and adrenergic synapses of the brain.

Adrenergic nerves use noradrenaline as the neurotransmitter in their synapses.

The effect of drugs on the nervous system

By the end of this section, you should be able to...

● explain how the effects of drugs can be caused by their influence on synaptic transmission

The more we understand about synapses, the better we will understand many of the diseases that affect the peripheral and central nervous systems. Understanding synapses also helps us decipher how many commonly-used drugs have their effect.

The effects of drugs on the nervous system

Fig A shows the main ways in which drugs affect synapses.

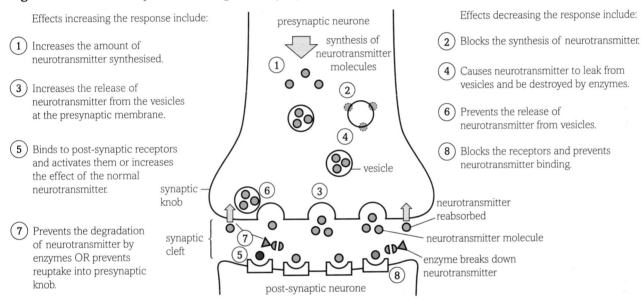

Effects increasing the response include:

(1) Increases the amount of neurotransmitter synthesised.

(3) Increases the release of neurotransmitter from the vesicles at the presynaptic membrane.

(5) Binds to post-synaptic receptors and activates them or increases the effect of the normal neurotransmitter.

(7) Prevents the degradation of neurotransmitter by enzymes OR prevents reuptake into presynaptic knob.

Effects decreasing the response include:

(2) Blocks the synthesis of neurotransmitter.

(4) Causes neurotransmitter to leak from vesicles and be destroyed by enzymes.

(6) Prevents the release of neurotransmitter from vesicles.

(8) Blocks the receptors and prevents neurotransmitter binding.

fig A The main ways in which drugs affect synapses.

This information enables us to understand how many different substances affect our bodies. Here are three specific examples:

* **Nicotine** mimics the effect of acetylcholine and binds to specific acetylcholine receptors in post-synaptic membranes known as nicotinic receptors. It triggers an action potential in the post-synaptic neurone, but then the receptor remains unresponsive to more stimulation for some time. Nicotine causes raised heart rate and blood pressure. It also triggers the release of another type of neurotransmitter in the brain called dopamine. This is associated with pleasure sensations. At low doses nicotine has a stimulating effect, but at high doses it blocks the acetylcholine receptors and can kill. It is a highly addictive drug. One cigarette contains 1–6 mg nicotine. Even at a low dose, nicotine has a big effect on your acetylcholine synapses, and dopamine release is also large. People trying to give up cigarettes, with their lethal load of carcinogens and other toxins, often need to replace the nicotine to help them do so.

fig B Nicotine is a very addictive drug found in cigarettes. It affects the acetylcholine synapses and the release of the neurotransmitter dopamine in the brain. People who are trying to give up smoking may wear a nicotine patch on their skin.

* **Lidocaine** is a drug used as a local anaesthetic. It is commonly used by dentists before drilling or removing a tooth. If you have ever had an injection at the dentist you will probably have been injected with lidocaine. Lidocaine molecules block voltage-gated sodium channels, preventing the production of an action potential in sensory nerves and so preventing you from feeling pain. It is also used to prevent some heart arrhythmias where it works in the same way – it blocks sodium channels, raising the depolarisation threshold. In this way it reduces or prevents early or extra action potentials from the pacemaker region that can cause arrhythmias.

fig C Lidocaine is so effective at blocking sensory nerve impulses and so preventing pain that you can have a tooth removed without feeling it.

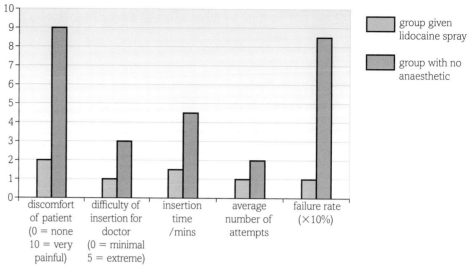

fig D Inserting a tube into the nose of patients in hospital accident and emergency departments is not easy and can be very painful. Research by C.P. Chan and F.L. Lau in 2010 shows that using a topical spray of lidocaine, and giving plenty of time for it to work, made the process much less painful for the patient, and faster and much more successful for the doctor.

- **Cobra venom** (α-cobratoxin) is a substance made by several species of cobra that is toxic and often fatal in snake bites. It binds reversibly to acetylcholine receptors in post-synaptic membranes and neuromuscular junctions. It prevents the transmission of impulses across synapses, including the neuromuscular junctions between motor neurones and muscles. As a result, muscles are not stimulated to contract and gradually the person affected becomes paralysed. When the toxin reaches the muscles involved in breathing it causes death. In very low concentrations however it can relax the muscles of the trachea and bronchi in severe asthma attacks, and so save lives.

Questions

1 Strychnine inactivates cholinesterase as the post-synaptic membrane. Botulinus toxin affects the presynaptic membrane, preventing the release of acetylcholine. Explain how electron micrographs and these toxins could be used as evidence for the current model of synaptic transmission.

2 Make a table to summarise the main ways in which drugs may affect synapses.

3 Use the data in **fig D**, taken from a sample of over 200 patients randomly assigned lidocaine or a placebo before having a nasogastric tube inserted, to answer the following questions:
   (a) The first two scores refer to the pain recorded by the patient during the procedure and the difficulty recorded by the doctor in getting the nasogastric tube in place. What was the percentage reduction in pain for the patients and difficulty for the doctors?
   (b) What was the mean time taken to insert the tube with and without lidocaine?
   (c) What percentage of procedures failed completely with and without lidocaine?
   (d) How does lidocaine work and state why it has such a marked effect on the success of inserting a nasogastric tube?

Key definitions

Nicotine is a drug found in cigarettes that mimics the effect of acetylcholine and binds to specific acetylcholine receptors in post-synaptic membranes known as nicotinic receptors.

Lidocaine is a drug used as a local anaesthetic that works by blocking the voltage-gated sodium channels in post-synaptic membranes in sensory neurones, preventing the production of an action potential.

Cobra venom is a substance made by several species of cobra that binds reversibly to acetylcholine receptors in post-synaptic membranes in motor neurones, preventing the production of a post-synaptic action potential.

By the end of this section, you should be able to...

- describe the structure of the human retina
- explain the role of rhodopsin in initiating action potentials
- explain how the distribution of human rod and cone cells maintain vision in different light intensities

The coordinated activity of any organism does not depend solely on the passage of action potentials along nerve fibres or successful reflex arcs. There must be a continuous input of information from both the outside world and the internal environment, if the animal is to be able to modify its behaviour and survive as situations change. Sensory receptors play a vital role in providing an animal with information about both its internal and external environment.

Simple sensory receptors are just neurones with a dendrite that is sensitive to one particular stimulus. When the dendrite receives a stimulus, chemical events occur that result in an action potential in the nerve fibre of the neurone. This type of cell is known as a primary receptor. A secondary receptor is slightly more complicated. It consists of one or more completely specialised cells (not neurones) that are sensitive to a particular type of stimulus. These cells synapse with a normal sensory neurone, which carries the impulse to the central nervous system. The retinal cells in the retina of the eye are an example of these secondary receptors.

There are various different types of receptors and a variety of ways of categorising them.

As animals become increasingly complex, so do their sensory systems. While many sensory receptors are always found as isolated entities, in higher animals sensory receptors are often found in systems known as sense organs.

How do sensory receptors work?

Receptor cells, like nerve fibres, have a resting potential that depends on maintaining the charge of the cell interior negative in relation to the outside, by using membrane sodium pumps. When a receptor cell receives a stimulus, sodium ions move rapidly across the cell membrane along concentration and electrochemical gradients, and this sets up a generator potential. A small stimulus results in a small generator potential and a large stimulus results in a large generator potential – generator potentials *do not* obey the all-or-nothing law. If the generator potential produced is large enough to reach the threshold of the sensory neurone, an action potential will result in that neurone. If it is not, there will be no action potential – the action potential *does* obey the all-or-nothing law.

The following process is common in one form or another to most sensory receptors:

$$\text{stimulus} \rightarrow \begin{array}{c}\text{local change}\\\text{in permeability}\end{array} \rightarrow \begin{array}{c}\text{generator}\\\text{potential}\end{array} \rightarrow \begin{array}{c}\text{action}\\\text{potential}\end{array}$$

In sense organs such as the eye, several receptor cells will often synapse with a single sensory neurone. If the generator potential from an individual receptor cell is insufficient to set up a synapse, the potentials from several may add together or summate and trigger an action potential. This is known as convergence and it is a useful adaptation for increasing the sensitivity of a sensory system to low-level stimuli. It is an important feature of the light-sensitive cells of the retina of the eye.

A weak stimulus results in a low frequency of action potentials along the sensory neurone. A strong stimulus results in a rapid stream of action potentials being fired along the sensory neurone. As a result, although the axon obeys the all-or-nothing law in terms of each individual action potential, a graded response is still possible, giving information about the strength of the stimulus. In the eye, this means that we are not only aware of the difference between light and dark, but of all the varying degrees of light and shade (see **fig A**).

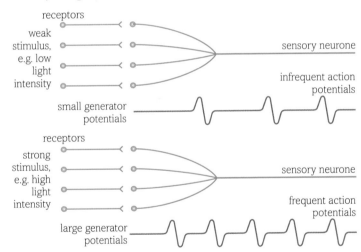

fig A The interactions between sensory receptors and sensory neurones give high levels of sensitivity to different strength stimuli, as seen in the eye and other sense organs.

The human eye

Sensitivity to light is found in some of the simplest organisms. In humans, this has developed far beyond a simple sensitivity to the absence or presence of light into a remarkably sophisticated optical system, allowing clear and focused vision in a wide variety of circumstances. Our eyes are sensitive to electromagnetic radiation with a wavelength between 400 and 700 nm. The ranges of other species often vary from this. The structure of the human eye can be seen in **fig B**.

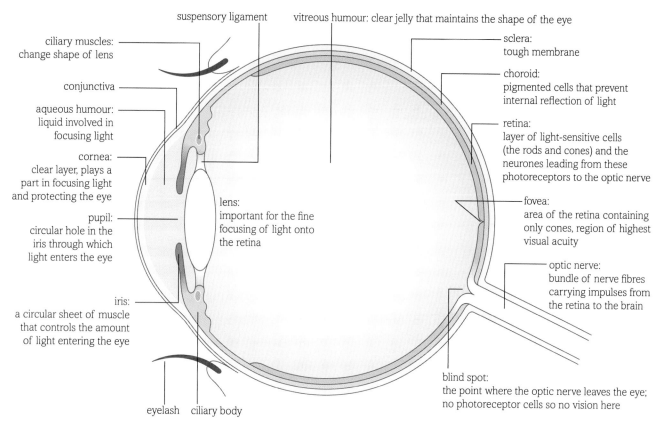

ciliary muscles:
change shape of lens

conjunctiva

aqueous humour:
liquid involved in
focusing light

cornea:
clear layer, plays a
part in focusing light
and protecting the eye

pupil:
circular hole in the
iris through which
light enters the eye

iris:
a circular sheet of muscle
that controls the amount
of light entering the eye

eyelash   ciliary body

suspensory ligament

vitreous humour: clear jelly that maintains the shape of the eye

sclera:
tough membrane

choroid:
pigmented cells that prevent
internal reflection of light

retina:
layer of light-sensitive cells
(the rods and cones) and the
neurones leading from these
photoreceptors to the optic nerve

fovea:
area of the retina containing
only cones, region of highest
visual acuity

optic nerve:
bundle of nerve fibres
carrying impulses from
the retina to the brain

lens:
important for the fine
focusing of light onto
the retina

blind spot:
the point where the optic nerve leaves the eye;
no photoreceptor cells so no vision here

fig B The main structures of the human eye.

The role of the retina

Focusing the light onto the retina of the eye is only the first step. The retina must then perceive the light and provide the information the **brain** needs to make sense of the image. The retina contains over a hundred million light-sensitive cells (photoreceptors), along with the neurones with which they synapse. There are two main types of photoreceptors in the retina, known as the **rods** and the **cones**. Both types are secondary exteroceptors – they signal changes in the external environment.

You have around 120 million rods in each eye, which are spread evenly across the retina except at the fovea where there are none. They provide greyscale vision only and are used mainly for seeing in low light intensities or at night – they are very sensitive to light. The rods are not very tightly packed together and several of them synapse with the same sensory neurone. Although this means they do not give a particularly clear picture, it does make them extremely sensitive both to low light levels and to movements in the visual field, because several small generator potentials can trigger an action potential to the CNS by summation (see **fig C**).

Cones, on the other hand, are found tightly packed together in the fovea. There are only around 6 million of them. They are used mainly for vision in bright light and they also provide colour vision. As a result of their tight packing in the fovea, along with the fact that each cone usually has its own sensory neurone, cones provide great visual acuity in bright light. It is only when light is focused directly on the fovea that an image is clearly in focus.

The arrangement of the retina is surprising – it appears to be 'back-to-front'. The 'outer segments' are next to the choroid, and the neurones are at the interior edge of the eyeball. The light has to pass through the synapses and the inner segments before reaching the outer segments containing the visual pigments. The reason for this somewhat unexpected arrangement is the origin of the retinal cells in the embryo and the way in which the eye is formed during embryonic development. This is why there is a blind spot where all the neurones pass through the layers of the eye to go into the brain (see **fig D**).

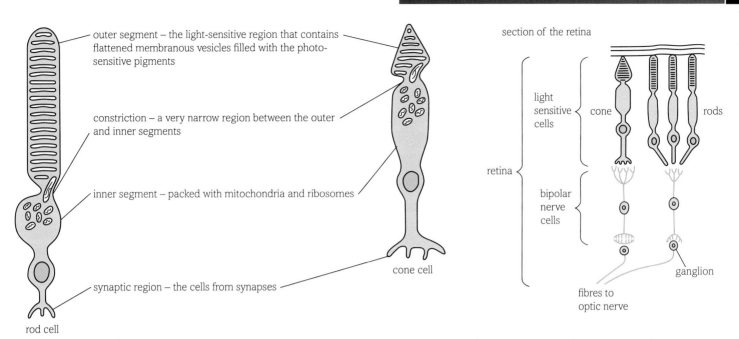

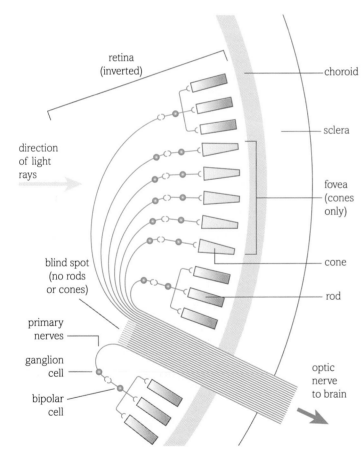

fig C The two types of receptor cells found in the human retina, along with their different arrangement of synapses, give a visual system that combines great sensitivity to low levels of light with high visual acuity and clarity of colour vision.

fig D The structure of the retina.

How do the rods and cones work?

The rods and cones work in a similar way that is based on the reactions of a visual pigment with light. In the rods this visual pigment is **rhodopsin (visual purple)**, which is formed from two components, opsin and retinal. Opsin is a lipoprotein and

retinal is a light-absorbing derivative of vitamin A. Retinal exists as two different isomers, *cis*-retinal and *trans*-retinal. In the dark, it is all in the *cis*-form. When a photon of light (the smallest unit of light) hits a molecule of rhodopsin, it converts the *cis*-retinal into *trans*-retinal. This changes the shape of the retinal, and puts a strain on the bonding between the opsin and retinal. As a result, the rhodopsin breaks up into opsin and retinal. This breaking up of the molecule is referred to as bleaching.

The membranes of most neurones are relatively impermeable to sodium ions, but rod cell membranes are normally very permeable to them. Sodium ions move into the rod cell through sodium ion channels, and the sodium/potassium pump moves them out again. When rhodopsin is bleached, it triggers a cascade reaction that results in the closing of the sodium ion channels, so the rod cell membrane becomes much less permeable to sodium ions and fewer sodium ions diffuse into the cell. The sodium pump continues to work at the same rate, pumping sodium ions out of the rod cell, so the interior becomes more negative than usual. This hyperpolarisation is known as the generator potential in the rod. The size of the generator potential depends on the amount of light hitting the rod, and therefore the amount of rhodopsin bleaching that takes place. If it is large enough to reach the threshold, or if several rods are stimulated at once, neurotransmitter substances are released into the synapse with the bipolar cell, which synapses on both the photoreceptor and the nerve fibre. An action potential is then set up in the bipolar cell that passes across the synapse to cause an action potential in the sensory neurone. All the sensory neurones leave the eye at the same point to form the optic nerve leading to the brain.

Once the visual pigment has been bleached, the rod cannot be stimulated again until the rhodopsin is resynthesised. It takes ATP produced by the many mitochondria in the inner segment of the rod to convert the *trans*-retinal back to *cis*-retinal and rejoin it to the opsin to form rhodopsin again. In normal daylight the rods are almost entirely bleached and can no longer respond to dim light – the eye is said to be light-adapted. After about 30 minutes in complete darkness the rhodopsin will be almost fully reformed, the eye is sensitive to dim light and is said to be dark-adapted. This explains why you become almost blind when you walk from a sunny garden into a house. The bright light has completely bleached the rhodopsin you need to see in the dimmer interior light. As the rhodopsin reforms in your rods, your vision returns as your eyes become dark-adapted again.

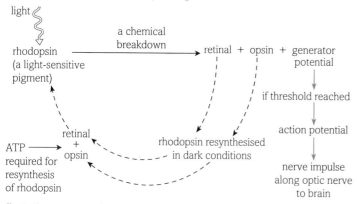

fig E The reactions of rhodopsin in the light.

Did you know?

Investigating the graded response of the rods

The rods of your eye do not give you an all-or-nothing response. You are aware of varying levels of light intensity. Yet you know that the axons that fire off potentials through the optic nerve operate on an all-or-nothing response. Scientists have developed a model based on a graded response in the receptor cells. This response is amplified through a cascade system that causes hyperpolarisation in the receptor cells. A single photon of light bleaches a single molecule of rhodopsin, but through the cascade that is set in place, several thousand sodium ion channels can be closed.

The brighter the light, the more rhodopsin is bleached and so the more sodium ion channels are closed. This gives a bigger depolarisation of the cell which lasts longer, so the cell can respond in a measurably different way to very small differences in light intensity. This can be demonstrated experimentally, as in **fig F**. The membrane potential is recorded from the inner segment of the rod cell and associated bipolar cell. The rod cell is then stimulated with a flash of light. When the outer segment of the rod cell is exposed to light, the inner segment hyperpolarises.

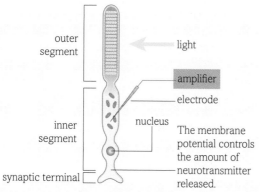

fig F A practical technique for investigating the change in the membrane potentials of rod cells in response to light.

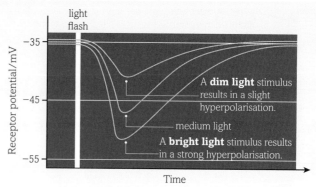

fig G The graded response of the rods to light levels and the interpretation of the recorded signals by the brain.

The cones and colour vision

The cones work in a very similar way to rods, except that their visual pigment is known as iodopsin. There appear to be three types of iodopsin, each sensitive to one of the primary colours of light. Iodopsin needs to be hit with more light energy than rhodopsin in order to break down, and so it is not sensitive to low light intensities. The cones provide colour vision because the brain interprets the numbers of different types of cones stimulated as different colours. This complements the low light vision and sensitivity to movement provided by the rods.

Questions

1. (a) Describe how a sensory receptor works.
   (b) Explain how convergence increases the sensitivity of a system to low level stimuli.

2. (a) Name the light-sensitive pigment in the rod cells and describe the effect that light has on the pigment.
   (b) Explain how this change in the visual pigment is translated into a visual image in the brain.

3. Explain how the different neural connections of the rods and the cones, which you can see in **fig C**, account for the different visual acuity they give.

4. Account for the following:
   (a) Rods transmit information in dimmer light than cones.
   (b) Rods are more sensitive to movement than cones.
   (c) Cones give a clearer image than rods.

Key definitions

The **brain** is the area of the CNS where information can be processed and from where instructions can be issued as required to give fully coordinated responses to a whole range of situations.

Rods are photoreceptors found in the retina which contain the visual pigment rhodopsin. They respond to low light intensities, give black and white vision and are very sensitive to movement.

Cones are photoreceptors found in the fovea of the retina which contain the visual pigment iodopsin. They respond to bright light, give great clarity of vision and colour vision.

Rhodopsin (visual purple) is the visual pigment in the rods.

By the end of this section, you should be able to...

- describe the location and main functions of the medulla oblongata, the cerebellum, the cerebrum and the hypothalamus
- describe the structure of the spinal cord

A nervous system made up of receptors, nerves and effectors allows an animal to respond to basic stimuli from the environment, but large, complex animals need more than this. Evolution has resulted in the central nervous system (CNS), which includes:

- the brain, where information can be processed and from where instructions can be issued as required to give fully coordinated responses to a whole range of situations
- the **spinal cord**, which carries the nerve fibres into and out of the brain and also coordinates many unconscious reflex actions.

The formation of the brain

The mammalian brain forms as a swelling in the hollow neural tube at the front or anterior end of a vertebrate embryo that folds back on itself. The basic brain pattern has three distinct areas – the forebrain, midbrain and hindbrain (**fig A**).

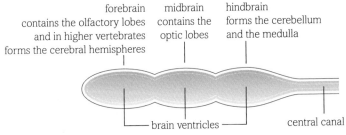

forebrain
contains the olfactory lobes and in higher vertebrates forms the cerebral hemispheres

midbrain
contains the optic lobes

hindbrain
forms the cerebellum and the medulla

brain ventricles

central canal

fig A The human brain has evolved considerably from the basic vertebrate pattern.

In some vertebrates this brain remains fairly simple, with areas very specific to particular functions such as sight or smell. In mammals, including humans, the brain becomes a remarkably complex structure. The original simple arrangement of the brain into three areas is very difficult to see, because a part called the **cerebrum** (made up of the two **cerebral hemispheres**) is folded back over the entire brain.

The brain is made up of a combination of **grey matter**, made up of neurone cell bodies, and **white matter**, consisting of nerve fibres. There are areas of the human brain with very specific functions concerned with the major senses and control of basic bodily functions. There are also many regions of the brain for which we do not yet clearly understand the precise functions and interrelationships with other areas of the brain. Scientists have estimated that there are around a hundred thousand million neurones working together in the human brain and that each neurone synapses with up to 10 000 other neurones. The brain contains centres or nuclei made up of cell bodies that make intercommunication between millions of cells possible. The great nerve tracts from the spinal cord cross over as they enter and leave the brain, so that the left-hand side of the brain receives information from and controls the right-hand side of the body, and vice versa. The brain is a very complex system to try to understand!

Some of the major areas of the brain

The cerebrum consists of the two cerebral hemispheres and they are the site of many of the higher functions of the brain. It is the biggest and most highly developed area of the human brain, making up 65–67% of the mass of brain tissue. This is the site of our abilities to see, think, learn and feel emotions. Our motor functions (all of our conscious movements) are also controlled by the cerebrum. The outer layer of the cerebrum is the cerebral cortex. This layer is only 2–4 mm thick, but it is made up almost entirely of grey matter – nerve cell bodies, dendrites and synapses. The cerebral cortex is also deeply folded to give a huge surface area. The left and right cerebral hemispheres are connected by a band of axons (white matter) known as the **corpus callosum**. They are subdivided into a number of lobes that have been found to be associated with particular functions. For example, the frontal lobe is associated with the higher brain functions such as emotional responses, planning ahead, reasoning and decision making.

There are a number of other areas of the brain that are known to be linked to specific aspects of the way your body works. Many of these are involved in the unconscious responses of your body which maintain the processes of life.

Other regions of the brain

Inside the brain there are other structures that have important functions:

- The hypothalamus coordinates the autonomic (unconscious) nervous system (see **Section 9.1.3**). It plays a major part in thermoregulation (the regulation of the core temperature of the body) and osmoregulation (the regulation of the osmotic potential of the body fluids) – see **Section 9.3.2** and **9.3.4**. It monitors the chemistry of the blood and controls the hormone secretions of the pituitary gland. It also controls many basic drives – thirst, hunger, aggression and reproductive behaviour.
- The **cerebellum** coordinates smooth movements. It uses information from the muscles and the ears to control balance and maintain posture.
- The **medulla oblongata** (often referred to as the medulla) is the most primitive part of the brain. It contains reflex centres that control functions such as the breathing rate, heart rate, blood pressure, coughing, sneezing, swallowing, saliva production and peristalsis. It is this region that may maintain the basic life responses, even when the higher areas of the brain have been effectively destroyed.

Fig B gives you a more detailed illustration of the various areas of the brain.

To coordinate the functions of the body, the brain needs a way of communicating and this is one of the major functions of the spinal cord.

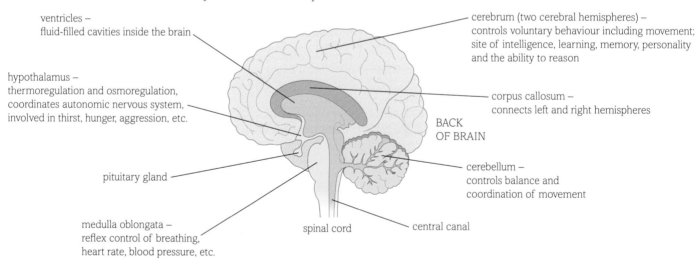

ventricles – fluid-filled cavities inside the brain

hypothalamus – thermoregulation and osmoregulation, coordinates autonomic nervous system, involved in thirst, hunger, aggression, etc.

pituitary gland

medulla oblongata – reflex control of breathing, heart rate, blood pressure, etc.

spinal cord

central canal

cerebrum (two cerebral hemispheres) – controls voluntary behaviour including movement; site of intelligence, learning, memory, personality and the ability to reason

corpus callosum – connects left and right hemispheres

BACK OF BRAIN

cerebellum – controls balance and coordination of movement

fig B The brain is an organ of remarkable complexity and so far our knowledge of it remains limited, although we are discovering more all the time.

The structure and functions of the spinal cord

The spinal cord is a tube made up of a core of grey matter surrounded by white matter, which runs out from the base of the brain (the medulla oblongata) through the vertebra. It is approximately 43–45 cm long. Impulses from sensory receptors travel along sensory nerve fibres into the spinal cord through the dorsal roots, and then travel in sensory fibres up the cord to the brain. Instructions from the brain travel as impulses down motor fibres in the spinal cord and out in motor neurones through the ventral roots to the effector organs.

The spinal cord is also an important coordination centre in its own right. Many simple organisms have little CNS – and certainly no complex brain. Their actions take place without conscious thought, as a result of **reflex responses**. Many of the actions of more complex animals are also the result of unconscious reflex actions. Well-known examples of human reflexes include moving a hand or foot away rapidly from something hot or sharp, swallowing as food moves down the back of the throat, blinking if an object approaches the eyes and the contracting and dilating of the pupils in response to changing light levels.

These unconditioned reflexes are controlled by the simplest type of nerve pathway in the body, known as a reflex arc. In vertebrates including mammals, this involves a receptor, a motor neurone and a sensory neurone at the very least. Part of the pathway occurs in the CNS – often the spinal cord. The reflex arc may involve simply a sensory and motor neurone, but often there is a small third relay neurone situated in the CNS. The function of the reflex arc is to bring about an appropriate response to a particular stimulus as rapidly as possible without the time delay that occurs when the conscious centres become involved. There are two main types of reflexes: spinal reflexes (e.g. hand withdrawing from heat or sharp object) and cranial reflexes (e.g. blinking, pupil reflexes), which both involve unconscious areas of the brain. Sensory neurones will also relay information to the conscious areas of the brain, so you know what has happened.

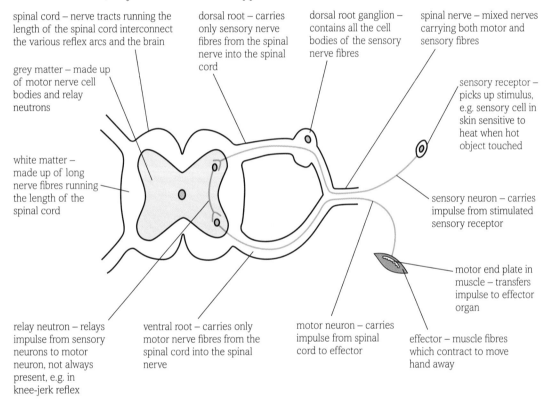

spinal cord – nerve tracts running the length of the spinal cord interconnect the various reflex arcs and the brain

dorsal root – carries only sensory nerve fibres from the spinal nerve into the spinal cord

dorsal root ganglion – contains all the cell bodies of the sensory nerve fibres

spinal nerve – mixed nerves carrying both motor and sensory fibres

grey matter – made up of motor nerve cell bodies and relay neutrons

sensory receptor – picks up stimulus, e.g. sensory cell in skin sensitive to heat when hot object touched

white matter – made up of long nerve fibres running the length of the spinal cord

sensory neuron – carries impulse from stimulated sensory receptor

motor end plate in muscle – transfers impulse to effector organ

relay neutron – relays impulse from sensory neurons to motor neuron, not always present, e.g. in knee-jerk reflex

ventral root – carries only motor nerve fibres from the spinal cord into the spinal nerve

motor neuron – carries impulse from spinal cord to effector

effector – muscle fibres which contract to move hand away

fig C A spinal reflex showing the structure of the spinal cord and how this is related to its function.

Questions

1. The degree of folding in the cerebral hemispheres varies among species, with more folding in humans than other primates for example. Dolphins have much folding but the layer of tissue is thinner. Give an explanation for these differences.

2. Make a table to summarise the functions of the main areas of the human brain.

3. Damage to both the cerebrum and the cerebellum can affect the way a person moves. Explain why, and state ways in which damage to either area might show itself.

4. (a) Compare and contrast the structure of the brain and the spinal cord.
   (b) Produce a flow diagram for the response if a person treads on a sharp stone in bare feet.

Key definitions

The **spinal cord** is the area of the CNS that carries the nerve fibres into and out of the brain and also coordinates many unconscious reflex actions.

The **cerebrum** is the area of the brain responsible for conscious thought, personality, control of movement and much more.

The **cerebral hemispheres** are the two parts of the cerebrum, joined by the corpus callosum.

Grey matter consists of the cell bodies of neurones in the CNS.

White matter consists of the nerve fibres of neurones in the CNS.

The **corpus callosum** is the band of axons (white matter) that join the left and right cerebral hemispheres of the brain.

The **hypothalamus** is the area of the brain that coordinates the autonomic (unconscious) nervous system.

The **cerebellum** is the area of the brain that coordinates smooth movements. It uses information from the muscles and the ears to control balance and maintain posture.

The **medulla oblongata (medulla)** is the most primitive part of the brain that controls reflex centres controlling functions such as the breathing rate, heart rate, blood pressure, coughing, sneezing, swallowing, saliva production and peristalsis.

Reflex responses are rapid responses that take place with no conscious thought involved.

By the end of this section, you should be able to...

● describe the division of the peripheral nervous system into the autonomic and voluntary systems

● explain why the autonomic system is divided into the sympathetic and parasympathetic systems, which act antagonistically

Changes in the internal or external environment picked up by sensory receptors must be carried to the CNS and instructions from the CNS must be carried to the effector organs for nervous coordination to work. This is the role of the peripheral nervous system.

The peripheral nervous system

The peripheral nerves are divided into two systems, which you have met already. The sensory or afferent nerves carry impulses from the receptors about changes in both the external and internal environment into the CNS. The motor or efferent nerves carry impulses out from the CNS to the effectors in the body. All the sensory nerves of the peripheral system function in much the same way. The motor nerves can be sub-divided into two main types:

• The **voluntary nervous system** involves motor neurones that are under voluntary or conscious control involving the cerebrum. Voluntary motor neurones function as a result of conscious thought. When you consider an action, such as picking up a drink or switching on the computer, the instructions that need to be issued to the muscles will be carried along voluntary nerve fibres.

• The **autonomic nervous system** (the involuntary nervous system) involves motor neurones that are not under the control of the conscious areas of the brain. They control bodily functions that are normally involuntary. Examples include control of the heart and breathing rate, the movements and secretions of the gut, sweating, the dilation or constriction of the iris of the eye in response to changing light levels and the dilation or constriction of the blood vessels in response to changing demands for blood.

The autonomic nervous system itself can be sub-divided into the **sympathetic nervous system** and the **parasympathetic nervous system**.

Most of the body organs are supplied by both the parasympathetic and the sympathetic nervous system. The differences between them are both anatomical and functional.

Structural differences between the parasympathetic and sympathetic systems

In both the parasympathetic and the sympathetic autonomic nervous systems, myelinated preganglionic fibres leave the CNS and synapse in a ganglion (a collection of cell bodies) with unmyelinated post-ganglionic fibres.

In the sympathetic system, the ganglia are very close to the CNS, so the preganglionic fibres are short and the postganglionic fibres are long.

In the parasympathetic system the situation is reversed. The ganglia are near to or in the effector organ, so the preganglionic fibres are very long and the postganglionic fibres are very short. These structural differences can be seen clearly in **fig A**.

Functional differences between the parasympathetic and sympathetic autonomic nervous systems

There are two basic differences between these two autonomic nervous systems:

• The sympathetic nervous system produces noradrenaline at the synapses and usually produces a rapid response in the target organ system. This sympathetic autonomic nervous system is sometimes referred to as the 'fight-or-flight' system. When you are active, or under physical or psychological stress, the sympathetic nervous system will dominate, stimulating the organs of the body to cope with the stress imposed.

• The parasympathetic nervous system often has a slower, damping down or inhibitory effect on organ systems and the neurotransmitter produced at the synapses is acetylcholine. The parasympathetic system maintains normal functioning of the body and restores calm after a stressful situation. It is sometimes referred to as the 'rest and digest' or 'feed and breed' system, in contrast to the action of the sympathetic system.

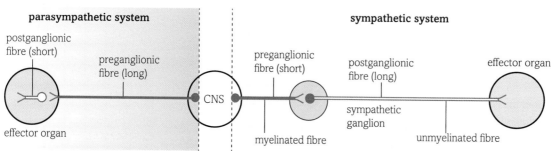

fig A Functional differences between parasympathetic and sympathetic automatic nervous systems.

	Sympathetic system	Parasympathetic system
Eyes	dilates pupil	constricts pupil
Salivary glands	inhibits flow of saliva	stimulates flow of saliva
Lacrimal glands	–	stimulates flow of tears
Lungs	dilates bronchi	constricts bronchi
Heart	accelerates heartbeat	slows heartbeat
Liver	stimulates conversion of glycogen to glucose	stimulates release of bile
Stomach	inhibits peristalsis and secretion	stimulates peristalsis
Kidneys	stimulates secretion of adrenaline and noradrenlaine	–
Intestines	inhibits peristalsis and anal sphincter contraction	stimulates peristalsis and contracts anal sphincter
Bladder	inhibits bladder contraction	contracts bladder

table A The opposing effects of the sympathetic and parasympathetic nerves on body systems.

The sympathetic and parasympathetic nervous systems act antagonistically, rather like the accelerator and brake of a car. So, for example, the sympathetic system speeds up the breathing rate and the heart rate, whilst the parasympathetic slows them down (see **table A**). However, as with the accelerator and brake, it is often not a case of all or nothing. The interplay of these complementary systems allows for fine control, allowing the body to match its responses exactly to the demands placed upon it.

Several bodily functions, which we might consider to be under voluntary control – opening the bowel and bladder sphincters, for example – appear in **table A**, as under the control of the autonomic system. However, the nervous system is extremely complex and sophisticated and many areas of the body are supplied with voluntary nerves as well as involuntary nerves. Most of us have control over our bladder and bowels, and can control our breathing rate if we wish to. It is relatively easy even to control the heart rate to some degree, and mystics the world over have shown control over many other normally involuntary activities. Scientists still do not understand all of the ways in which the voluntary and autonomic nervous systems, and the CNS itself, interact to control both our body functions and our behaviour.

Questions

1 What are the advantages of having both autonomic and voluntary motor nerves?

2 (a) What are the parasympathetic and sympathetic nervous systems?
  (b) Why are they sometimes described as antagonistic and why is this now regarded as a rather limited description?

Key definitions

The **voluntary nervous system** involves motor neurones that are under voluntary or conscious control involving the cerebrum.

The **autonomic nervous system** is the involuntary nervous system. Autonomic motor neurones control bodily functions that are not normally dealt with by the conscious area of the brain.

The **sympathetic nervous system** involves autonomic motor neurones with very short myelinated preganglionic fibres that leave the CNS and synapse in a ganglion very close to the CNS. They have long unmyelinated postganglionic fibres. They produce noradrenaline at the synapses and often stimulate a rapid response, activating an organ system.

The **parasympathetic nervous system** involves autonomic motor neurones with very long myelinated preganglionic fibres that leave the CNS and synapse in a ganglion very close to the effector organ. They have very short unmyelinated post-ganglionic fibres. They produce acetylcholine at the synapses and often have a relatively slow, inhibitory effect on an organ system.

THE PARALYSED MAN WHO WALKED AGAIN

Neurones make and lose connections all the time in the brain, as learning takes place and memories are formed and lost. Yet if the spinal cord is severed, the neurones simply do not grow and rejoin, and the affected individual will be paralysed below the break. Until recently, that is…

WORLD FIRST AS MAN WHOSE CORD WAS SEVERED WALKS

Fireman paralysed by knife attack recovers after UK scientists use brain cavity cells to re-grow nerve cells in his spine

- **Darek Fidyka is the first person in the world to recover from chronic injuries.**
- **Fireman was paralysed from the waist down after severing spinal cord.**
- **He received treatment pioneered by Wroclaw Medical University in Poland and University College London (UCL).**
- **Cells taken from his brain cavity were injected into the spine and regrew to repair broken link, enabling him to begin walking again.**

A fireman paralysed from the waist down after his spinal cord was completely severed has regained the ability to walk thanks to British scientists.

Darek Fidyka, 40, is believed to be the first person in the world to recover from such chronic injuries, in an achievement hailed as more impressive than putting a man on the Moon.

Mr Fidyka's spinal cord was sliced in half during a stabbing four years ago, leaving him paralysed below the waist.

It is the first time the procedure has been shown to work on a human. Professor Geoffrey Raisman of UCL said: 'We believe that this procedure is the breakthrough which, as it is further developed, will result in a historic change in the currently hopeless outlook for people disabled by spinal cord injury.'

THE PROCEDURE MORE IMPRESSIVE THAN PUTTING A MAN ON THE MOON

Scientists from Wroclaw Medical University in Poland and the University College London pioneered the breakthrough procedure. … The procedure involved transplanting olfactory ensheathing cells – or OECs – from the olfactory bulb in the brain cavity to the spinal cord. OECs assist the repair of damaged nerves in the nose that transmit smell messages to the brain.

The cells were transplanted into the spinal cord, using 100 micro injections across the site of the injury. A small piece of nerve tissue, which was taken from the ankle, was grafted onto the 'gap' in the spinal cord to act as a scaffold for the spinal neurons to extend, as guided by the OECs.

This enabled the ends of severed nerve fibres to grow and join together – something that was previously thought to be impossible. Three months after the operation, Mr Fidyka began to gain muscle in his left leg and regained sensations in it, such as hot and cold, and pins and needles. A year later, he is able to walk again, with the aid of parallel bars.

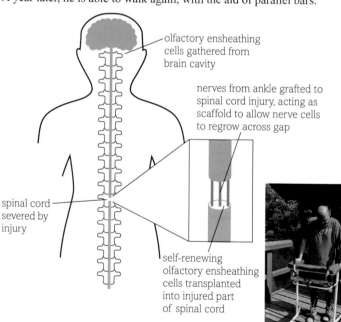

- olfactory ensheathing cells gathered from brain cavity
- nerves from ankle grafted to spinal cord injury, acting as scaffold to allow nerve cells to regrow across gap
- spinal cord severed by injury
- self-renewing olfactory ensheathing cells transplanted into injured part of spinal cord

fig A Darek Fidyka has been able to walk again using a frame and drive an adapted car after receiving pioneering treatment which has repaired his severed spine.

Where else will I encounter these themes?

| Book 1 | 5.1 | 5.2 | 6.1 | 6.2 | 6.3 | 7.1 | 7.2 |

This article comes from the MailOnline, and was written by Ben Spencer, science reporter for the *Daily Mail* and Sarah Griffiths for MailOnline. It was published in October 2014.

1. This is a sensational story – even the doctors and scientists involved in the treatment seemed excited by what happened to Darek Fidyka. Explain how this online version of a popular newspaper sets out to:

   a. Convey the ground-breaking nature of the scientific development.

   b. Give a clear explanation of the procedure.

2. Comment on whether you think the article succeeds in communicating both the excitement and the science, justifying your opinion.

3. a. Explore online and find other descriptions of the same medical breakthrough. Describe four points that are common to all the accounts you look at, and three points that occur in some but not all of them.

   b. Why do you think there are discrepancies in the way the story is reported in different places?

> Being able to communicate scientific ideas to a wide audience is an important skill, particularly when the science has implications for decision making in society.

Now let us examine the biology. You already know about the structure and functions of the cell membrane, diffusion, active transport, the structure and function of the mammalian nervous system and about nervous transmission. This will help you answer the questions below.

4. Describe the structure of the spinal cord and use it to explain how, in a spinal injury:

   a. The amount of paralysis depends on the site of the injury.

   b. Function above the injury is usually unaffected.

   c. There may be a loss of motor function, the loss of sensory function or both.

   d. Autonomic functions such as bladder and bowel control may be lost, as well as consciously controlled movements.

5. Some news articles at the time of the story of Darek Fidyka's recovery described the cells used in the process as 'cells from the nose' or 'stem cells'. Neither of these descriptions was accurate. Explain why not.

6. Sometimes people with partial spinal cord injuries can recover to a greater or lesser extent. When the spinal cord is completely severed, there is usually no recovery. From what you know of nerve cells and using what you have learned from the case of Darek Fidyka, suggest reasons for this.

Activity

Spinal cord injury

Spinal cord injuries can and do happen to anyone of any age, although they are relatively rare in children.

(a) Investigate the main causes of spinal injuries in the UK.

(b) Find out more about research into repairing spinal injuries.

(c) Use your findings to help you write a letter to the research councils arguing for protected funding for spinal injury research.

● Adapted from http://www.dailymail.co.uk/sciencetech/article-2800988/world-man-spinal-cord-severed-walks-paralysed-fireman-recovers-thanks-uk-research.html

Exam-style questions

1   Which of these neurones will transmit an impulse fastest?
  A   myelinated axon 30 µm diameter
  B   myelinated axon 2 µm diameter
  C   unmyelinated axon 30 µm diameter
  D   unmyelinated axon 20 µm diameter   [1]
  [Total: 1]

2   Which of these is a drug that interferes with the action of acetylcholine?
  A   cobra venom
  B   lidocaine
  C   MDMA
  D   rhodopsin   [1]
  [Total: 1]

3   The diagram below shows a human brain seen from the side.

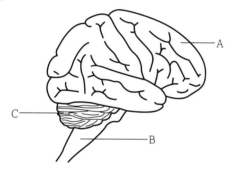

  (a)   Name the parts labelled **B** and **C**.   [2]
  (b)   Give **two** functions of the part labelled **A**.   [2]
  [Total: 4]

4   (a)   When an action potential arrives at a synaptic knob, acetylcholine is released.
          Describe how acetylcholine is released into the synaptic cleft.   [3]
    (b)   The graph below shows a recording of an action potential produced after the binding of acetylcholine to receptors on a post-synaptic membrane.

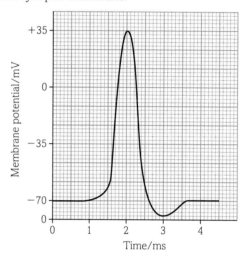

      (i)   Use the graph to state the time at which the sodium channels open to allow an increased flow of sodium ions into the neurone.   [1]
      (ii)   Use the graph to state the time at which the hyperpolarisation is at its greatest.   [1]
      (iii) Calculate the number of action potentials that could occur in one second if the stimulus is maintained. Show your working.   [2]

    (c)   When a transmitter substance called gamma-aminobutyric acid (GABA) is released at a synapse, it causes chloride ion (Cl^-) channels and potassium ion (K^+) channels to open in the post-synaptic membrane. This results in chloride ions moving into the post-synaptic neurone and potassium ions moving out.

          Explain why an action potential is less likely to develop when GABA is released at the same time as acetylcholine.   [2]
  [Total: 9]

5 Myelinated and non-myelinated neurones carry impulses at various speeds. The speed of a nerve impulse along an axon is known as the conduction velocity.

(a) The graph below shows the conduction velocities of myelinated and non-myelinated neurones of different axon diameters.

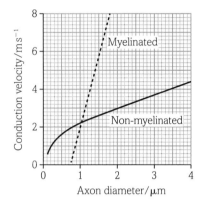

Compare and contrast the conduction velocities of myelinated and non-myelinated neurones. [3]

(b) Explain how the presence of myelin affects the conduction velocity of nerve impulses along an axon. [3]

(c) An investigation was carried out to determine the conduction velocity of a nerve impulse along an axon of a myelinated neurone. A stimulus was applied at a point on the axon that was 25 mm from the synapse.
It took 3.4 milliseconds (ms) for an action potential to arrive at the synapse.
(i) Calculate the conduction velocity in **metres per second**. Show your working. [2]
(ii) Use the information in the graph to give an estimate for the diameter of this axon. [1]

(d) Some snakes produce a toxin that has a similar structure to acetylcholine. When a person has been bitten by a snake, this toxin blocks nerve pathways.

Explain how this toxin could stop post-synaptic neurones from being stimulated. [2]

[Total: 11]

6 (a) (i) Draw a diagram of a synapse. On your drawing label the following structures: pre-synaptic membrane, post-synaptic membrane, synaptic vesicles and mitochondria. [3]
     (ii) Explain the functions of the mitochondria and the synaptic vesicles. [3]

(b) The effects of nicotine from cigarette smoke can be felt in the body within 15 to 30 seconds of inhaling it. Describe how nicotine in cigarette smoke can enter the body and reach the synapse. Explain the effect of nicotine on the synapse. [6]

[Total: 12]

7 Rhodopsin is the light-sensitive pigment contained in rod cells. The diagram below shows a rod cell from the retina of a mammal.

(a) Use the letter **R** to label on the diagram where rhodopsin is found in the rod cell. [1]

(b) When a person enters a dimly-lit room, after being in bright sunlight, objects in the room only gradually become more visible. Give an explanation for this. [3]

[Total: 4]

TOPIC 9
Control systems

CHAPTER
9.3 > Homeostasis

Introduction

Gemsbok are a species of oryx, a type of antelope native to South Africa. These animals have evolved an amazing temperature tolerance that helps them survive in the heat of the African savannah. During the day their core body temperature may rise as high as 45 °C, but at night it can fall as low as 35 °C. This reduces their need to pant or sweat and so greatly reduces their water loss. They also have a specialised blood flow. Blood passes close to the cool nasal passages before it reaches the brain. This cools the blood so that, in spite of the fluctuations of the body temperature, the temperature of the brain of a gemsbok stays much the same day and night.

In this chapter you will learn about homeostasis and how your body maintains a stable dynamic equilibrium. You will consider how the basic control of the heart rate given by the sino-atrial node is fine-tuned by the autonomic nervous system and hormones, so that your heart can respond to both exercise and stress. You will also look at the production of urea in the liver as a waste product of the deamination of excess amino acids. You will then consider the detailed structure of the kidney and discover the mechanisms by which this extraordinary organ controls the concentration of the blood. This includes how, as a result of the counter-current multiplier system in the kidney, the urine you produce may be extremely dilute or more concentrated than your blood plasma. You will investigate how the negative feedback mechanism involving the chemoreceptors and baroreceptors in the brain affects the release of the antidiuretic hormone (ADH), and the effect this hormone has on the distal tubules and the collecting ducts of the kidney.

Finally, you will consider how ectotherms and endotherms regulate their body temperatures. You will consider the similarities and differences between the two groups, and look at adaptations that allow both types of animals to regulate their core temperature. In the case of most endotherms, the core temperature is maintained within a very narrow range.

All the maths you need

- Recognise and use expressions in decimal and standard form (*e.g. when comparing composition of plasma and glomerular filtrate*)
- Construct and interpret frequency tables and diagrams, bar charts and histograms (*e.g. interpret diagrams to show urine production and concentration after drinking water*)
- Translate information between graphical, numerical and algebraic forms (*e.g. interpret data on the relationship between heart rate, stroke volume and exercise*)

What have I studied before?

- The principles of the nervous system
- The production of metabolic waste including urea
- The cell membrane and receptors within the membrane
- Diffusion and osmosis
- Transport across membranes
- The production of ATP in cellular respiration
- Homeostasis as the maintenance of a state of dynamic equilibrium and the importance of negative and positive feedback control in maintaining that equilibrium
- The importance of maintaining pH, temperature and water potential in the body
- The division of the peripheral nervous system into autonomic and voluntary systems, and the further division of the autonomic nervous system into the sympathetic and parasympathetic systems that act antagonistically
- The principles of mammalian hormone production by the endocrine glands
- The structure of the mammalian heart and their relationship to its functions

What will I study later?

- How organisms are adapted to their habitat and this affects their distribution in an ecosystem
- That human activities have an effect on the environment and as a result ecosystems may become drier or wetter, affecting the organisms that live there

What will I study in this chapter?

- The control of the heart rate in mammals by the autonomic nervous system, including the role of baroreceptors and chemoreceptors, the cardiac centre in the medulla oblongata, and the sympathetic and parasympathetic nervous system
- How osmoregulation is brought about in the mammalian body
- The gross and microscopic structure of the kidney
- How urea is produced in the liver and removed from the blood in the kidney by ultrafiltration
- That solutes are selectively re-absorbed in the proximal tubule of the kidney and how the loop of Henle acts as a counter-current multiplier to increase the absorption of water
- How the pituitary gland and osmoreceptors in the hypothalamus combined with the action of ADH bring about negative feedback control of mammalian plasma concentration
- How the kidneys of some desert animals are adapted for life in a very dry environment
- That endotherms maintain their body temperature through metabolic processes, but ectotherms rely on the external environment
- How endotherms use both behavioural and physiological mechanisms to maintain their body temperature

By the end of this section, you should be able to...

● describe homeostasis as the maintenance of a state of dynamic equilibrium through the response of the body to internal and external stimuli

● explain how the autonomic nervous system controls the heart rate in a variety of ways

In **Section 9.1.1** you began to look at homeostasis in mammals. Homeostasis is the maintenance of a state of dynamic equilibrium through the response of the body to internal and external stimuli.

Now you are going to consider some specific examples of different homeostatic systems: the control of the heart rate, maintenance of the pH of the body fluids, osmoregulation and thermoregulation.

Responding to demand

The fine control of the heart rate in mammals demonstrates clearly the importance of having different receptors to internal changes, the ways in which the parasympathetic and sympathetic branches of the autonomic nervous system work in a complementary way in the body, and the interactions between nervous and hormonal control.

In **Book 1 Section 4.3.6** you learned how the intrinsic rhythm of the heart is controlled by impulses initiated in the sinoatrial node (SAN) and then spreads through the atrioventricular node (AVN) and the bundle of His to give a regular, coordinated heartbeat. This intrinsic rhythm, however, is not sufficient to cope with changes in demand. For example, during exercise, more oxygen must be carried to the rapidly respiring muscle tissues and the waste carbon dioxide and lactate that accumulate in the muscle fibres need to be removed. Then, once the exercise stops, your body needs to return to normal. The response of the heart to changes in demand is the result of a number of negative feedback systems.

When your body demands more glucose and oxygen, the heart can respond in two ways. The rate at which the heart beats can increase and the volume of blood pumped at each heartbeat, the **cardiac volume** (otherwise known as stroke volume), can be increased by a more efficient contraction of the ventricles. The combination of these two factors gives a measure called the **cardiac output**:

cardiac output = cardiac volume × heart rate
$dm^3\ min^{-1}$　　　　　dm^3　　　$beats\ min^{-1}$

In a normal individual at rest, the heart beats about 70 times a minute and pumps between 4 and 6 dm^3 of blood per minute. In a trained athlete the resting heart beats more slowly – around 60 beats per minute. When a fit individual anticipates exercise, the heart rate begins to increase before the exercise begins. The cardiac volume increases more slowly, as it becomes clear from the changes in the body that the exercise is going to continue. Cardiac output during exercise can be increased to around 30 $dm^3\ min^{-1}$.

Nervous control of the heart

Most of the nervous control of your heart is by the autonomic (involuntary) nervous system, so you do not have to think about it. The **cardiac control centre**, situated in the medulla oblongata of the brain, is key to the control of changes in the heart rate and the volume of blood pumped with each heartbeat in response to changes in the internal environment.

Chemical, stretch and pressure receptors in the lining of the blood vessels and the chambers of the heart send nerve impulses to the cardiac centre. The cardiac centre responds by sending impulses to the heart along parasympathetic and sympathetic nerves, giving a fine level of control.

Nerve impulses travelling down the sympathetic nerve from the cardiac centre in the brain to the heart release noradrenaline to stimulate the SAN. This increases the frequency of the signals from the pacemaker region, so that the heart beats more quickly. Branches of this sympathetic nerve also pass into the ventricles, so they also increase the force of contraction. In contrast, nerve impulses in the corresponding parasympathetic nerve release acetylcholine, inhibiting the SAN and slowing the heart down.

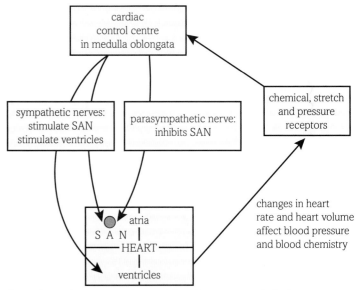

fig A The cardiac centre in the medulla oblongata controls the heart rate via parasympathetic and sympathetic nerve stimulation.

The balance of impulses passing to the heart from the cardiac centre is affected by inputs from a number of different sensory receptors in the main arteries and in the heart itself.

The role of baroreceptors

Baroreceptors found in the sinuses of the carotid arteries in the neck and on the aorta are important in the feedback control of the heart rate during exercise. At rest they send a steady stream of signals back through sensory neurones to the cardiac centre in the brain. When exercise starts, the blood vessels dilate (vasodilation) in response to the hormone adrenaline, which is released in anticipation of exercise, and the blood pressure falls a little. This reduces the stretch on the baroreceptors and they almost stop responding, but when stimulation from the baroreceptors is reduced, the cardiac control centre immediately sends signals along the sympathetic nerve to stimulate the heart rate and increase the blood pressure again by vasoconstriction. When exercise stops, blood pressure in the arteries increases as the heart continues to pump harder and faster than it needs to, and so the baroreceptors are stretched. They respond by sending more sensory nerve impulses to the cardiac centre that in turn sends impulses through the parasympathetic system to slow down the heart rate and cause a widening of the blood vessels. These two actions lower the blood pressure again.

The role of chemoreceptors in the aorta

The walls of the aorta and carotid arteries contain **chemoreceptors** as well as baroreceptors. These are sensitive to the levels of carbon dioxide in the blood. As carbon dioxide levels go up, the pH of the blood goes down and this is detected by the aortic and carotid chemoreceptors. They send impulses along sensory neurones to the cardiac control centre in the medulla, and this increases the impulses travelling down the sympathetic nerve to the heart. As a result the heart rate increases, giving an increased blood flow to the lungs and more carbon dioxide is removed from the blood. As blood carbon dioxide levels fall, the blood pH rises. The chemoreceptors respond to this by reducing the number of impulses to the cardiac centre. This reduces the number of impulses in the sympathetic nerve to the heart and reduces the acceleration of the heart rate so it returns to the intrinsic rhythm. The chemoreceptors are also involved in the control of the breathing rate.

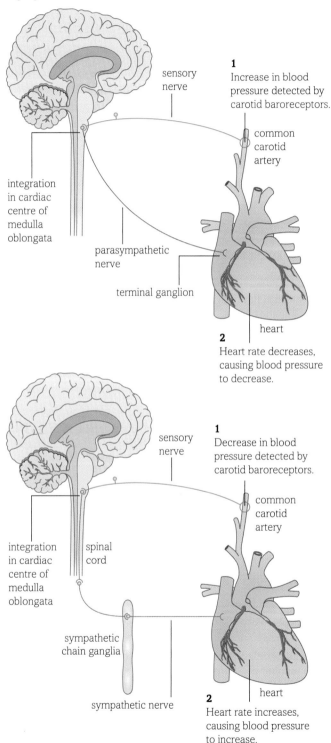

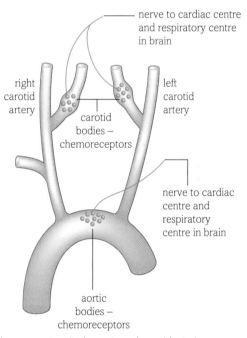

fig C The chemoreceptors in the aorta and carotid arteries.

We also have a certain amount of conscious control over our heart rate, so nerves from the conscious areas of our brains can also stimulate or inhibit the SAN. Some people can slow their heart rate right down just by concentrating on it!

Hormonal control of the heart

You are probably familiar with the way your heart beats faster when you are nervous, frightened or excited, and when you are anticipating exercise. This happens even if you are sitting still at

fig B A negative feedback system for controlling the heart through the baroreceptors – one of the complex interactions that enable the output of the heart to match the demands of the body.

the time, so the change is not in response to exercise. When you are stressed, the sympathetic nerve stimulates the adrenal medulla to release the hormone adrenaline. Adrenaline is very similar to the noradrenaline released in the synapses of the sympathetic nervous system. It is carried around the body in the blood and binds to receptors in the target organs, including the SAN. Adrenaline stimulates the cardiac centre in the brain, increasing the impulses in the sympathetic neurones supplying the heart, and it also has a direct effect on the SAN, increasing the frequency of excitation and so increasing the heart rate, supplying you with extra oxygen and glucose for the muscles and brain, in case you need to run away or stand and fight!

Additional responses

The response of the body to exercise or stress is complex. When you exercise, impulses from the cardiac centre travel to other effectors as well as the heart. When many impulses travel along the sympathetic nerve to the heart to speed it up, fewer impulses are sent along sympathetic nerves to many blood vessels, causing the smooth muscles lining the vessels to contract, narrowing or closing the vessels. In this way the blood flow is diverted from temporarily less important areas to provide more blood for the heart and the muscles to use. As you can see from **table A**, the blood supply to the brain is fairly constant, but the amounts of blood flowing elsewhere in the system vary considerably. The changes seen in **fig D** and **table A** are all the result of a combination of nervous and hormonal responses affecting the heart, the blood vessels and a range of body systems.

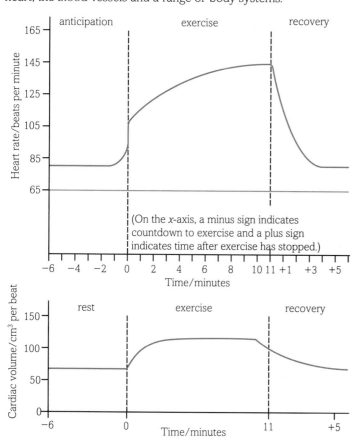

fig D The impact of exercise on the heart rate and on the cardiac volume.

Structure	At rest		Vigorous exercise	
	cm³ min⁻¹	% of total	cm³ min⁻¹	% of total
heart	190	3.3	740	3.9
liver	1340	23.5	590	3.1
adrenal glands	24	0.4	24	0.1
brain	690	12.1	740	3.9
lung tissue	100	1.8	200	1.0
kidneys	1050	18.4	590	3.1
skeletal muscles	740	13.0	12 450	65.9
skin	310	5.4	1850	9.8
other parts	1256	22.0	1716	9.1
total blood flow	5700		18 900	

table A Redistribution of blood flow in response to exercise.

Questions

1 Draw a clear flow diagram to show the negative feedback system involving the baroreceptors that is used in the response of the heart to exercise.

2 Using data from **fig D**, produce a graph to show what you would expect to happen to cardiac output (dm⁻³ per minute) during a period of rest, followed by 10 minutes of exercise and another 10 minutes of recovery time.

3 (a) Using data from **table A**, draw bar charts or pie charts to show the difference in the percentage blood flow to different areas of the body at rest and during exercise.
  (b) Explain why these differences occur.

4 Summarise how nervous and hormonal controls ensure the output of the heart is matched to the demands of the body.

Key definitions

The **cardiac volume** is the volume of blood pumped at each heartbeat.

The **cardiac output** is a measure of the volume of blood pumped by the heart per minute, worked out by the combination of cardiac volume and heart rate.

The **cardiac control centre** in the medulla oblongata receives input from a number of different receptors and controls changes to the heart rate and the cardiac volume through parasympathetic and sympathetic nerves.

Baroreceptors are mechanoreceptors in the aorta and carotid arteries that are sensitive to pressure changes.

Chemoreceptors are sensory nerve cells or organs that respond to chemical stimuli.

By the end of this section, you should be able to...

- describe the gross and microscopic structure of the mammalian kidney

- explain how urea is produced in the liver from excess amino acids and how it is removed from the blood stream by ultrafiltration

- explain how solutes are selectively reabsorbed in the proximal tubules and how the loop of Henle acts as a countercurrent multiplier to increase the reabsorption of water

- explain how the kidney of a kangaroo rat is adapted for life in a dry environment

Water moves into and out of cells by osmosis. If the concentration of water and solutes inside and outside of a cell is not balanced, water may enter the cells by osmosis, causing the cells to swell and burst, or leave the cells by osmosis so the cytoplasm becomes shrunken, concentrated and unable to function (see **Book 1 Section 4.1.3**). **Osmoregulation**, the maintenance of the osmotic potential in the tissues of a living organism within narrow limits by controlling water and salt concentrations, is vital for life.

Animals that live on the land have to drink all the water that they need, so they must conserve water. The cells of a land-living mammal are surrounded by tissue fluid that comes from the blood capillaries (see **Book 1 Section 4.3.9**). By controlling the water potential of the blood (both the water content and the solute concentration), the body can control the water potential of the tissue fluid and so protect the cells from osmotic damage.

In mammals, the main organ involved in the homeostatic control of the water balance of the body is the kidney. The liver is also involved in homeostasis, for example in the breakdown of excess amino acids and the removal of toxins. You will look at the liver first, because urea and other toxins from the liver are removed from the body by the kidneys.

The liver, protein metabolism and homeostasis

The liver has in the region of 500 different functions in the body, many of them involved in homeostasis. It plays an important role in the **deamination** of excess amino acids in protein metabolism. Your body cannot store protein or amino acids and without the action of the liver, any excess protein you eat would be excreted and wasted. The hepatocytes (liver cells) deaminate excess amino acids. They remove the amino group and convert it first into ammonia, which is very toxic, and then to less toxic urea,

which can be excreted by the kidneys (remind yourself of the structure of amino acids in **Book 1 Section 1.2.4**). The ammonia produced in the deamination of proteins is converted into urea by a series of enzyme-controlled reactions known as the **ornithine cycle**. The remainder of the amino acid can then be used in cellular respiration or converted into lipids for storage.

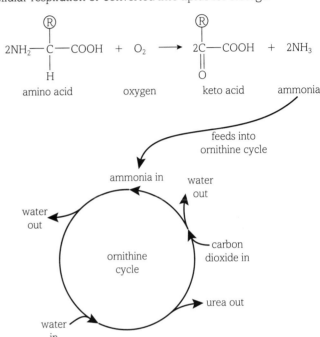

fig A The deamination of excess amino acids and the production of urea in the ornithine cycle of the liver is important for preventing the waste of excess protein from the diet.

Osmoregulation in mammals

Osmoregulation in mammals is largely brought about by the kidneys, a pair of organs capable of producing urine, which is more concentrated than (hypertonic to) the body fluids. This makes it possible to conserve water, an ability that has allowed mammals to spread into most of the land environments of the Earth.

In humans, as in other mammals, the kidneys are a pair of dark reddish brown organs attached to the back of the abdominal cavity. They are surrounded by a thick layer of fat, which helps to protect them from mechanical damage. They control the water potential of the blood that passes through them, removing substances that would affect the water balance as well as getting rid of urea. Blood from the body is passed through the kidneys and the urea, along with excess salts and water, is removed and forms urine. The urine is stored in the bladder and released from the body at intervals.

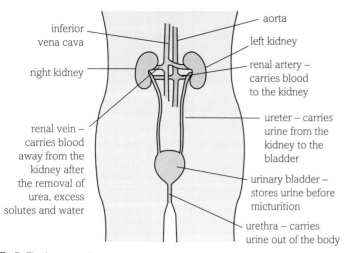

fig B The human urinary system.

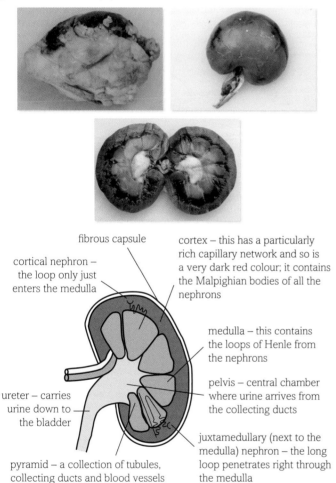

fig C The gross structure of the kidney as seen with the naked eye. The basic structure and position of the two main types of kidney tubules have been superimposed.

The structure and functions of the kidney

The mammalian kidney has two main roles in the body. One is excretion, the removal of urea from the body. The other is osmoregulation. In humans, around 120 cm³ of blood per minute passes through the kidneys, a rate that means all of the blood in your body travels through the kidneys and is filtered and balanced approximately once every hour. Around 180 dm³ of blood is filtered every day!

The kidney carries out three main functions in its osmoregulatory role. These are **ultrafiltration**, **selective reabsorption** and **tubular secretion**.

Each kidney is made up microscopic tubules called **nephrons**, which are 2–4 cm long. There are about 1.5 million nephrons in each kidney, all engaged in filtering and balancing the blood. There are two main types of nephron:

- Cortical nephrons are found mainly in the renal cortex. They have a loop of Henle (a U-shaped tubule) that only just reaches into the medulla. About 85% of human nephrons are cortical.

- Juxtamedullary nephrons have long loops of Henle that penetrate right through the medulla. They are particularly efficient at producing concentrated urine.

The balance of these different types of nephrons varies in different organisms.

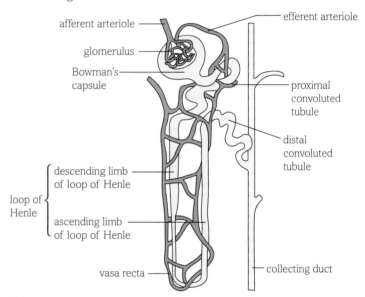

fig D The structure and functions of a nephron.

Ultrafiltration

The first stage in the osmoregulation of the blood is ultrafiltration, which is involved in the formation of the tissue fluid in the body see (**Book 1 Section 4.3.9**). Ultrafiltration in the kidney tubules occurs due to a combination of very high blood pressure in the glomerular capillaries, and the structure of the Bowman's capsule and glomerulus. The glomerulus and the Bowman's capsule together make up the Malpighian body (see **fig E**).

High blood pressure develops in the glomerular capillaries, because the diameter of the blood vessel coming into the glomerulus is greater than that of the blood vessel leaving. The high pressure squeezes the blood out through the pores in the capillary wall – imagine water passing along a hosepipe with holes in it. The size of the pores means that almost all the contents of the plasma can pass out of the capillary – only the blood cells and the largest plasma proteins cannot get through the pores.

The cells of the Bowman's capsule next to the capillaries act as an additional filter. The wall of the capsule is made up of special cells called podocytes. They have extensions called pedicels that wrap around the capillaries, forming slits that ensure any cells, platelets or large plasma proteins that have left the capillary do not get through into the tubule itself. The filtrate that enters the capsule contains glucose, salt, urea and many other substances in the same concentrations as they are in the blood plasma (see **table A**).

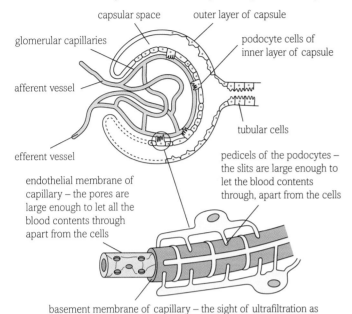

fig E The site of ultrafiltration in the Malpighian body.

Substance	Approximate concentration g/dm³	
	In plasma	In filtrate
water	900.0	900.0
protein	80.0	0.0
inorganic ions	7.2	7.2
glucose	1.0	1.0
amino acids	0.5	0.5
urea	0.3	0.3

table A Comparison of the composition of human plasma and glomerular filtrate.

If all of the filtrate produced in the Malpighian bodies over a 24 hour period was then passed out of the body, we would produce around 200 dm³ of urine a day and would have to drink continually to replace it! In fact, the average daily urine production is 1–2 dm³, so ultrafiltration is only the first step in the process. Most of the filtrate is later reabsorbed into the blood.

Selective reabsorption

Ultrafiltration is passive and indiscriminate. It removes urea from the blood, but it also removes a lot of water along with glucose, salt and other substances that are present in the plasma. Glucose is needed for cellular respiration and is never, under normal circumstances, excreted. Most of the water, salt and other inorganic ions passed into the tubule during ultrafiltration are also needed by the body. After the ultrafiltrate has entered the nephron, the main function of the kidney tubule is to return most of what has been removed from the blood back to its rightful place.

The proximal convoluted tubule

Over 80% of the glomerular filtrate is reabsorbed back into the blood in the first or **proximal convoluted tubule**. The cells lining this tubule are covered with microvilli, which greatly increase the surface area through which substances can be absorbed. The cells also have large numbers of mitochondria, indicating that they are involved in active processes.

All of the glucose, amino acids, vitamins and most hormones are moved back into the blood by active transport. About 85% of the sodium chloride and water is reabsorbed as well. The sodium ions are actively transported, and the chloride ions and water follow passively down concentration gradients. Once the substances are removed from the tubule cells into the intracellular spaces, they then pass by diffusion into the extensive capillary network that surrounds the tubules. The blood is constantly moving through the capillaries, maintaining a concentration gradient for diffusion. By the time the filtrate reaches the loop of Henle it is isotonic with the tissue fluid that surrounds the tubule. The amount of reabsorption that occurs in the proximal tubule is always the same – the fine tuning of the water balance takes place further along the nephron in the loop of Henle.

The loop of Henle

The loops of Henle are found in the medulla of the kidney in close contact with the network of capillaries known as the vasa recta. They create a water potential gradient between the filtrate and the medullary tissue fluid that enables water to be reabsorbed from the distal convoluted tubule and collecting duct. It is this water potential gradient that allows mammals to produce urine that is more concentrated than their own blood.

The creation of the high concentration of sodium and chloride ions in the tissue fluid of the medulla is due to the flow of fluid in opposite directions in the adjacent limbs of the loop of Henle, combined with the different permeabilities of the different sections to water and a region of active transport. This creates a **countercurrent multiplier**, a biological system that uses active transport to set up and maintain concentration gradients. This is key to the ability of the kidney to concentrate the fluid (urine) in the distal proximal tubule and collecting duct. These processes are summarised in **fig F**.

The changes that take place in the descending limb of the loop of Henle depend on high concentrations of sodium and chloride ions in the tissue of the medulla, and these are the result of events in the ascending limb of the loop. The system is explained below:

- *The descending limb*: the descending limb is freely permeable to water but is not permeable to sodium and chloride ions. No active transport takes place here. The fluid entering this limb is isotonic with the blood. As it travels down the limb deeper into the medulla, the external concentration of sodium and chloride ions in the tissue fluid of the medulla and the blood in the vasa recta is higher and higher. As a result, water moves out of the descending limb into the tissue fluid by osmosis down a concentration gradient. It then moves into the blood of the vasa recta, again down a water potential gradient. By the time the fluid reaches the hairpin bend at the bottom of the loop it is very concentrated and hypertonic to the arterial blood.

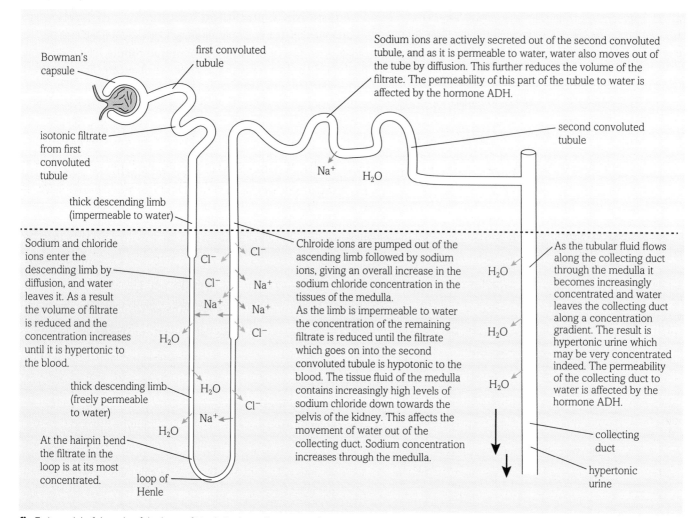

Sodium ions are actively secreted out of the second convoluted tubule, and as it is permeable to water, water also moves out of the tube by diffusion. This further reduces the volume of the filtrate. The permeability of this part of the tubule to water is affected by the hormone ADH.

Sodium and chloride ions enter the descending limb by diffusion, and water leaves it. As a result the volume of filtrate is reduced and the concentration increases until it is hypertonic to the blood.

Chloride ions are pumped out of the ascending limb followed by sodium ions, giving an overall increase in the sodium chloride concentration in the tissues of the medulla. As the limb is impermeable to water the concentration of the remaining filtrate is reduced until the filtrate which goes on into the second convoluted tubule is hypotonic to the blood. The tissue fluid of the medulla contains increasingly high levels of sodium chloride down towards the pelvis of the kidney. This affects the movement of water out of the collecting duct. Sodium concentration increases through the medulla.

As the tubular fluid flows along the collecting duct through the medulla it becomes increasingly concentrated and water leaves the collecting duct along a concentration gradient. The result is hypertonic urine which may be very concentrated indeed. The permeability of the collecting duct to water is affected by the hormone ADH.

At the hairpin bend the filtrate in the loop is at its most concentrated.

fig F A model of the role of the loop of Henle in the reabsorption of water and the production of concentrated urine in the kidney.

- *The ascending limb*: the first section of the ascending limb is very permeable to sodium and chloride ions but not permeable to water. No active transport takes place in this section. Sodium and chloride ions move out of the very concentrated fluid in the loop of Henle into the tissue fluid of the medulla down concentration gradients. The second, thicker section of the ascending limb is also impermeable to water, but sodium and chloride ions are actively pumped out of the tubule into the tissue fluid of the medulla and the blood of the vasa recta. This gives the tissues of the medulla the very high sodium and chloride ion concentration that causes the water to pass out of the descending limb by osmosis. However, the ascending limb is impermeable to water, so water cannot follow the chloride and sodium ions out down the concentration gradient. As a result, the fluid left in the ascending limb becomes less concentrated.

Distal convoluted tubule

The **distal convoluted tubule** is permeable to water, but the permeability varies with the levels of **antidiuretic hormone (ADH)**. It is here, and in the collecting duct, that the balancing of the water needs of the body takes place (see **Section 9.3.3**). If there is not enough salt in the body, sodium may be actively pumped out of the tubule with chloride ions following. Water also leaves by diffusion, if the walls of the tubule are permeable.

The collecting duct

The permeability of the **collecting duct** is also strongly affected by the hormone ADH. The water moves out of the collecting duct down a water potential gradient as it passes through the medulla, with its high levels of sodium and chloride ions, and the urine becoming steadily more concentrated. Because the concentration of sodium ions in the surrounding fluid increases through the medulla towards the pyramids, water may be removed from the collecting duct all the way along. This allows for the creation of very hypertonic urine when it is necessary to conserve water for the cells of the body.

The urine

The urine is the fluid that is produced by the kidney tubules. It is collected first in the pelvis of the organ. It then passes along the ureters to the bladder, where it is stored until the bladder is sufficiently stretched to stimulate micturition. The urine passes out of the body along a tube called the urethra. The volume of urine produced is very variable, depending on both what is taken into the body and the activity levels.

The urine contains variable amounts of water and salts, depending on the diet and the demands of the body, along with relatively large quantities of urea. The colour of the urine varies from almost colourless to deep yellow/brown, depending on its concentration. The colour can also be affected by certain foods, for example, eating a lot of beetroot can cause pink urine. Substances such as glucose or protein should never appear in the urine. If they do, this indicates problems elsewhere in the body, the pancreas for example, or shows that the kidneys themselves are not working properly.

Desert animals spend most or all of their lives short of water. In some extreme examples, such as the kangaroo rats (*Dipodomys* spp.), they never drink. A combination of behavioural, anatomical and physiological adaptations enable them to survive. Kangaroo rats spend much of their time in burrows below the surface of the desert, where the temperature is both cooler and more stable. This reduces the resources they use to maintain a stable body temperature (see **Section 9.3.4**, Thermoregulation). They generate up to 90% of the water they need by the oxidative reactions in their cells – as you know from **Book 1**, many metabolic processes involve condensation reactions, which produce water. In contrast, humans can only get 12% of the water they need from their metabolic reactions. Kangaroo rats get the rest of the water they need from water contained in their food.

In spite of not drinking, kangaroo rats must still remove waste products such as urea, so they need to produce tiny amounts of very concentrated urine. In fact, they produce urine with a concentration of more than 6000 mosmol/kgH$_2$O. In contrast, humans can concentrate their urine to around 1400 mosmol/kgH$_2$O and camels to 2800 mosmol/kgH$_2$O. Scientists are still unsure exactly how the kidneys of kangaroo rats produce such concentrated urine, but they have a number of adaptations that seem to contribute to this ability:

- a relatively large proportion of juxtamedullary nephrons;

- relatively long loops of Henle – although the kangaroo rat actually has relatively shorter loops of Henle than some other animals that cannot produce such concentrated urine. Recent research suggests that although the loops of Henle themselves may not be longer than those in other species, the region of the thin descending loop that is permeable to water may be longer in members of this species and this may contribute to their ability to form a very high concentration of ions in the medulla, which in turn makes it possible to produce very highly concentrated urine;

- higher numbers of infoldings in the cell membranes of the epithelial cells lining the tubules, which give increased surface area for diffusion of inorganic ions and water, making steep concentration gradients possible;

fig G A number of adaptations, both anatomical, physiological and behavioural, enable kangaroo rats to live in the desert without drinking.

- high numbers of mitochondria, and mitochondria with densely arranged cristae for maximum cellular respiration, are found in the epithelial cells of the nephrons of kangaroo rats, providing the energy for the active pumping of inorganic ions into or out of the tubules as required.

It appears a combination of these factors and others enable kangaroo rats to retain the maximum amount of water and produce very small quantities of extremely concentrated urine.

Questions

1 What is osmoregulation and why is it important in mammals?

2 How are the homeostatic functions of the liver linked to the osmoregulatory functions of the kidney?

3 Explain how the kidney filters the blood and produces urine that may be more concentrated than the blood.

4 Discuss the anatomical and physiological adaptations of kangaroo rats that enable them to survive in the desert without drinking water.

Key definitions

Osmoregulation is the maintenance of a constant osmotic potential in the tissues of a living organism by controlling water and salt concentrations.

Deamination is the removal of the amino group from excess amino acids in the ornithine cycle in the liver. The amino group is converted into ammonia and then to urea, which can be excreted by the kidneys.

The **ornithine cycle** is the series of enzyme-controlled reactions that convert ammonia from excess amino acids into urea in the liver.

Ultrafiltration is the process by which fluid is forced out of the capillaries in the glomerulus of the kidney into the kidney tubule through the epithelial walls of the capillary and the capsule.

Selective reabsorption is the process by which substances needed by the body are reabsorbed from the kidney tubules into the blood.

Tubular secretion is the process by which inorganic ions are secreted into or out of the kidney tubules as needed to maintain the osmotic balance of the blood.

Nephrons are microscopic tubules that make up most of the structure of the kidney.

The **proximal convoluted tubule** is the first region of the nephron after the Bowman's capsule, where over 80% of the glomerular filtrate is absorbed back into the blood.

A **countercurrent multiplier** is a system that produces a concentration gradient in a living organism, using energy from cellular respiration.

The **distal convoluted tubule** is the section of the nephron after the loop of Henle that leads into the collecting duct, where balancing the water needs of the body takes place.

Antidiuretic hormone (ADH) is a hormone produced in the hypothalamus and stored in the posterior pituitary that increases the permeability of the distal convoluted tubule and the collecting duct to water.

The **collecting duct** takes urine from the distal convoluted tubule to be collected in the pelvis of the kidney. It is the region of the kidney where most of the water balancing needed for osmoregulation takes place.

By the end of this section, you should be able to...

- explain how the pituitary gland and osmoreceptors in the hypothalamus, combined with the action of antidiuretic hormone (ADH), bring about negative feedback control of mammalian plasma concentration

The kidney is involved in the balance of both water and solutes in the body. Urea is produced continuously by metabolic processes and the kidney plays a key role in removing it from the body. However, levels of other important substances vary according to the situation of the individual. A long walk on a hot, sunny day, a salty meal or drinking several pints of liquid all threaten the equilibrium of the body. How is the functioning of the kidney controlled to bring about homeostasis?

Osmoregulation

The osmotic potential of the blood is maintained within narrow boundaries by balancing the water and salts taken in by eating and drinking with the water and salts lost by sweating, defaecation and in the urine. It is the concentration of the urine that is most important in this dynamic equilibrium, and this is controlled by a negative feedback system involving antidiuretic hormone (ADH).

ADH is produced by the hypothalamus and secreted into the posterior lobe of the pituitary where it is stored (see **Sections 9.1.2** and **9.1.3**). ADH increases the permeability of the distal convoluted tubule and the collecting duct to water.

Mechanism of ADH action

The mechanism by which ADH increases the permeability of the walls of the distal convoluted tubule and the collecting duct to water is very elegant. ADH does not cross the membrane of the tubule cells. It binds to specific receptors, triggering reactions that result in the formation of cAMP as the second messenger (see **Section 9.1.1**). The cAMP sets up a series of reactions that cause vesicles within the cells lining the tubules to move to, and fuse with, the cell membranes. The vesicles contain water channels, which are inserted into the membrane, making it permeable to water. Water then moves through the channels out of the tubules and into the surrounding blood capillaries by osmosis.

The amount of ADH released controls the number of channels that are inserted, so the permeability of the tubules can be very closely controlled to match the water demands of the body. When ADH levels fall, levels of cAMP also drop and the water channels are withdrawn from the membranes and repackaged in vesicles. This makes the tubule impermeable to water once again – and the channels are stored ready for reuse when needed.

ADH and negative feedback control

When water is in short supply or you sweat a lot as a result of exercise or eat a very salty meal, the concentration of inorganic ions in the blood rises so its water potential becomes more negative. If this continued, the osmotic balance of the tissue fluids would become disturbed, causing cell damage (see **Book 1 Section 4.1.3**). This is prevented by a negative feedback system involving ADH.

An increasingly negative water potential in the blood is detected by **osmoreceptors** in the hypothalamus. They send nerve impulses to the posterior pituitary, which in turn releases stored ADH into the blood. The ADH is picked up by receptors in the cells of the kidney tubules. ADH increases the permeability of the distal convoluted tubule and the collecting duct to water. As a result, water leaves the tubules by osmosis into the surrounding capillary network. This means more water is returned from the filtrate to the blood, and a small volume of concentrated urine is produced.

When large amounts of liquid are taken in, the blood becomes more dilute – its water potential becomes less negative. Again the change is detected by the osmoreceptors of the hypothalamus, and in this case the release of ADH by the pituitary is inhibited. The walls of the distal convoluted tubule and the collecting duct remain impermeable to water and so little or no reabsorption takes place. Therefore, the concentration of the blood is maintained and large amounts of very dilute urine are produced (see **fig A**).

It is very easy to test the effectiveness of this system – simply drink about a litre of water over a short period of time and wait for the results!

Extra feedback

The release of ADH is also stimulated or inhibited by changes in the blood pressure. These changes are detected by the baroreceptors in the aortic and carotid arteries, which also help control the heart rate (see **Section 9.3.1**). A rise in blood pressure (often a sign of an increase in blood volume) will suppress the release of ADH and so increase the volume of water lost in the urine. This in turn reduces the blood volume and so the blood pressure falls.

A fall in blood pressure, which may indicate a loss of blood volume, causes an increase in the release of ADH from the pituitary and the conservation of water by the kidneys. Water is returned to the blood and a small amount of concentrated urine is produced. This is part of the normal dynamic equilibrium of the body, but it also plays an important role if you lose a lot of blood for any reason.

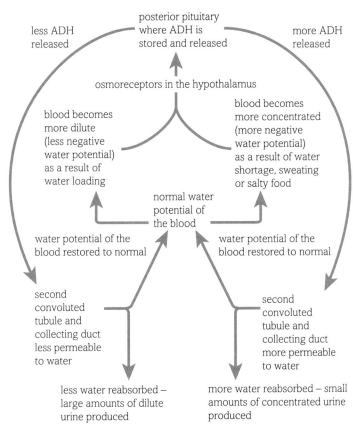

fig A A negative feedback system involving ADH maintains the osmotic potential of the blood within very narrow limits.

Diabetes insipidus

The most common form of diabetes, diabetes mellitus, is the result of insufficient insulin being produced. The name literally means 'sweet fountain', because large volumes of urine containing sugar are produced.

There is, however, another form of diabetes – a relatively rare condition known as **diabetes insipidus**. The name means 'dilute fountain' and affected individuals continuously produce large volumes of very dilute urine. It is caused when an individual does not produce any ADH, or their kidneys do not respond to ADH. Without ADH, the distal convoluted tubules and collecting ducts are permanently impermeable to water. The patient feels extremely thirsty and has to drink large quantities of liquid to avoid severe dehydration. Diabetes insipidus is treated in different ways, depending on the severity of the condition and the cause, either with drugs that replace the ADH or with drugs that enable the kidneys to produce a more concentrated urine.

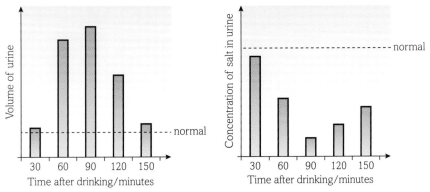

fig B These graphs show the effect of drinking a given volume of distilled water on both the volume and salt concentration of the urine. Urine was collected at 30 minute intervals from the time that the water was drunk. This shows the sensitivity of the response to an imposed change – the water load is removed, but without an unwanted loss of salt.

ADH is not the only hormone that affects the kidney tubules and controls the amount of urine formed in response to changes in the body. There is another major homeostatic system involving the kidney that helps control the salt concentration of the body. It involves the hormone aldosterone, which is produced by the cortex of the adrenal glands. Aldosterone causes the active uptake of sodium ions from the filtrate in the kidney tubules into the plasma in the capillaries. Water follows by osmosis, increasing the blood volume and as a result increasing the blood pressure.

If sodium ions are lost from the blood, for example after a lot of sweating, then the water potential of the blood becomes less negative and water tends to be lost from the blood into the tissue fluid and cells. This causes a slight drop in blood pressure, which is detected by a group of cells in the kidney itself. When the blood pressure drops, these cells produce an enzyme called rennin. Rennin acts on a protein in the blood to produce the hormone angiotensin. Angiotensin stimulates the release of aldosterone from the adrenal glands, triggering the active uptake of sodium ions from the filtrate to the blood. Once aldosterone is being produced, the adrenal glands are also stimulated by adreno-corticotrophic hormone from the pituitary gland.

This is a complex system which helps to fine-tune the concentration of sodium ions and the volume of the blood.

Questions

1   Produce a flowchart to show the role of ADH in osmoregulation:
    (a) if you drink two litres of water
    (b) if you have a fever and sweat more than normal.

2   (a) How does ADH increase the permeability of the distal convoluted tubule and the collecting duct to water?
    (b) Explain why treatments for diabetes insipidus often involve a drug that mimics ADH.

3   ADH and aldosterone are both important in controlling the volume and ion balance of the blood. Discuss how their actions complement each other.

Key definitions

Osmoreceptors are sensory receptors in the hypothalamus that detect a change in the concentration of inorganic ions, and therefore changes in the osmotic potential of the blood.

Diabetes insipidus is a relatively rare condition caused by lack of ADH in the body or an inability of the kidney to respond to ADH. The main symptoms are constant production of large volumes of dilute urine and constant thirst.

4 ▶ Thermoregulation

By the end of this section, you should be able to...

- explain that endotherms regulate their temperature through metabolic processes, but ectotherms must rely on the external environment

- explain how an endotherm is able to regulate its temperature through behaviour and also physiologically through the autonomic nervous system, including the role of thermoreceptors, the hypothalamus and the skin

The chemical reactions in cells only take place within a relatively narrow range of temperatures. This is largely due to the sensitivity of the enzymes that control the reactions. For example, once temperatures rise above 40 °C, most enzymes are denatured as their protein structure is destroyed. Many organisms have evolved ways to control their internal body temperature in one way or another – **thermoregulation** is important to their survival and is a major aspect of homeostasis.

How do living organisms lose or gain heat?

The surface temperature of an animal or plant can change rapidly, but it is the internal core temperature that is relevant to enzyme activity. Living organisms are continually cooling down or warming up their surroundings; most of the ways in which they do this are affected by the size of the animal. Small animals have a large surface area-to-volume ratio, so they transfer energy more rapidly than larger organisms.

fig A Animals employ various methods, such as fluffing their feathers, to maintain their body temperature as external conditions change.

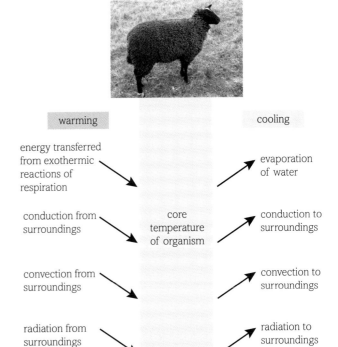

warming		cooling
energy transferred from exothermic reactions of respiration		evaporation of water
conduction from surroundings	core temperature of organism	conduction to surroundings
convection from surroundings		convection to surroundings
radiation from surroundings		radiation to surroundings

fig B The warming and cooling experienced by an organism. It is the balance of these that determines the core temperature. Organisms use a variety of means to shift the balance and allow them to get warmer or cooler as needed.

There are several ways in which organisms warm up or cool down:

- They *warm up* as a by-product of metabolism. Chemical inefficiency means that energy is wasted, which warms the core of the organism.

- They *cool down* by the evaporation of water from the body surfaces. A certain amount of cooling always takes place in this way from the mouth and respiratory surfaces of land-dwelling animals, and this cannot be controlled. Sweating and behavioural patterns such as wallowing can increase this cooling.

- Energy may be transferred to or from the environment by radiation (the transfer of energy in the form of electromagnetic waves). Infrared radiation is the most important form of energy absorbed and radiated by animals.

- Energy may be transferred to or from the environment by convection (the transfer of energy by currents of air or water). Convection currents are set up around relatively hot objects, so adaptations to prevent cooling by these currents are common in animals.

• Energy may be transferred to or from the environment by conduction (the transfer of heat by the collision of molecules). Conduction is particularly important between organisms and the ground or water, as air does not conduct well. Neither does fat, so adipose (fatty) tissue is a very valuable insulator, preventing energy exchanges with the environment.

How do animals control their body temperature?

Not all animals need to control their body temperature. Protoctists and simple animals living in water have no means of temperature regulation. For organisms living in large water masses such as the sea, this raises no difficulties, because the temperature of their environment is very stable. Similar organisms living in small ponds have more problems. They either develop tolerance to temperature fluctuations or have survival strategies for adverse conditions, such as forming cysts and emerging when the temperature is suitable again.

However, larger animals in a wide variety of land habitats need to regulate their temperatures, either to avoid damage to their cells or to enable them to have an active way of life. Animals are often classified according to the main mechanisms of temperature control as either **endotherms** or **ectotherms**.

Endotherms

An endotherm relies on its own metabolic processes to provide at least some warming and usually has a body temperature higher than the ambient temperature. They regulate their body temperature in a number of different ways. They are adapted to conserve their body temperature when necessary and also take advantage of warmth from the environment when possible. This means that there are few environments where they cannot survive, and the main groups of endotherms, the mammals and birds, are found in a very wide variety of environmental niches. They can cope with high external temperatures and can also live in areas with very low ambient temperatures where ectotherms would not be able to stay active. In order to maintain their body temperature against adverse environmental conditions, the metabolic rate of endotherms has to be high. On average, endotherms have a metabolic rate five times higher than an ectotherm of the same size. This means they have to eat considerably more food to supply their metabolic needs.

Ectotherms

Ectotherms rely heavily on the external environment to control their body temprature. They usually have behavioural and structural modifications that take advantage of the environment in order to maintain a reasonably steady temperature. When cold they may bask in the sun, press themselves to warm surfaces or have special areas of skin to erect in order to maximise their absorption of the radiation from the sun. They may move into the shade, or into water or mud in order to cool down if necessary.

Endotherms also modify their behaviour to aid thermoregulation. There is more about this in **Section 9.3.5**. By using a variety of strategies, some ectotherms maintain a body temperature that is almost as stable as that of an endotherm, although they are always more vulnerable to fluctuations. Because their metabolic rate is relatively low, ectotherms need less food, which is an advantage in many environments where food may be in short supply.

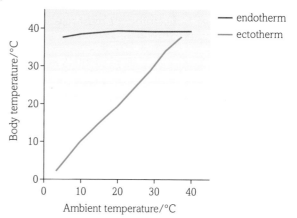

fig C Graph to show the core body temperature of an endotherm and an ectotherm over a range of external temperatures.

Thermoregulation in endotherms

We will consider mammals as our main example of endotherms. Temperature control in humans is a good example of homeostasis, as we are endotherms that regulate our body temperature within a very narrow range. The main source of warming is from our metabolism, but humans survive in almost every area of the world and have effective ways of both losing and conserving energy.

We have a wide variety of behavioural mechanisms allowing us to warm up or cool down as needed. The major difference between humans and other mammals with respect to temperature regulation is, of course, that we can manipulate our environment to help us to survive. People build houses, light fires and wear clothes to help prevent cooling when it is cold. We make air-conditioning and build swimming pools to keep ourselves comfortable in high temperatures. In spite of these behavioural adaptations, we still maintain all our physiological adaptations for temperature regulation and they play the major role in the maintenance of our core body temperature. All we can do with our technology is reduce the extremes with which we have to deal.

The skin

The major homeostatic organ involved in thermoregulation in most endotherms is the skin. Some cooling takes place as water evaporates from the surfaces of the mouth and nose, and this cannot be prevented because of their moist surfaces. The skin, however, has an enormous surface area that can be modified either for cooling or keeping warm (see **fig D**).

The skin is the largest single organ. It covers the entire surface of the body in a waterproof layer, providing protection both against mechanical damage and from the ultraviolet radiation of the sun. The surface area of your skin is about 2 m² and your skin makes up about 16% of your body weight!

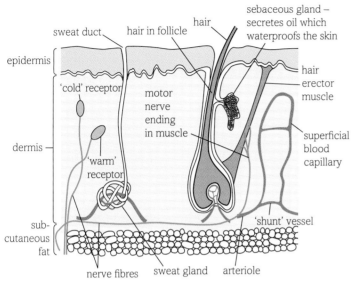

fig D The human skin. The same major structures are present in the skin of all other mammals too. Many of the features are involved in thermoregulation.

Keeping cool

The skin plays a major homeostatic role in maintaining the core body temperature when you are exercising or the external temperature is high. It helps prevent overheating in a number of ways. A rich supply of capillaries runs near to the surface of the skin. Cooling by radiation, convection and conduction to the environment takes place from the blood flowing through the skin. This cooling is controlled via the **arteriovenous shunt**. When you are exercising or the external temperature rises, the shunt is closed, which allows more blood to flow through the capillaries at the surface of the skin and **vasodilation** occurs. As a result the skin appears red and flushed, and more energy is transferred to the environment by radiation.

When you are exercising or in a hot environment, the erector pili muscles, which are attached to hair follicles, are relaxed and the body hairs lie flat against the body, minimising any insulating air layer that is trapped next to the skin. This has little or no impact on cooling in human beings, because we have so little body hair, but it is important in other mammals.

The rate of sweat production in the sweat glands increases when the core temperature starts to increase. As more sweat is released onto the skin surface, cooling takes place as the water evaporates. Almost 1 dm³ of sweat is usually produced and lost in a day, but this can rise to 12 dm³ under very hot conditions. This means it is very important to keep drinking plenty of water when you are getting hot, so that the tissues remain hydrated and sweat can be produced.

Subcutaneous fat acts as insulation, reducing cooling. People who are very physically active, such as elite athletes, tend to have very little subcutaneous fat, as they use up all the energy from their food. This reduces the insulation and increases the amount of energy that can be lost by conduction from the surface of the skin. Very overweight people overheat easily – a disadvantage when they start to exercise, as it can put them off!

Keeping warm

Homeostatic mechanisms also act when the core temperature starts to fall. Some of the energy conservation measures are the exact opposite of the cooling responses of the skin. So the arteriovenous shunt in the blood supply to the skin opens, reducing the blood flow through the capillaries. This is called **vasoconstriction**, and it reduces energy lost from the surface of the skin.

Sweat production is reduced and so cooling by evaporation is reduced too. The erector pili muscles are contracted, pulling the hairs upright. In humans this is visible as 'goose-pimples' and has little effect on temperature regulation, but in hairy mammals or feathered birds it traps an insulating layer of air that helps to reduce cooling.

The metabolic rate of the body also speeds up, warming the body. This takes place particularly in the liver and the muscles. Shivering, involuntary contractions of the skeletal muscles, also helps with metabolic warming. The energy released raises the body temperature. This is particularly important in the successful emergence of animals from hibernation and for temperature control in very young human babies.

Animals living in cold areas develop thick layers of subcutaneous fat that act as an effective insulator against cooling.

In a warm environment	In a cold environment
• vasodilation occurs • sphincter muscles around arterioles leading to superficial capillaries are not stimulated to contract and therefore relax • more blood can flow into these capillaries, dilating them with the pressure; less blood flows through deeper shunt vessels • more blood flows close to the body surface • as more blood flows close to the body surface, the temperature gradient between the body surface and the environment becomes steeper, so cooling by conduction and radiation is increased.	• vasoconstriction occurs • sphincter muscles around arterioles leading to superficial capillaries contract • this constricts the passage into these capillaries and more blood flows through deeper shunt vessels • less blood flows close to the body surface • as most blood is diverted further from the body surface, the temperature gradient between the body surface and the environment is less steep, so cooling by conduction and radiation is reduced.

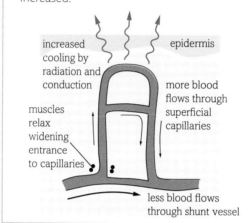 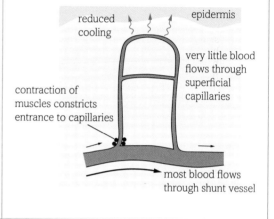

fig E The role of the superficial blood vessels in thermoregulation.

Control of the core (blood) temperature

In a homeostatic feedback system there need to be receptors sensitive to changes in the system. In the case of thermoregulation there are two types of receptors. Receptors in the brain directly monitor the temperature of the blood, while receptors in the skin detect changes in the external temperature. This allows for great sensitivity, not only to actual changes in the core temperature but to potential changes too. The temperature receptors in the brain are sited in the hypothalamus, which effectively acts as the thermostat of the body. As a result of sensitive feedback mechanisms, the temperature of the human body is usually controlled within a 1°C range.

When the temperature of the blood flowing through the hypothalamus increases, the **thermoregulatory centre** is activated and sends out impulses along autonomic motor nerves to effectors that increase the blood flow through the skin and increase sweating. The erector pili muscles are relaxed so that the hairs lie flat, and any shivering stops. The metabolic rate may be reduced to lower the amount of warming in the body. The response is the same, whether the increase in core temperature comes from warming due to internal factors during exercise or a fever, or external factors such as a warm room.

If the temperature of the blood flowing through the hypothalamus drops, the thermoregulatory centre reacts by sending nerve impulses through the autonomic nervous system to the skin. These cause a reduction in the blood flow through the capillaries in the skin, along with a reduction in the production of sweat and contraction of the erector pili muscles to raise the hairs. Impulses in autonomic motor neurones from the thermoregulatory centre also stimulate involuntary muscle contractions (known as shivering) and raise the amount of metabolic warming. Thus the core temperature is usually maintained within very narrow limits (see **fig F**).

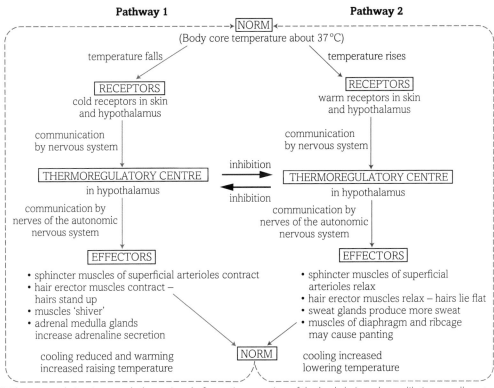

fig F Thermoregulation in an endotherm. Level of sweating, erection of the body hair and vasodilation are all controlled by negative feedback systems and each has an impact on the core temperature.

Questions

1 Why is thermoregulation such an important aspect of homeostasis?

2 What is (a) an endotherm and (b) an ectotherm?

3 Produce a table to show how a mammal regulates its body temperature when:
   (a) the external temperature drops
   (b) the external temperature rises.

4 Some endotherms give birth to immature, bald babies that are kept in nests or dens lined with moss, fur or feathers. Other endotherms produce relatively mature offspring that are often very fluffy.
   (a) What is the main thermoregulatory problem for young endotherms?
   (b) How do the adaptations described above help them to survive?

Key definitions

Thermoregulation is a homeostatic mechanism that enables organisms to control their internal body temperature within set limits.

Endotherms are animals that warm their bodies through metabolic processes at least in part and usually have a body temperature higher than the ambient temperature.

Ectotherms are animals that are largely dependent on the external environment for their body temperature.

An **arteriovenous shunt** is a system which closes to allow blood to flow through the major capillary networks near the surface of the skin, or opens to allow blood along a 'shortcut' between the arterioles and venules, so it does not flow through the capillaries near the surface of the skin.

Vasodilation is the widening of the blood vessels by relaxation of their muscle walls, increasing blood flow

Vasoconstriction is the narrowing of the blood vessels by contraction of their muscle walls, reducing blood flow.

The **thermoregulatory centre** comprises temperature receptors sited in the hypothalamus in the brain that act as the thermostat of the body and fire when the temperature of the blood flowing through the hypothalamus increases or decreases.

By the end of this section, you should be able to...

● explain how an endotherm is able to regulate its temperature through behaviour, and also physiologically through the autonomic nervous system, including the role of thermoreceptors, the hypothalamus and the skin

Cooling and its control is an important element in homeostatic temperature regulation. The rate of cooling is related to the surface area-to-volume ratio of an animal, and this affects the type of endotherms found in different environments. Endotherms also show a wide variety of behavioural and physiological adaptations that help them regulate their body temperatures.

Cold environments

In a cold environment, animals are usually larger than their equivalent in a warmer climate. This reduces their surface area-to-volume ratio and so reduces their rate of cooling. Animals also develop thick layers of fat beneath the skin as insulation against cooling. They also tend to have small extremities, such as ears, as parts that stick out are particularly vulnerable to cooling. Finally, some mammals and birds living in cold external conditions have developed countercurrent exchange systems in their limbs that allow body temperature to be conserved. The arterial blood is cooled as it flows to the extremity and the venous blood is warmed as it returns to the body, minimising loss to the environment (see **fig A**).

arterial blood flows into the foot through the artery deep in the centre of the leg

thermal energy is transferred from the incoming arterial blood to the outgoing venous blood in a counter-current exchange system

veins returning blood to the body run very close to the artery

capillaries in the foot

fig A A countercurrent exchange system enables ducks to walk on ice without their core body temperature falling below a critical level.

Some endotherms cannot generate enough metabolic warming for them to survive in cold conditions. These are generally small animals living in Arctic or temperate regions, where the winter temperatures are much lower than the summer ones. This problem can be overcome through **hibernation**, which usually involves both behavioural and physiological adaptations. Before hibernation, animals usually dig or find a sheltered place and line it with leaves, fur or other similar materials. They also eat more than usual, which results in a thick layer of insulating fat that prevents cooling and acts as a food store during hibernation. In hibernation, the animal goes into a very deep sleep. Its metabolic rate slows right down and the core body temperature is greatly lowered and then maintained at this lower level, making substantial energy savings. A few endotherms, including bats, allow their body temperature to follow that of the external environment. This saves even more energy, but if the temperature drops too low the tissues might freeze, killing the animal.

Animals usually go into hibernation as a result of both low temperatures and a shortening of the day length. Warm temperatures and lengthening days bring them out of hibernation again. The metabolic rate needs to speed up very quickly, so the animal can feed and to make sure it does not become easy prey for predators. Stores of brown fat (fat with a particularly rich blood supply) are saved during hibernation and used up quickly to produce metabolic warming at the end of hibernation.

Hot environments

In a hot environment, cooling is not a problem, so animals do not need to be particularly large in order to keep warm. In fact, one of the problems for animals in very hot areas is to cool sufficiently to prevent overheating. This is why the extremities of many of these animals are large and thin with a rich blood supply. The large, thin ears and many skin folds of an elephant are a good example. Energy can be transferred to the environment relatively easily through adaptations such as these.

Some endotherms, such as camels, simply cannot cool down enough to maintain a steady internal temperature, and they have evolved an ability to tolerate much larger fluctuations in body temperature than most organisms. Living in the desert, camels' major priority is to conserve water. They do this very effectively, largely because they do not sweat. This saves lots of water but it also removes a major way of cooling down. As a result, the body temperature of a camel climbs to around 40 °C during the day, but then falls to a low of 34 °C at night. By tolerating these fluctuations in temperature, the camel allows osmoregulation to occur successfully and also prevents temperature damage to its cells. The low night temperature ensures that the temperature does not climb too high during the day.

Cooling the brain

In most mammals, the brain is particularly susceptible to damage as a result of overheating. A countercurrent exchange system enables them to prevent this damage (see **fig B**). This type of physiological adaptation for cooling is valuable in many mammals, but it is of particular importance and is extremely well-developed in animals such as the camel and the oryx, which live in conditions of heat stress.

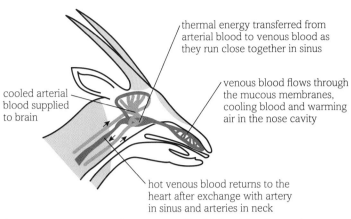

thermal energy transferred from arterial blood to venous blood as they run close together in sinus

cooled arterial blood supplied to brain

venous blood flows through the mucous membranes, cooling blood and warming air in the nose cavity

hot venous blood returns to the heart after exchange with artery in sinus and arteries in neck

fig B A countercurrent exchange system like this involving the nasal passages can prevent brain damage due to overheating and so make more environments available as habitats.

Control of body temperature by behaviour

Physiological controls are not the only thermoregulatory method available to animals. Modification of the behaviour can lead to major changes in temperature. Endotherms use behaviour to control their body temperature to some extent, but ectotherms are particularly good at modifying their behaviour to control their body temperature – in fact, it is their major homeostatic mechanism. There are several major and very obvious behavioural modifications that are used by both endotherms and ectotherms to help maintain body temperatures within viable limits:

- Basking – desert lizards are among many groups of animals that will bask in the sun when the body temperature is tending to fall. By orientating the body differently to the sun, or by erecting special areas of tissue that have evolved to absorb radiation from the sun, the body temperature may be raised to a level that makes rapid activity and, in birds, even flight.

- Sheltering – when the core temperature is rising too high, many animals, both endotherms and ectotherms, shelter from the direct warmth of the sun in burrows, holes or crevices in rocks. They may also attempt extra cooling by conduction, pressing the body against the cool earth.

- Evaporation – animals can increase evaporation and therefore cooling from the body by panting and so exposing the moist tissues of the mouth, by licking the body surface or wallowing in mud or water.

- Moving into or out of the sun – one of the most common behavioural means of regulating body temperature. It is used by both endotherms and ectotherms as part of their homeostatic control system.

The Cape ground squirrel, *Xerus inauris,* found across southern Africa, shows several of these behavioural adaptations, which enable it to feed in the middle of the day, unlike many other small mammals. It uses its flattened tail to keep its body shaded from direct sunlight. This keeps the body temperature down so that it can feed for long periods of the day. Only at the peak of the daily temperatures do the animals retreat to cool underground burrows, demonstrating another behavioural way of controlling the body temperature.

fig C The Cape ground squirrel continues to feed in the heat of the day in the Kalahari, by using its tail as a portable parasol providing shade.

One extreme physiological adaptation for avoiding hot, dry conditions is **aestivation**. This is very similar to the hibernation used by some animals to overcome particularly cold conditions. When animals aestivate, their metabolic rate slows right down, they become completely inactive and remain torpid (dormant) for months at a time. They often bury themselves in mud, which dries out completely, or hide in crevices in rocks. The African snail, *Helix desertorum*, has been known to aestivate for up to five years! It is not only molluscs that aestivate – animals from frogs and crocodiles to lungfish and snakes have been known to use this survival mechanism to avoid severe hot, dry conditions. It is rare in mammals, but long-eared hedgehogs use the technique to survive desert summer temperatures.

So thermoregulation is an important part of homeostasis in animals, including ourselves. And when conditions are too hot or too cold for animals to control their temperature, some of them at least have mechanisms by which they can avoid the problem completely.

Questions

1 Countercurrent exchange mechanisms are physiological adaptations that can prevent some of the problems caused by extreme temperatures. Explain how they work, with examples.

2 Both endothermic and ectothermic animals are found in a wide range of habitats, both hot and cold. Describe some of the ways in which endotherms regulate their temperatures in both extremes.

Key definitions

Hibernation is the state when an animal goes into a very deep sleep to avoid very cold conditions. The metabolic rate slows and the core body temperature is greatly lowered, making substantial energy savings.

Aestivation is an extreme physiological adaptation for avoiding hot, dry conditions, similar to hibernation. The metabolic rate slows right down and the animals become completely inactive and torpid for months, buried in dried mud or in rock crevices.

THINKING BIGGER

THE COLD KILLER

People, like other warm-blooded animals, fight a constant battle to maintain a constant body temperature, regardless of the temperature of their surroundings. At the beginning and towards the end of our lives we can find this very difficult.

HYPOTHERMIA

Hypothermia is diagnosed as a lowering of the normal core temperature of the human body from 37 °C to below 35 °C, but even before that the body will begin to malfunction. In hypothermia, the skin turns greyish-blue and pale, with a puffy face and blue lips. The skin is cold to the touch, even in places where it is almost always warm such as the armpits. The sufferer will become drowsy and confused, with very slow reflexes and slurred speech, but they are very unlikely to complain of feeling cold – that sensation is long past. The breathing will be slow and shallow, the heart rate falls and the blood pressure drops. If the temperature continues to fall, the person will become unconscious and eventually die. People recover completely from hypothermia as long as they are warmed up gently – sudden heating can cause a severe fall in blood pressure and death.

The ageing trap

As we get older, our thermoregulatory systems become less efficient. Older people tend to be less physically active due to health problems or simply the choice of a more sedentary lifestyle, so they generate less metabolic warmth. Older people are more likely to be lonely after bereavement or poorer than average, making it more likely that they will not cook and eat the proper meals which would help keep them warm. The poverty suffered by some older people means they do not always heat their homes adequately. All of these factors mean that older people have a much raised risk of developing hypothermia. Every winter in the UK several hundred old people die of hypothermia, and thousands more die of illnesses which they might well have shrugged off in the summer.

Not just the old...

Although hypothermia is a particular problem for the elderly, it can affect anyone.

fig A Young people can die out on mountains and wild country such as Dartmoor, if the weather changes fast and they are not properly prepared. The confusion that sets in with the onset of hypothermia means sensible route planning or survival strategies disappear.

In severe frosty or snowy weather, hypothermia kills people stranded in cars, particularly if they try to walk to get help when the blood is moved from the body core to the muscles and is cooled in the cold body tissues. This makes the temperature drop faster and often brings unconsciousness before help is found. If people have been drinking alcohol, hypothermia sets in even faster, as the vasodilation of the skin caused by the drug means cooling happens faster than ever. The best way to survive if stranded in the cold is to put on all the clothing you have with you – most importantly cover your head, as that is often the most exposed part of your body, stay in the car or find a small enclosed space, curl up so that you have the minimum surface area exposed to the cold and if there is more than one of you huddle together. It really is important to do this, because cold kills the old – and the young, and the teenagers…

Where else will I encounter these themes?

Book 1 | 5.1 | 5.2 | 6.1 | 6.2 | 6.3 | 7.1 | 7.2

This extract is from a different biology textbook.

1. Make a bullet pointed list of:
   a. The main symptoms of hypothermia.
   b. How to avoid hypothermia if you find yourself stranded in cold conditions.
2. How effective is writing like this at giving you the information you need to deal with hypothermia in either an elderly neighbour, or if you and a group of friends become lost or stranded in wintry conditions on a walk in the country or after a night out?
3. This content was used at the beginning of a chapter on homeostasis, including thermoregulation, to give some context for the biology to come. Discuss the advantages and disadvantages of using an article like this to stimulate interest at the beginning of a topic in a textbook.

Now let us examine the biology. You already know about the effect of temperature on reaction rates and enzyme structures, the main metabolic reactions of the body, the nervous and hormonal controls of the body and thermoregulation in endotherms and ectotherms. This will help you answer the questions below.

4. Reflect on the symptoms of hypothermia and try to explain what is happening biologically and why the body reacts as it does.
5. Give biological explanations for the following.
   a. If someone is suffering from hypothermia, they need to be warmed up gently rather than fast.
   b. You should never give someone suffering from hypothermia alcohol or a very hot drink, although a warm drink is fine.

Activity

Premature babies

A baby in the uterus does not have to control its own temperature – that is done by the maternal control systems. If a baby is born prematurely, it struggles to regulate its temperature.

(a) Discover why premature babies (born earlier than the 37th week of pregnancy) cannot easily regulate their own body temperature.

(b) Find out how, in areas such as Europe and North America, incubators overcome many of these problems.

(c) Investigate kangaroo care, a way of looking after premature babies that is effective and easily used everywhere, including areas with poor access to technology such as Africa.

With this information, plan an advertising campaign to raise funding for a global education programme on how to increase the survival rates of premature babies.

Think about the level of science you should present in your campaign, keeping in mind the target audience. However much scientific detail you choose to include, make sure that it is all factually correct!

● *Heinemann Advanced Science: Biology* by Ann Fullick

Exam-style questions

1 There are a number of structures involved in spreading the wave of depolarisation over the atria and ventricles.
(1) bundle of His
(2) ventricular muscle cells
(3) sino-atrial node
(4) atrio-ventricular node

In what order are these structures depolarised?
A (3)–(4)–(2)–(1)
B (3)–(4)–(1)–(2)
C (4)–(3)–(1)–(2)
D (4)–(2)–(3)–(1) [1]
[Total: 1]

2 A diet cola tastes sweet because it contains an artificial sweetener, aspartame, rather than glucose syrup or sucrose. Aspartame is very sweet so very little needs to be present and the diet cola has a very high water potential. If you consume a large bottle of this drink your body will respond in a certain way. Which of these best describes the way your body responds?
A More ADH is produced, increasing the volume of urine.
B Less ADH is produced, decreasing the volume of urine.
C Less ADH is produced, increasing the volume of urine.
D More ADH is produced, decreasing the volume of urine. [1]
[Total: 1]

3 Describe the roles of each of the following in the production of concentrated urine in a mammal.
(a) The loop of Henle. [5]
(b) Antidiuretic hormone. [4]
[Total: 9]

4 (a) At high environmental temperatures, the rate of sweating in humans increases.
Explain how sweating is involved in the regulation of body temperature. [2]

(b) In an investigation, a healthy volunteer measured his body temperature.

After 5 minutes, he got into a bath of water at a temperature of 18 °C.

He stayed in the bath for 10 minutes, then got out and sat on a chair.

During the investigation, he recorded his body temperature at regular time intervals.

The results of this investigation are shown in the following table.

Time/min	Activity	Body temperature/°C
0	Started investigation	37.0
5	Got into bath	36.9
10	Lying in bath	36.7
15	Got out of bath	36.5
20	Sitting on a chair	36.8
25	Sitting on a chair	37.0

(i) Describe the changes in body temperature that occurred during this investigation. [3]
(ii) Suggest explanations for the changes in body temperature that occurred between the following time intervals:
5–10 minutes
15–25 minutes [3]
[Total: 8]

5 (a) Name the region of the human brain involved in control of heart rate. [1]

(b) Heart rate increases during exercise. Explain the mechanisms involved in controlling this increase in heart rate. [4]
[Total: 5]

6 (a) The diagram below represents a nephron (kidney tubule).

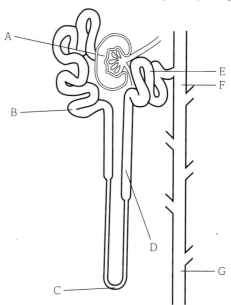

(i) Name the parts labelled **A** and **B**. [1]
(ii) All the glucose in region **A** is reabsorbed back into the bloodstream as the fluid in the nephron passes from region **A** to region **B**. Explain how this glucose reabsorption occurs. [2]

(b) The graph below shows the concentration of solutes in
the fluid in the nephron in each of the labelled regions
shown in the diagram. The graph shows the concentration
of solutes when there is a high level of ADH (antidiuretic
hormone) in the blood and when there is a low level of
ADH in the blood.

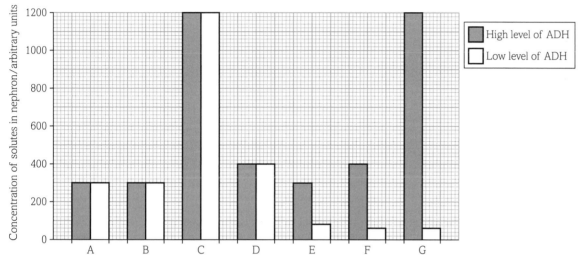

(i) Calculate the percentage decrease in the concentration
of solutes between regions **A** and **G** when there is a
low level of ADH in the blood. Show your working. [3]

(ii) The concentration of solutes in the fluid changes as
it passes from region **A** to region **G**. Compare and
contrast the changes that occur when the level of ADH
in the blood is high with changes that occur when the
level of ADH is low.   [3]

(iii) Use the information in the graph to explain how a rise
in the level of ADH results in the production of a more
concentrated urine.   [3]

[Total: 12]

TOPIC 10
Ecosystems

10.1 ▶ The nature of ecosystems

Introduction

In the UK, we have many species that are rare and becoming rarer, many species that are rare but are becoming more common, and many other species that are simply not well understood. Increasingly, research teams are recruiting the help of 'citizen scientists'. These are ordinary people who get involved in projects as diverse as the Royal Society of Biology's Flying Ant Day, through to monitoring the remaining populations of natterjack toads or Glanville fritillary butterflies, or trying to record the elusive New Forest cicada. Apps, web resources and smartphone images mean scientists can be sure that their citizen scientists are making correct identifications, because if the identification is wrong or uncertain, the information collected is of no value to anyone. Similarly, if the habitat is damaged or destroyed during a survey, this will mean any data collected is not of value.

In this chapter you will learn about the nature of ecosystems. You will define the term and look at the variations in size. You will learn what is meant by a trophic level and the different ways that the organisms at each level feed. You will look at different ways of representing ecosystem structures, and you will consider the advantages and disadvantages of each. You will discuss the transfer of energy and biomass through an ecosystem.

You will discover a variety of techniques that can be used to measure the abundance and distribution of organisms in an ecosystem, and gain practical experience of as many of them as possible. You will learn to select the best ecological technique to use according to the ecosystem and organisms you are studying, as slow-moving animals and plants require different techniques to faster moving, flying or swimming organisms. Finally, you will learn several statistical techniques that can be used to analyse the data you collect.

All the maths you need

- Construct and interpret frequency tables and diagrams, bar charts and histograms (*e.g. interpret diagrams to show secondary succession*)
- Translate information between graphical, numerical and algebraic forms (*e.g. interpret data increase in biodiversity during colonisation*)
- Determine the intercept of a graph (*e.g. when looking at data on two different populations*)
- Understand the principles of sampling as applied to scientific data (*e.g. when analysing data from field investigations*)
- Select and use a statistical test (*e.g. the Student's t-test or the Spearman's rank correlation coefficient*)

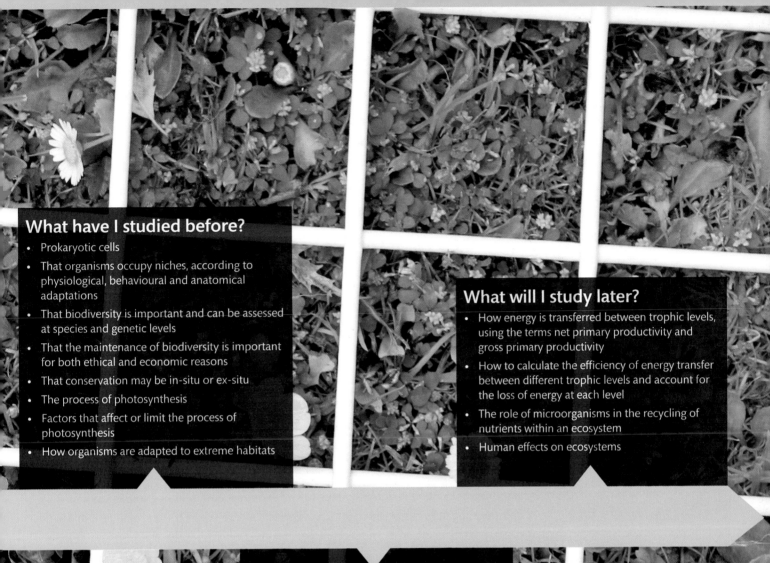

What have I studied before?

- Prokaryotic cells
- That organisms occupy niches, according to physiological, behavioural and anatomical adaptations
- That biodiversity is important and can be assessed at species and genetic levels
- That the maintenance of biodiversity is important for both ethical and economic reasons
- That conservation may be in-situ or ex-situ
- The process of photosynthesis
- Factors that affect or limit the process of photosynthesis
- How organisms are adapted to extreme habitats

What will I study later?

- How energy is transferred between trophic levels, using the terms net primary productivity and gross primary productivity
- How to calculate the efficiency of energy transfer between different trophic levels and account for the loss of energy at each level
- The role of microorganisms in the recycling of nutrients within an ecosystem
- Human effects on ecosystems

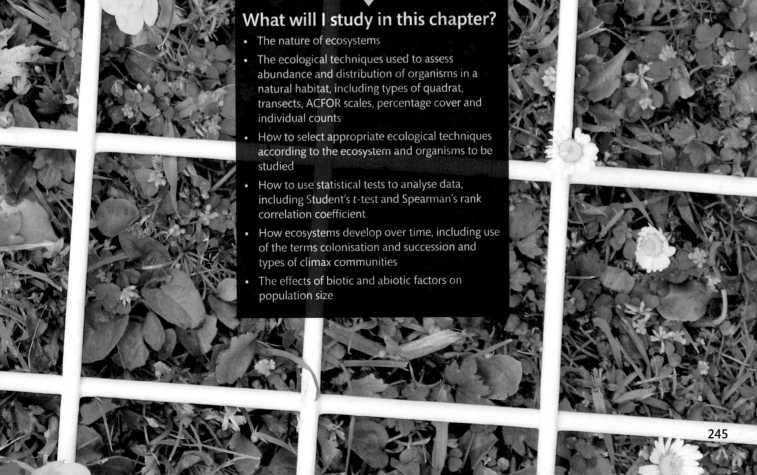

What will I study in this chapter?

- The nature of ecosystems
- The ecological techniques used to assess abundance and distribution of organisms in a natural habitat, including types of quadrat, transects, ACFOR scales, percentage cover and individual counts
- How to select appropriate ecological techniques according to the ecosystem and organisms to be studied
- How to use statistical tests to analyse data, including Student's t-test and Spearman's rank correlation coefficient
- How ecosystems develop over time, including use of the terms colonisation and succession and types of climax communities
- The effects of biotic and abiotic factors on population size

10.1 ▶ 1 ▶ What is ecology?

By the end of this section, you should be able to...

● explain what is meant by the term ecosystem and how ecosystems range in size

Your study of adaptation, biodiversity and endemism in **Book 1 Topic 3** will have shown you that real ecology is a long way from the emotive picture of the 'environment' often portrayed in the media. Ecology is the study of the interactions between organisms and their environment. These interrelationships determine the distribution and abundance of organisms within a particular environment. The word ecology comes from the Greek '*oikos*', meaning 'house'. Put simply, ecology is the study of living things in their home environment.

What is an ecosystem?

An **ecosystem** is a life-supporting environment. It includes all of the living organisms, which interact together, the nutrients that cycle throughout the system, and the physical and chemical environment in which the organisms are living. An ecosystem consists of a network of habitats and the communities of organisms associated with them.

Useful terms

Like any other area of scientific study, ecology has its own very specific terms. Some of these are reminders from your **Book 1** studies. Others will be important in this part of your course.

● A **habitat** is the place where an organism lives, such as a stream, a tropical rainforest or a sand dune. You can think of the habitat of an organism as its address. Many organisms live only in a small part of a habitat – a single fig on a tree may be home to a fig wasp for example. Such habitats are referred to as **microhabitats**.

● A population is a group of organisms of the same species, living and breeding together in a habitat. The three-spined sticklebacks in a particular pond or the dust mites in your mattress are examples.

● A **community** is all the populations of all the different species of organisms living in a habitat at any one time. For example, in a habitat such as a rock pool, the community consists of populations of different seaweeds, sea anemones, small fish such as gobies, shrimps and crabs and other crustacea, molluscs such as mussels and barnacles, and many other species. More details about a particular community may be given in the name, such as the soil community, or the animal community in the soil.

● In **Book 1 Section 3.2.1** you learned that the niche of an organism can be described as the role of the organism in the community, or its way of life. If the habitat is the address of the organism, the niche describes its profession. Several organisms can share the same habitat, occupying different niches. So for example, the fox occupies the food niche in a woodland as the top predator, squirrels occupy the large tree-dwelling omnivore niche, and rabbits fill the large ground and burrow-living herbivore niche.

● **Abiotic factors** are the non-living elements of the habitat of an organism. They include those related to the climate, such as the amount of sunlight, temperature and rainfall, and those related to the soil (edaphic factors), including the drainage and the pH. In aquatic habitats the oxygen availability in the water is very important. Abiotic factors have a big effect on the success of an organism in a particular habitat.

● **Biotic factors** are the living elements of a habitat that affect the ability of a group of organisms to survive there. For example, the presence of suitable prey species will affect the numbers of predators in a habitat.

fig A There are many ecosystems in the biosphere, from the Arctic wastes to lush tropical forests, oceans and caves.

fig B These minute protists live in a microhabitat on the palps of a freshwater shrimp. The magnification is × 500.

Biomes – the major ecosystems

The **biosphere** could be considered as the largest ecosystem on Earth. However, it is so large that it is very difficult to study as a whole. So it is divided into smaller parts, distinguished by their similar climates and plant communities. These major ecosystems or **biomes** are shown in **fig C**. Biomes are generally subdivided into smaller ecosystems again for ease of study.

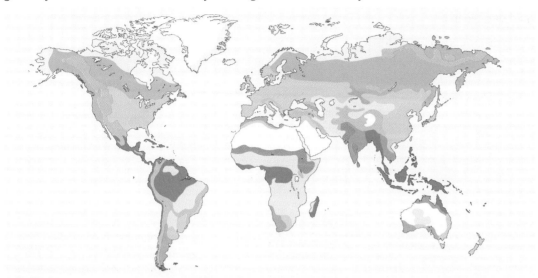

fig C Map of the world showing the major land biomes.

Colour	Name of biome	Description of biome	Level of biodiversity
	tropical rainforest	high humidity (rain all year), warm and plenty of sunlight	very high
	tropical seasonal forest	drier than tropical rainforest, warm, sunny	high
	savannah	dry tropical grassland	medium
	tropical woodland	wetter than savannah, grassland with thornwoods, bushes and trees	more than savannah
	desert	very little rainfall, often extremes of temperature between day and night	very low
	temperate grassland	warm dry temperate areas, e.g. prairies, steppes and pampas	medium
	temperate shrublands	hot dry summers and cool wet winters	medium
	temperate forests	warm moist regions, including deciduous and conifers	less than tropical rainforest
	taiga	evergreen forests in cold subarctic and subalpine regions	low
	tundra	very cold, arctic and high mountian regions	very low
	high mountain	very cold, high altitude	very low
	polar ice	very cold, little available water	very low

Key definitions

An **ecosystem** is an environment including all the living organisms interacting within it, the cycling of nutrients and the physical and chemical environment in which the organisms are living.

A **habitat** is the place where an organism lives.

A **microhabitat** is a small area of a habitat.

A **community** is all the populations of all the different species of organisms living in a habitat at any one time.

Abiotic factors are the non-living elements of the habitat of an organism.

Biotic factors are the living elements of a habitat that affect the ability of a group of organisms to survive there.

The **biosphere** is all of the areas of the surface of the Earth where living organisms survive.

Biomes are the major ecosystems of the world.

Questions

1   How does the habitat of an organism differ from its niche? Give examples to illustrate your answer.

2   Choose three of the major biomes of the Earth:
   (a) Find out about the climate of the biome, including the range of temperatures experienced and the level of rainfall.
   (b) Find out about two plants and two animals found in each biome you have chosen and their adaptations to the conditions.
   (c) Using **fig C**, link the water availability and temperatures in the biomes you have chosen to the level of biodiversity found.

247

The major biomes of the Earth have developed over thousands or even millions of years from bare rock into the ecosystems of today. This has been brought about by **succession**, a process by which communities of animals and plants colonise an area and then over time are replaced by other, usually more varied, communities.

Primary succession

Primary succession starts with an empty inorganic surface, such as bare rock or a sand dune. This type of succession can be seen after a volcanic eruption or landslide, or after the emergence of a new volcanic island. The first organisms are opportunists or pioneer species such as algae, mosses and fungi. These organisms can penetrate the rock surface, helping to break it into small grains, and trap organic material that will break down to form humus. The inorganic rock grains and the organic humus are the start of the formation of soil.

Once there is soil, other species such as grasses and ferns can establish root systems. The action of their roots, and the humus they form when they die and decay, add to the soil. As the soil layer develops, more water and nutrients are retained and become available for plant roots, and so less hardy species can survive. Gradually larger plants can be supported and the diversity of species increases. As plant biodiversity increases, so the diversity of animals that can be supported also increases. Eventually a **climax community** is reached, where the biodiversity and range of species are generally constant. A climax community is self-sustaining and is usually the most productive group of organisms that the environment can support. Primary succession must have been of prime importance in the formation of the biosphere, but today it is found in only a few places, such as the island of Surtsey.

In 1916, F.E. Clements proposed the idea that climate is the major factor in determining the make-up of the climax community in a particular place. He said that for any given climate, there was only one possible climax community, and that this should be known as the **climatic climax community**. This view has been modified over time as scientists learn more about ecosystems and recognise that many factors interact to determine any given climax (see later in this chapter). So a climatic climax community is now defined as one that remains generally constant over time.

In the modern landscape there are many examples of another type of climax community – a **plagioclimax**. These are constant and self-sustaining, but they are not truly natural. They are a final community that is partly the result of human intervention. Humans have changed the landscape, for example by clearing woodland and grazing animals, and this has changed the ultimate climax community. Examples include chalk grassland and lowland heaths. Probably all of the modern British countryside consists of plagioclimax communities rather than climatic climaxes. If the limiting factors are removed, if people move away, for example, a climatic climax community will eventually develop.

Secondary succession

Secondary succession is the development of an ecosystem from existing soil that is clear of vegetation. It occurs as rivers shift their courses, after fires and floods, and after disturbances caused by humans. The sequence of events is very similar to that seen in primary succession, but because the soil is already formed and contains seeds, roots and soil organisms, the numbers of plants and animals present right from the beginning of the succession are much higher. Simply digging a patch of earth and leaving it is sufficient to observe the beginnings of a secondary succession.

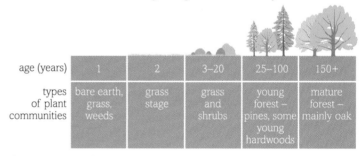

age (years)	1	2	3–20	25–100	150+
types of plant communities	bare earth, grass, weeds	grass stage	grass and shrubs	young forest – pines, some young hardwoods	mature forest – mainly oak

fig A The stages of a secondary succession from bare earth to oak woodland. The timescale is very approximate.

The time it takes to go from an area of bare earth to a climax community varies enormously. It depends on many different factors, including temperature, rainfall and the underlying soil fertility. However, a succession of different types of plants and animals is always seen. The climax community formed will depend not only on the climatic factors, but also on the plants, animals and microorganisms that are within or able to colonise the area. This can mean that a secondary climax community differs from the original primary climax community, as we see when virgin rainforest is cut down and the area left to regenerate naturally.

Observing succession is not always easy, because it is a process that occurs over a long time. However, sand dunes can show a complete record of the stages of the succession. For example, at Gibraltar Point on the east coast of England the oldest dunes (those furthest from the sea) are in the late stages of the succession. Nearest to the sea are the youngest dunes, in the very earliest stages of colonisation (see **fig B**).

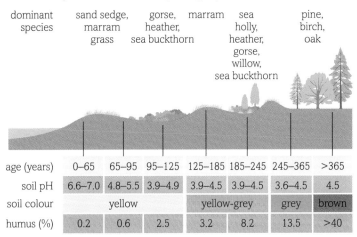

dominant species	sand sedge, marram grass	gorse, heather, sea buckthorn	marram	sea holly, heather, gorse, willow, sea buckthorn		pine, birch, oak
age (years)	0–65	65–95	95–125	125–185	185–245	245–365	>365
soil pH	6.6–7.0	4.8–5.5	3.9–4.9	3.9–4.5	3.9–4.5	3.6–4.5	4.5
soil colour	yellow			yellow-grey		grey	brown
humus (%)	0.2	0.6	2.5	3.2	8.2	13.5	>40

fig B The gradual change from sand to a more mature soil, the maturing topography and the differences in the plants and animals making up the communities populating each habitat can readily be observed on the sand dunes at Gibraltar Point.

Surtsey – a case study in succession

On 14 November 1963, a volcanic eruption burst through the surface of the sea off the coast of Iceland. The column of steam and ash could be seen for miles and the hot magma cooled rapidly to form a new island. By 1967, when the eruptions stopped, the island covered 2.7 km², a huge area of bare volcanic rock.

While the island was forming, a group of researchers got together, determined to make the most of the rare opportunity to observe the primary succession of completely new land. The Surtsey Research Society has had an overview of all the research carried out on the island since then. The colonisation and succession sequences of the island have been carefully recorded and photographed. Access to the island has been carefully controlled and no tourism has been allowed, so the succession of plant and animal life has been as natural as possible.

The earliest colonisers were moulds, bacteria and fungi, followed by flowering plants. This surprised the ecologists. Initially all the plants died over winter but after permanent plant communities established, gull colonies followed. The bird droppings added fertility to the soli. Many plant species flourished including willow trees and orchids, and many different species of invertebrates and birds now live there too.

There have been a number of fascinating discoveries, some showing that earlier predictions about primary succession do not always occur and that the process varies, depending on the availability of colonising organisms.

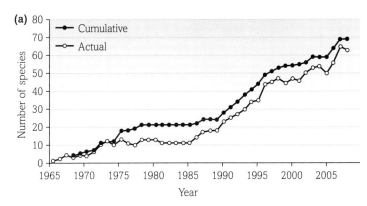

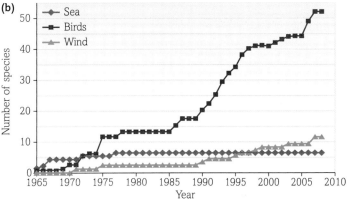

fig C Graph (a) shows the increase in the biodiversity of plants with the passage of time, while graph (b) shows the different dispersal routes that have brought plant species to the island of Surtsey.

Questions

1 What is meant by the term ecological succession?

2 What is meant by a climax community and do such communities always look the same?

3 Describe the difference between primary and secondary succession.

4 Use the data in **fig C** to answer these questions:
  (a) Discuss the information you get from **fig C(a)** (the curve showing the actual number of species) and **fig C(b)** about the natural succession process on Surtsey.
  (b) Why is there a difference between the actual and cumulative number of plant species observed on Surtsey?
  (c) If the island develops as the scientists predict, how would you expect these curves to develop over the next 20 years?

Key definitions

Succession is the process by which the communities of organisms colonising an area change over time.

A **climax community** is a self-sustaining community with relatively constant biodiversity and species range. It is the most productive group of organisms that a given environment can support long term.

A **climatic climax community** is the only climax community possible in a given climate.

A **plagioclimax** is a climax community that is at least in part the result of human intervention.

By the end of this section, you should be able to...

● explain the effects of abiotic factors on population size

It is important to develop an understanding of how different factors affect living things and determine the distribution and population size of organisms in a particular habitat. The community of organisms in a habitat, whether in the early stages of colonisation or in a climax community, is controlled by both abiotic (non-living) and biotic (living) factors. The interaction of these factors results in different ecological niches, which change as the factors change. Understanding the abiotic and biotic factors that affect living organisms can help us predict the effect of changes on an ecosystem.

Abiotic factors can vary a great deal within a habitat to produce **microclimates**. These provide different niches and so determine the distribution and abundance of different populations within the habitat. For example, logs may be placed for seating on a grass area. Although the area will be largely dry and well lit, under a log it will be damp and shady, allowing very different organisms to grow and thrive compared with those in the grassy areas.

Light

The amount of light (energy input) in a habitat has a direct effect on the numbers of organisms found there. Plants are dependent on light for photosynthesis, so any plant populations that are going to thrive in habitats with low light levels must be able to cope with this factor. For example, some plants reproduce early to avoid the shade caused by larger plants. Other plants are able to photosynthesise and reproduce successfully in low light levels, often by having extra chlorophyll or a different ratio of photosynthetic pigments that are sensitive to lower light levels. This enables these plants to thrive in a niche where other plants would die. Another strategy is to have very large leaves to collect light.

Animals are affected by light levels indirectly as a result of the distribution of food plants. Seasonal light changes also affect reproductive patterns in many animals and the 24-hour cycle of light and dark imposes order on the circadian rhythms that control much of animal physiology and behaviour.

Temperature

Every organism has a range of temperatures within which it can grow and successfully reproduce. Reproduction does not occur above or below that range, even if the individual organism survives. It is the extremes of temperature that determine where an organism can live, not the average.

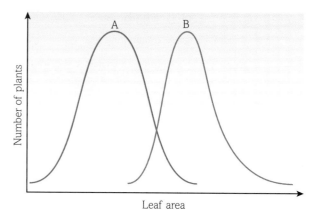

fig A Some plants cope with a shady situation by growing larger leaves. Curve A shows the range of leaf area in a nettle population grown in the open. Curve B shows that the leaves of nettles grown in the shade have on average a considerably larger area and are less likely to show the full range of sizes in the normal distribution curve.

The temperature of the environment particularly affects the rate of enzyme-controlled reactions in plants and ectothermic animals. In some areas of the world, the daytime temperatures can exceed the temperature at which endotherms can normally control their body temperatures (see **Section 9.3.5**). Many animals have evolved behaviours and physiological adaptations that enable them to cope. Organisms without these adaptations do not survive.

Wind and water currents

Wind has a direct effect on organisms in a habitat. Wind increases water loss from the body and cooling and so adds to the environmental stress on an organism. Fewer species can survive in areas with strong prevailing winds, while occasional gales and hurricanes can devastate populations. Whole woodlands may be destroyed and the communities of plant and animal life within them lost (see **fig B**).

fig B The damage seen here was caused by very high winds which swept across the south of the UK in the autumn of 1987. Within 20 years, recovery was almost complete.

In water currents, organisms have to flow with the current, be strong swimmers or be able to attach to a surface and resist the force of the water. Currents are most damaging to populations when the strength increases suddenly, such as when flooding occurs.

Water availability

In a terrestrial environment, the availability of water is affected by several factors including the amount of precipitation (rain, sleet, snow or hail), the rate of evaporation, and edaphic factors such as the rate of loss by drainage through the soil. Water is vital for living organisms, so where the supply is limited it will cause severe problems.

If the water stress becomes too severe organisms will die unless, like camels, kangaroo rats and cacti, they have special adaptations to enable them to survive and reproduce in very dry conditions. Equally, an increase in the availability of water can lead to a huge change in a habitat and to a massive increase in population size of some organisms. For example, in deserts after a little rain has fallen the seeds of many desert plants germinate, grow and flower in a very short space of time in the phenomenon known as the 'flowering of the desert'. This in turn provides a food bonanza for insects and other animals that normally just manage to survive in the harsh conditions, so there is a population explosion all round!

Oxygen availability

Oxygen can be in short supply in both water and soil. When water is cold or fast flowing, sufficient oxygen dissolves in it to support life. If the temperature of the water rises, or it becomes still and stagnant, then the oxygen content will drop, often affecting the survival of populations within it. Soil is also usually a well-aerated habitat. The spaces between the soil particles contain air, so there is plenty of oxygen for the respiration of plant roots. However, in waterlogged soil the air spaces are filled with water and the plant roots may be deprived of oxygen. The plants may die as a result. Some plants, like mangroves, have special adaptations such as aerial roots that allow them to thrive in waterlogged conditions (see **Book 1 Section 4.2.5**).

Edaphic factors: soil structure and mineral content

Edaphic factors relate to the structure of the soil and can affect the various populations associated with it. Sand has a loose, shifting structure that allows very little to grow on it. Plant populations that are linked by massive root and rhizome networks, such as marram grass, can and do survive on sand. They not only reproduce successfully but also bind the sand together, which makes it more suited for colonisation by other species. Marram is also well adapted to survive the physiological drought conditions that occur on the seashore. The leaves curl round on themselves with the stomata on the inside, creating a microenvironment that reduces water loss. Marram grass fills the sandy soil, salt-resistant, dune-binding niche perfectly.

Soils that contain a high proportion of sand are light, easily worked and easily warmed. However, they are also very easily drained. Water passes through them rapidly, carrying with it minerals that may be needed by plants. This **leaching** of minerals reduces the population density of plants that can grow in the soil. Conversely, it is difficult for water to drain through soils that are made up predominantly of tiny clay particles. They are heavy, take longer to warm up, are hard to work and are easily waterlogged. Mineral leaching is not a problem in soil of this type, but the populations which it will support are still limited. The ideal soil, loam, has particles of a wide range of sizes. It is heavier and less prone to leaching than sandy soils, yet easier to warm and work than clay. Different types of plants have evolved to grow well in different soil types – and will not grow as well in other soils.

fig C The edaphic features of different types of soils produce different growing conditions that support very different plants. For example, sundews only grow in the damp, acid soil of bogs.

Learning tip

As part of your study of this topic, you will carry out **Core Practical 16: Investigate the effect of one abiotic factor on the distribution or morphology of one species taking into account the safe and ethical use of organisms**. Make sure you have a good understanding of this practical as your understanding of the experimental method may be assessed in your examination.

Questions

1. (a) What are abiotic factors? Give some examples.
   (b) Why do abiotic factors have such a major impact on the distribution of all organisms in a habitat?

2. Abiotic factors interact to make up the conditions of a particular habitat. Describe an example of the way in which the impact of one abiotic factor may be influenced by another.

3. Choose one abiotic factor and investigate organisms that are adapted to survive in the extremes of these conditions.

Key definitions

A **microclimate** is a small area with a distinct climate that is different to the surrounding areas.

Edaphic factors relate to the structure of the soil.

Leaching describes the loss of minerals from soil as water passes through rapidly.

4 ▸ The effect of biotic factors on populations

By the end of this section, you should be able to...

● explain the effects of biotic factors on population size

Biotic factors are all of the living elements in a habitat, including plants, algae, fungi, herbivores, predators, parasites and disease-causing organisms.

Predation

It is easy to see how one species can affect the abundance of another. Horses grazing a field reduce the reproduction of the grass by eating the potentially flower-forming parts. A predator such as a fox will reduce the numbers of the prey species, reducing the numbers of the local rabbit population.

A mathematical model that describes the relationships between predator and prey populations predicts that the populations will oscillate in a repeating cycle. The reasoning underlying this model is straightforward. As a prey population increases there is more food for the predators and so, after an interval, the predator population grows too. The predators will increase to the point where they are eating more prey than are replaced by reproduction, so the numbers of prey will fall. This will reduce the food supply of the predators, so they will not produce as many offspring, and their numbers will fall, allowing the abundance of prey to increase again and so on (see **fig A**).

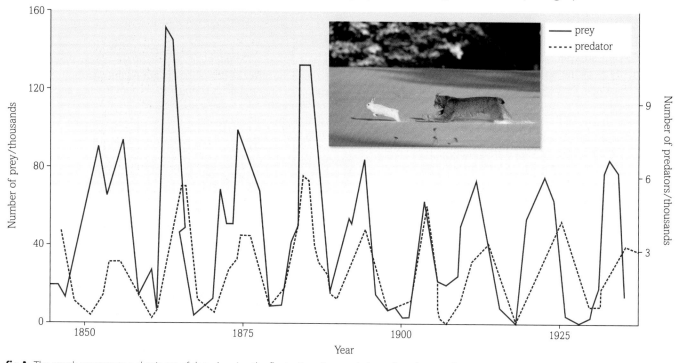

fig A The graph represents a classic set of data showing the fluctuations in populations of predator and prey.

The situation in a natural habitat is always more complex than a model. Other research shows that the prey population follows a similar pattern, even in areas where there are no predators. The prey are responding to cycles in their food availability that appear to be related to climatic variations and changes in insect pest populations. This is why it is important to study all the factors in an ecosystem. You will be looking at this in more detail on the following pages.

Finding a mate

Reproduction is a powerful driving force and the likelihood of finding a mate, or achieving pollination for plants, will help to determine the organisms that are found in any habitat. So if a single seed is dispersed to a new area and germinates, grows and survives, that species of plant is unlikely to become a permanent resident unless other plants of the same species also grow in the area, or the plant can reproduce successfully asexually. Similarly, a single individual of any animal species in an area does not mean that the species lives in the habitat. There must be males and females, so mates can be found. Availability of mates has a big effect on the abundance of any type of animal in an area.

Territory

Many species of animals show clear territorial behaviour. A territory is an area held and defended by an animal or group of animals against other organisms, which may be of the same or different species. Territories have different functions for different animals, but they are almost always used in some way to make sure that a breeding pair has sufficient resources to raise young. The type and size of territory will help to determine which species live in a particular community. In **Section 10.1.5** you will be looking at territories in more detail when you look at the role of competition in determining the distribution of animals living in a habitat.

fig B Gannets have small breeding territories because they feed out at sea, but without one a bird cannot nest and reproduce.

Parasitism and disease

Parasitism and disease are biotic factors that can have a devastating effect on individuals. Diseased animals will be weakened and often do not reproduce successfully. Sick predators cannot hunt well, and diseased prey animals are more likely to be caught. Some diseases are very infectious and can be spread without direct contact, such as avian (bird) flu, which can be spread in the faeces of an infected bird.

Parasites affect their hosts, usually by feeding off the living body of their host and so weakening it. Occasionally they can wipe out whole populations. For example, Dutch elm disease is a parasitic fungus spread by elm bark beetles that has destroyed most of the mature elm trees in the UK. In recent years, ash trees in the UK have become infected by the fungus *Hymenoscyphus fraxineus*. This causes a disease called ash dieback that threatens to destroy our ash population in a similar way.

Parasites and other infectious diseases spread more rapidly when there is a high population density, as individuals are in close proximity to each other. If almost all of the organisms in an area are of the same species, an infectious disease or parasite can have a devastating impact as it will affect most of the individuals directly or indirectly. In a community with greater biodiversity, although the effect on any infected individual will be as great, the effect of a disease or parasite on the whole community will be much less. Many species will be unaffected and there will be plenty of alternative food options.

Did you know?

Devil facial tumour disease affects and kills only Tasmanian devils, which are carnivorous marsupials endemic to the Australian state of Tasmania. Malignant tumours develop around the face and even in the mouths of affected individuals, spreading and killing them within about six months. This one disease could change the Tasmanian ecosystem forever by removing the top carnivore. Originally scientists suspected a virus was involved. Now it appears that this devastating biotic factor arose through a mutation in a single Tasmanian devil, probably in the mid-1990s, that caused a transmissible cancer. Tissues from all the tumours investigated have been shown to have this same mutation. Tasmanian devils bite and savage each other when they feed, mate or fight, and bits of the deadly tumours are passed on through bites on the face. The impact on the devil population has been huge, with numbers down from around 150 000 to under 50 000. Now that scientists understand how the disease is transmitted, there is some hope that a programme of vaccinations might be developed in time to save the species.

fig C The facial tumours that affect Tasmanian devils kill them by making it impossible to eat and drink.

Questions

1 Animals mark out and defend their territories in a number of ways. Find out how three different animals mark and protect their territories, including at least one species of bird, one fish and one mammal.

2 Why is disease likely to have a greater effect on an ecosystem with little biodiversity than on a more diverse community?

3 Different regions of **fig A** can be used to argue for or against the idea of repeating predator/prey cycles. Discuss this statement with reference to the data.

By the end of this section, you should be able to...

● explain the effects of abiotic and biotic factors on population size

When scientists are considering the ecology of an area and the impact of changes on the populations of living organisms, it is difficult to distinguish clearly between the biotic and abiotic factors that shape an ecosystem, as they are often interlinked, and it is rare that one factor works alone. In a natural habitat the factors determining distribution and abundance of organisms are always complex. Here are several examples that demonstrate the importance of studying all the factors in an ecosystem, and taking them into account when considering the numbers of organisms in a population and their distribution in a habitat.

Case study 1: Snowshoe hares and lynxes

In **Section 10.1.4**, **fig A**, you saw a classic graph showing the relationship between numbers of prey and predators. Charles Krebs and his team at the University of British Columbia have investigated snowshoe hare (prey) populations in Yukon territory, Canada. Their hypothesis was that both food supply and predation interact to affect the hare population and give rise to the pattern of the lynx (predator)/hare abundance. They chose nine $1\,km^2$ areas of undisturbed forest and measured the populations of hares over 11 years (the average length of the population cycle). In two of the areas, extra food was supplied to the hares all year round. Another two areas were fenced off with mesh that let hares through but not lynxes, and extra food was provided in one of these areas, but not the other. In two more areas the soil was fertilised to increase the food quality. The remaining three areas were untouched and left as controls. The data collected (see **fig A**) show that both food and predation affect the abundance of hares.

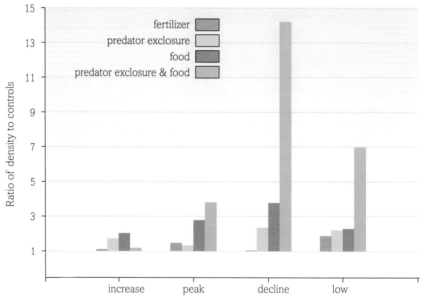

fig A Results of investigations into factors affecting snowshoe hare populations in Canada. Population censuses were carried out at intervals through the 11 year cycle. The results of four of these censuses are shown here, when the population was in an increase phase, a peak phase, a decline phase and a low phase.

Scientists often have to think 'outside the box' to explain their observations. The obvious answer is not always the right one, as you will see in the second case study.

Case study 2: Woodland birds

A regular census of breeding birds in a woodland in southern England showed major changes in the populations of breeding birds between 1950 and 1980 (see **fig B**). The woodpigeon population almost doubled quite suddenly between 1965 and 1970, but garden warblers disappeared altogether in 1971. The breeding blue tit population also increased fairly steadily.

The biggest change in the habitat of the wood itself was that regular felling stopped, so there were more mature and dead trees. Had the increase in pigeons caused the drop in garden warblers, or had the end of felling caused the changes in bird populations? While there appeared to be correlation between these events, it did not seem enough to explain the changes in the bird populations. Scientists felt they needed to look again to find the real causes of the changes.

It turned out that changes in biotic and abiotic factors far beyond the wood itself were affecting the bird populations. During the late 1960s, many farmers across the south of England started growing oilseed rape, and rape fields provide pigeons with an abundant supply of food through the winter. Many more pigeons survived to breed in the woodland due to a biotic factor beyond that habitat.

Garden warblers were found to be affected by abiotic factors thousands of miles away in West Africa. These small birds migrate to Africa for the winter, but lack of rain meant severe drought in their overwintering grounds, so the numbers surviving and making it back to the UK to breed fell dramatically.

Only the blue tits were affected by changes in the woodland habitat itself. The increase in old and dead trees meant an increase in the small holes that blue tits nest in, so the breeding population increased because more nest sites were available.

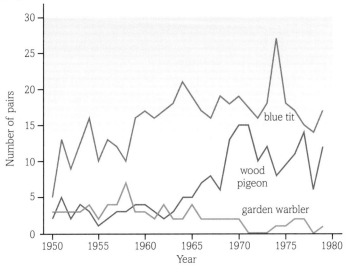

fig B Distant events and changes in abiotic or biotic factors can have a big impact on the distribution of organisms in any particular habitat. However, it is easy to see correlations and assume causation where there is none – as this data on bird populations in a UK woodland clearly shows.

Density-dependent and density-independent factors

The abiotic and biotic factors that affect the number of organisms occupying a particular niche may be density dependent or density independent. The effect of density-independent factors is the same regardless of population size. For example, extremes of temperature will usually have the same effect on all individuals, irrespective of how many individuals there are. These factors tend to limit the distribution of individuals and therefore of species. One of the effects of climate change as a result of global warming is the change in distribution of species as temperature extremes change.

The impact of a density-dependent factor will depend on how many organisms there are in a specific area. For example, disease and parasitism are factors that are strongly density dependent – the more individuals there are in a given area, the more likely it is that the disease or parasite can be transmitted between individuals. Breeding success in territorial animals and birds is also usually density dependent, as individuals without territories or with reduced territories are less able to breed. Only the strongest individuals are able to hold territories when space is a limited resource. Density-dependent factors are important in limiting the abundance of species.

Competition

Individual organisms often have to compete for density-dependent factors, and this can determine the size and density of a population. Competition occurs when two organisms compete for a resource that is in limited supply. The competition may be for abiotic resources such as sunlight, minerals or oxygen, or for biotic resources such as territories, nest sites or mates. Competition for mates, for example, has led to sexual selection, and to the evolution of sexual dimorphism, where the males and females of a species look very different.

fig C The competition between male animals for females can be very direct indeed.

Intraspecific competition

Intraspecific competition is competition between members of the same species for a limited resource. For example, meerkats are small mammals that live in family groups, and both males and females will defend their territory against other groups of meerkats wanting to forage there. The territory provides the food they need for their growing young, and so is a limiting resource.

As a result of intraspecific competition, some individuals may not survive, or may not reproduce, and so population growth slows. In contrast, if resources are plentiful, there is little or no competition and the numbers of individuals will increase.

Another example involves *Eleutherodactylus coqui*, a species of frog that lives in the tropical rainforests of Puerto Rico. These frogs feed on insects and are active at night, hiding during the day to avoid predators. They lay eggs in moist spots and the male guards them until tiny froglets hatch out. There is plenty of food – enough to support a larger population of frogs than exists. Scientists set about to discover the limiting resource. They investigated the idea that competition for shelter might be a factor limiting the population size, by dividing the study area into 100 m² plots. In some areas the frogs were provided with many small bamboo shelters, while in others the habitat was left unchanged. All of the shelters in the test plots were rapidly occupied and the population density increased accordingly, while the population density of the frogs in the control plots remained the same. This confirmed their hypothesis – a similar situation to the blue tits in UK woodland.

fig D Coqui frogs got their name from the sound the males make – ko-KEE. The 'ko' marks the territory and the 'kee' attracts females!

Interspecific competition

Interspecific competition occurs when different species within a community compete for the same resources. Competition will reduce the abundance of the competing species. If there is a greater density of one species or it has a faster reproduction rate, then competing species may become extinct in that area. For example, years ago sailors released goats on Abingdon Island, which is part of the Galápagos archipelago. Goats are relatively large, fast-breeding mammals, with appetites to match. Since their introduction, the growing goat population has consumed huge numbers of the island's plants, including the ones which the giant tortoises ate. The reptiles could not compete effectively because they reproduce much more slowly, and in the 1960s they became extinct on Abingdon. Isabela Island, which supports a higher proportion of endemic Galápagos species than any other, faces the same problem. Local scientists are working to eradicate the goats and remove the introduced competition.

fig E On Isabela Island many endemic species, including the giant tortoises, are threatened by competition from goats.

Questions

1 Explain the difference between intraspecific and interspecific competition and how they affect the distribution and abundance of organisms in a habitat.

2 Use data from **fig A** to answer these questions:
   (a) Describe the impact of the different experimental conditions on the density of the hare population.
   (b) Suggest an explanation for the difference between the effect of simply adding food or removing predators, compared with the enclosures where both food was added and predators excluded.
   (c) Suggest an explanation for the hare population dipping towards the end of the population cycle and back this up with evidence if possible. Why do you think the population where the grass was fertilised does not show this dip?

3 Explain how both abiotic and biotic factors can be density independent and density dependent, and how they affect the population numbers and distribution of organisms.

Key definitions

Intraspecific competition is competition between members of the same species for a limited resource.

Interspecific competition occurs when different species within a community compete for the same resources.

When scientists are considering the ecology of an area of natural habitat they will measure both the **abundance** and the **distribution** of the organisms in the area. Collecting this type of data can involve a number of different ecological techniques. The technique used depends on the type of ecosystem and the organisms that are being studied. For example, different techniques are needed to study the fish in an ocean from those to study the bluebells in a woodland.

Abundance and distribution

The abundance of an organism refers to the relative representation of a species in a particular ecosystem. It is more than simply the numbers of an organism. It is the numbers of an organism relative to the numbers of other organisms in the same habitat.

The distribution of an organism describes where a species of organism is found in the environment and how it is arranged. The distribution of organisms can change. Animals that migrate, for example, are found in different countries or even on different continents at different times of the year. Some organisms are always found in very high densities, for example, nematode worms in soil, ants in an ants' nest, people in a city or grass in a lawn. Other species are found widely scattered throughout a habitat. These are often the top predators. The distribution of an organism in a habitat usually falls into one of three main patterns:

- Uniform distribution. This occurs when resources are thinly but evenly spread, or when individuals of a species are antagonistic to each other. The territories may be very large or very small, for example nesting penguins, polar bears, hawks.

- Clumped distribution. This is the most common distribution seen with herds of animals or groups of plants and animals that have specific resource requirements and therefore clump in areas where those resources are found, for example a herd of sheep, a pod of dolphins or a stand of pines.

- Random distribution. This is the result of plentiful resources and no antagonism, for example dandelions in a lawn.

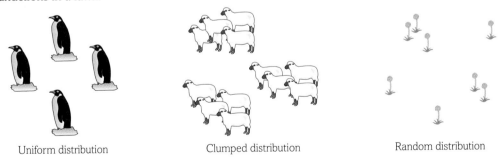

Uniform distribution   Clumped distribution   Random distribution

fig A Distribution patterns of different types of organisms.

Ecological techniques

In **Book 1 Chapter 3.3**, you saw the importance of biodiversity within a habitat. To calculate the biodiversity of an ecosystem, scientists often want to measure the abundance of an organism, or its distribution, or both. You need to assess both the number of species in the area and the size of their populations. You also need to identify them correctly.

Quadrats

The simplest way to sample an area is to use a **quadrat** (see **fig B**). This method is particularly useful for plants and for animals that do not move much. A frame quadrat is usually a square frame divided into sections that you lay on the ground to identify the sample area. Quadrats come in different sizes, from very large to very small, but we often use ones with sides of 50 cm (with an area of 0.25 m^2) or 25 cm (with an area of 0.0625 m^2), because they are easy to handle. This is known as a quantitative sampling technique as it allows us to quantify and measure such things as the number of individual organisms in the area (**individual counts**) or, using the divided grid, the area covered by the above-ground parts of a particular species (the **percentage cover**). A number of readings are taken to get a mean abundance or distribution. Frame quadrats are very important tools for the ecologist, but there are limitations including:

- limitations to the area you can sample
- the randomness (or otherwise!) of the sampling sites
- decisions made about whether to include or exclude organisms partly covered by the quadrat.

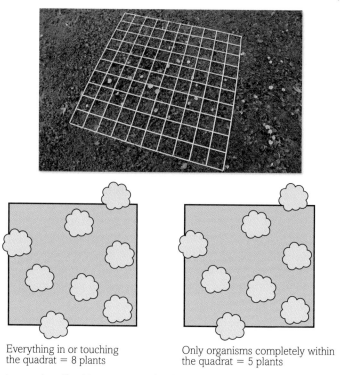

Everything in or touching
the quadrat = 8 plants

Only organisms completely within
the quadrat = 5 plants

fig B When you are using quadrats like this, to measure the abundance and distribution of organisms, you have to decide before you start whether organisms partly covered by the quadrat count as 'in' or 'out'. It does not matter which, as long as you decide and stick to it.

One way of measuring the abundance of organisms in a quadrat, or any other given area, is to use the **ACFOR scale**. This is a simple and rather subjective scale, but it is nevertheless useful in the field. It describes the abundance of a species in a given area as:

A = abundant

C = common

F = frequent

O = occasional

R = rare

Limitations of the ACFOR scale include:

- It is subjective – two people would probably never come up with exactly the same scores.
- There are no set definitions of the terms – how common is common?
- Species can easily be rated by how obvious they are rather than how abundant they are.

A point quadrat is a horizontal bar supported by two or three legs. At set intervals along the bar are holes through which a long pin is dropped. The species that the pins touch are recorded. They are often used to estimate the percentage cover by recording how many of the ten pins touch a particular species of plant. For example, if 5 out of 10 pins touch plants of species A, the percentage cover is 50%. A number of readings are taken and a mean percentage cover can be calculated.

Permanent quadrats are left in place all the time. This makes it possible to collect data through the seasons or from year to year reliably from the same place. For example, in a more than 40-year study being carried out on penguin breeding sites at Punta Tombo in Patagonia, researchers return to exactly the same (100 m^2) quadrat areas every year to count the number of active nests. This makes their data reliable, because it always refers to the same regions of the breeding colony.

Quadrats can be used to randomly sample areas by placing them at random coordinates, or simply throwing them in a given area. They can also be used to systematically sample a site.

Transects

Line transects are a widely used way of gathering data about an area more systematically. A transect is not random. You stretch a tape between two points, for example in a wildflower meadow, across sand-dunes (see **Section 10.1.2 fig B**) or down a mountainside, and record every individual plant (or animal) that touches the tape. An alternative is a **belt transect**, where two tapes are laid out and the ground between them is surveyed. Quadrats can be used for systematic sampling when they are laid out along a tape to form a belt transect. Transects can be used to investigate whether a change in the distribution of organisms is linked to a change in an abiotic factor. By sampling the abundance of organisms at regular intervals along the transect, and also sampling and recording abiotic factors such as soil pH or light intensity, correlations between them may be found.

More ecological techniques

There are many other ways of measuring the abundance of different types of organisms in a variety of habitats. Each method has advantages and disadvantages. Here are a few examples:

Light traps are often used to capture night-flying insects, as insects such as moths are strongly attracted to the light. Unfortunately, you do not know where the insects have travelled from, and some may have come many miles. This is useful for giving you a sense of the abundance of the organism – you may capture 3 or 300 – but less useful for measuring distribution, unless used in a wide variety of places at the same time of year.

fig C Moth traps can capture hundreds of moths, belonging to many different species, in one night.

Mark, release and recapture techniques allow you to build up a clearer picture of the animals that live in a particular area, or at least have it as part of their regular territory. This involves capturing animals on one occasion, marking them in a way that does not affect their survival chances and releasing them, for example, ringing birds. Subsequent captures are examined to find out how many marked animals have been recaptured, and these data can be used to estimate population numbers in the area.

The animals in tree canopies are often collected by beating the branches and collecting what falls out. However, this method misses many populations that live at the tops of the trees, including the birds. Some researchers spray an entire tree with anaesthetic gas or smoke (fogging), and then collect all the stunned organisms that fall out. Even this has its limitations – bark dwellers are not affected by the gas, because they are protected by the bark they live under.

Roadkill numbers can be used to estimate population size. The more dead badgers or hedgehogs there are on a road, the higher the numbers in the local population. Only a proportion of a population ends up as roadkill, so many dead animals indicates many live ones, although factors such as traffic density also have to be taken into account.

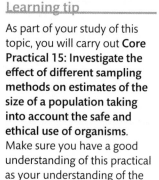

Learning tip

As part of your study of this topic, you will carry out **Core Practical 15: Investigate the effect of different sampling methods on estimates of the size of a population taking into account the safe and ethical use of organisms.** Make sure you have a good understanding of this practical as your understanding of the experimental method may be assessed in your examination.

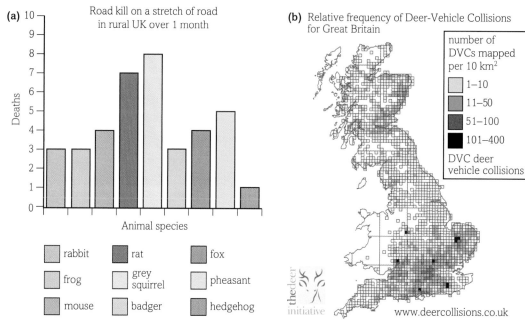

fig D Data from roadkills can be collected relatively easily and can give scientists an idea of both the species diversity and the population numbers in an area, as you can see from these two studies.

Questions

1 Why is a quadrat more useful for measuring the abundance and distribution of plants than of animals?

2 Compare the use of frame quadrats, point quadrats and permanent quadrats in a field investigation.

3 (a) What further details would you want to know about the collection of data for parts **(a)** and **(b)** of **fig D** before you tried to interpret the figures?

   (b) What conclusions can you draw about the British deer population from the data on vehicle collisions in **fig D(b)**?

Key definitions

The **abundance** of an organism refers to the relative representation of a species in a particular ecosystem.

The **distribution** of an organism describes where a species of organism is found in the environment and how it is arranged.

A **quadrat** is a square frame divided into sections that you lay on the ground to identify the sample area.

Individual counts measure the number of individual organisms in an area.

Percentage cover describes the area covered by the above ground parts of a particular species.

The **ACFOR scale** is a simple scale used to describe the abundance of a species in a given area.

Line transects are a way of gathering data more systematically. A tape is stretched between two points and every individual plant (or animal) that touches the tape is recorded.

A **belt transect** is when two tapes are laid out and the ground between them surveyed.

By the end of this section, you should be able to...

● use statistical tests to analyse data, including the Student's *t*-test and Spearman's rank correlation coefficient

In any field study you will collect a lot of data. It is not always easy to make sense of the data. Even when patterns appear to emerge, it is hard to know whether this is the result of a biological relationship or just down to chance and so we use statistical tests to help us decide. There are a number of statistical tests that are particularly useful and widely used by ecologists. These include **Spearman's rank correlation coefficient** and the **Student's *t*-test**.

What are you testing for?

Before you start to collect data in the field or lab, it is important to have a clear idea of what you are trying to prove or disprove – in other words, you need a clearly stated null hypothesis. This allows you to choose the right statistical test to help you analyse and evaluate your data.

After running your chosen statistical test, you will be left with a number known as the observed value. You need to look up this value in a published table of critical values to see which probability (p) value this corresponds to, so you can decide whether or not you should accept your null hypothesis. If your observed result fits with the critical value corresponding to the $p < 0.05$ level of significance given in the table, you can be 95% certain that the relationship between the factors you are studying is due to something other than chance.

Scientists use statistical computer programs to run the tests and give them a p value, without the need to use tables. However, it is important that you learn how to use these tests and do the calculations yourself.

Hypothesis testing

In statistical tests, by definition, you need something to test. In each situation you are investigating, you will make biological observations and collect sets of data. To carry out a statistical analysis of your results you need to produce a **null hypothesis**.

In general terms, the null hypothesis assumes that your current model explains your results.

When you have two or more sets of data, your null hypothesis will usually be that any differences between the data sets are simply due to chance. As scientists, we always want to find out something new. If statistical analysis shows a high probability that the null hypothesis is due to chance, it usually means we have not found something new and have confirmed an existing idea. Here are two examples:

1 You are investigating whether eating chocolate helps to prevent heart disease. You give one group of people a bar of chocolate every day for five years and another matched group get no chocolate. You record all the instances of heart disease in your two populations and then compare the results.

   The null hypothesis would be that eating chocolate has no effect on heart disease, and any difference in the instances of heart disease between the two groups is simply due to chance.

2 You are investigating the effect of grazing on the biodiversity of a meadow.

   The null hypothesis is that differences in biodiversity between grazed meadows and similar ungrazed meadows are due to chance.

In both cases, statistical analysis will show you the probability that any differences between the groups you observe are the result of chance. If it is chance, you accept the null hypothesis and you may need to study more data to create a new hypothesis to investigate. If the probability is that the differences are not due to chance, then you reject the null hypothesis, but may need to do a lot more analysis to work out exactly what is causing those differences!

Spearman's rank correlation coefficient

One method of determining whether there is correlation between two variables is to plot a scatter diagram. This is a useful technique but it does not always give a clear result. Using a correlation coefficient is a more precise way of measuring whether there is significant correlation between two variables. Spearman's rank correlation coefficient is calculated using a formula that gives a correlation coefficient r_s between −1 and +1. The size of the correlation coefficient indicates how strong the correlation is and the sign indicates whether the correlation is positive or negative:

• A value of +1 indicates that there is perfect positive correlation.

• A value of −1 indicates that there is perfect negative correlation.

• A value of 0 indicates that there is no correlation between the variables.

Using the test

Step 1: State the null hypothesis

The null hypothesis is that there is no correlation between the two variables, in which case r_s is equal to 0.

Step 2: Calculate the correlation coefficient, r_s (the observed value)

The formula for the Spearman's rank correlation coefficient is:

$$r_s = 1 - \frac{6\sum d^2}{n(n^2 - 1)}$$

where d is the difference in rank between each pair of variables and n is the number of pairs. To use it, find the difference in rank between each pair, square these differences and add them all together. Now substitute this value into the formula in place of $\sum d^2$.

Step 3: Decide whether to accept the null hypothesis

If your value for r_s suggests a correlation, that is, it is not 0, you need to see if the correlation is statistically significant by looking up your value in a table of critical values.

Find the probability (p) value that relates to your observed value and the number of pairs (n). Ignore whether r_s is positive or negative. If $p < 0.05$, the correlation is considered to be statistically significant, so you can reject your null hypothesis that there is no correlation.

WORKED EXAMPLE

Succession on Surtsey

Table A shows you how the numbers of species on Surtsey changed over time. This is an example of a data set where Spearman's rank correlation coefficient can be used.

Years after formation	0	5	10	15	20	25	30	35	40
Species total	0	4	10	13	11	18	34	46	54

table A

fig A Vegetation spreading on Surtsey.

The null hypothesis is that there is no correlation between the time in years after Surtsey was first formed and the number of species present. Rank the variables and calculate d and d^2 for each pair of data.

	x values (number of species)	Ranking of x values from lowest to highest	y values (years after formation in 1965)	Ranking of y values from lowest to highest	Difference in rank between pair of variables (d)	d^2
	0	1	0	1	0	0
	4	2	5	2	0	0
	10	3	10	3	0	0
	13	5	15	4	−1	1
	11	4	20	5	1	1
	18	6	25	6	0	0
	34	7	30	7	0	0
	46	8	35	8	0	0
	54	9	40	9	0	0
Sum of	$n = 9$					$\sum d^2 = 2$

table B Ranking number of species against years of formation of Surtsey.

Now substitute your data into the formula:

$$r_s = 1 - \frac{6\sum d^2}{n(n^2 - 1)} = 1 - \frac{6 \times 2}{9 \times 80} = 1 - \frac{12}{720} = 1 - 0.0167 = \mathbf{0.983}$$

You need to use a critical values table to see if you can accept your null hypothesis.

Number of pairs of values (n)	4	5	6	7	8	9	10	11	12
Critical values	1.00	0.90	0.83	0.79	0.74	0.68	0.65	0.61	0.59

table C Extract from a table of critical values for Spearman's rank correlation coefficient, at $p = 0.005$.

The critical values table shows that you need a figure greater than 0.68 to give $p < 0.05$. Your figure of 0.983 is considerably higher than this, relating to a p value of much less than 0.05, so the correlation is considered statistically significant. Therefore you can reject the null hypothesis that there is no correlation between the time since Surtsey was first formed and confirm that the longer the island of Surtsey has been in existence, the more species of plants are found growing on it.

The Student's *t*-test

The Student's *t*-test is used to determine whether the mean of a variable in one group differs significantly from the mean of the same variable in a different group; for example, comparing the blood pressure of males and females. Put simply, it tests whether two sets of data are significantly different from each other. The variable being considered must follow a normal distribution. The Student's *t*-test takes into account the degree of overlap between the two sets of data and allows you to judge whether any difference between the means of the two groups is statistically significant or just due to chance.

Step 1: State the null hypothesis

The null hypothesis is that there is no significant difference between the mean of the interval variable for the two categories.

Step 2: Calculate your observed value, t

You will need to calculate *t*, but first need to find the mean and the variance s^2 for both groups. *n* is the sample size (for example, the number of people in the group) and *x* is a value of the measured variable (for example, one person's blood pressure reading).

$$\bar{x} = \frac{\sum x}{n}$$

$$s^2 = \frac{\sum x^2 - \frac{(\sum x)^2}{n}}{n-1}$$

You can now use the following formula to calculate the value of *t*:

$$t = \frac{\bar{x}_1 - \bar{x}_2}{\sqrt{\frac{s_1^2}{n_1} + \frac{s_2^2}{n_2}}}$$

where

$\bar{x}_1$ = mean of the first set of data

$\bar{x}_2$ = mean of the second set of data

s_1^2 = variance of the first set of data

s_2^2 = variance of the second set of data

n_1 = number of data values in first set of data

n_2 = number of data values in second set of data

Step 3: Decide whether to accept the null hypothesis

To decide whether this value of *t* indicates a significant difference in variable between the two groups, you need to compare your observed value of *t* with a critical values table. You will need to know the degree of freedom, which for the Student's *t*-test is calculated as $df = n_1 + n_2 - 2$ (the total number of data values minus 2).

Find the probability (*p*) value that relates to your observed value and degrees of freedom; if $p < 0.05$, the difference between the means of the two groups is considered to be statistically significant and not due to chance.

fig B Mussels growing on the rocks vary in size, but is there a significant difference between the maximum sizes of the mussels on the two beaches investigated in this example?

WORKED EXAMPLE

Maximum mussel sizes

The Student's t-test is very useful in field work for helping you decide if there are genuine differences between populations, for example the size of mussels or dog whelks on different beaches or different areas of the same beach. In this example, a UK student compared the shell length (the interval variable) of the largest mussel in each of 10 quadrats on the rocks at the low water level on two different beaches (the nominal variable).

The null hypothesis is that there is no significant difference between the mean shell length of the largest mussels on the rocks at the low water mark on the different beaches.

	x_1 values (length of largest mussels from beach 1)/mm	$x_1{}^2$	x_2 values (length of largest mussels from beach 2)/mm	$x_2{}^2$
	76	5776	77	5929
	82	6724	71	5041
	65	4225	73	5329
	90	8100	69	4761
	71	5041	79	6241
	69	4761	75	5625
	75	5625	82	6724
	59	3481	76	5776
	84	7056	81	6561
	78	6084	85	7225
Sum of	749	56 873	768	59 212

table D Length of largest mussels from beach 1 and beach 2.

So

$$\bar{x}_1 = \frac{749}{10} = 74.9$$

$$\bar{x}_2 = \frac{768}{10} = 76.8$$

$$s_1^2 = \frac{56\,873.0 - (749.0)^2/10.0}{9.0} = \frac{56\,873.0 - 56\,100.1}{9.0} = \frac{772.9}{9.0} = 85.88$$

$$s_2^2 = \frac{59\,212.0 - (768.0)^2/10.0}{9.0} = \frac{59\,212.0 - 58\,982.4}{9.0} = \frac{229.6}{9.0} = 25.51$$

$$n_1 = 10$$

$$n_2 = 10$$

Now you can substitute these numbers to calculate the value of t:

$$t = \frac{\bar{x}_1 - \bar{x}_2}{\sqrt{\frac{s_1^2}{n_1} + \frac{s_2^2}{n_2}}} = \frac{74.9 - 76.8}{\sqrt{\frac{85.88}{10} + \frac{25.51}{10}}} = \frac{-1.9}{\sqrt{11.13}} = \frac{-1.9}{3.34} = -0.57 \text{ (to 2 decimal places)}$$

For the Student's t-test, the sign of the observed value does not matter – if you swapped the two data sets around we would get +0.57 – so you just need to take the absolute value and ignore the sign.

You now need to use a critical values table to see whether you can reject the null hypothesis that there is no significant difference between the maximum shell lengths of mussels on the two beaches. The degrees of freedom is $20 - 2 = 18$ (the total number of data values minus 2).

Degrees of freedom (df)	Levels of significance (p)				
	0.05	0.02	0.01	0.002	0.001
18	2.101	2.552	2.878	3.610	3.922

The critical value table shows that you need a figure greater than 2.101 to give $p < 0.05$ level of significance with 18 degrees of freedom.

The number we calculated was -0.57. This absolute value 0.57 is far less than the critical value of 2.101, indicating a p value much greater than 0.05. This means that there is no significant difference between the means of shell length on the two beaches. Therefore we accept the null hypothesis. There is no significant difference in the maximum shell sizes of the mussels measured at the low water mark on the two beaches.

Questions

1. In an area where some bog sites had been restored, ecologists wanted to evaluate if the process had been successful. They looked at the percentage cover of two different bog plants, *D. flexuosa* and *Agrostis* sp., on the same site in 26 quadrats and ranked the relative cover of the two species. They wanted to discover if there was any relationship between the populations of the two plants.
   (a) Suggest a null hypothesis for this investigation.
   (b) Suggest a suitable statistical test to use to analyse the data they collected and explain why you have chosen it.
   (c) How many degrees of freedom will there be for this data set?

2. (a) What is the Student's *t*-test?
   (b) Why is it not the right test to apply to the data described above?
   (c) Suggest data that might be collected from that site and analysed using the Student's *t*-test.

Key definitions

Spearman's rank correlation coefficient is a statistical tool used to test whether two variables are significantly correlated.

The **Student's *t*-test** is a statistical test that allows you to judge whether any difference between the means of the two sets of data is statistically significant.

The **null hypothesis** is the hypothesis that any differences between data sets are the result of chance.

Exam-style questions

1 A number of strategies could be used to analyse data from an ecological investigation. Choose one strategy from the following list to analyse each of the following four situations.

  A Spearman's rank test
  B student's t test
  C chi squared (χ^2) test
  D standard deviation

(a) Determine the concordancy of the height of *Patella* shells in one part of a rocky shore. [1]

(b) Investigate whether the height of *Patella* shells depends on whether they grow facing or away from the sea. [1]

(c) Compare the number of worms in two different fields. [1]

(d) Investigate how the percentage cover of marram grass changes with the increasing salt concentration in the sand. [1]

[Total: 4]

2 Salt marshes develop in sheltered areas of coastline where deposits of mud build up. The sea frequently floods these marshes with salt water and only plants with a tolerance to salt can survive. The frequency of flooding becomes less as the height of the land above low water increases due to the build up of organic matter and mud. The highest tides flood some areas only a few times a year.

The distribution of plant species in a salt marsh was investigated. The presence of plant species was recorded at ten sites along a transect. Site 1 was at the low water mark and site 10 was on higher ground 50 m inland.

The diagram below shows a profile of this salt marsh and the table shows the plant species recorded at each site.

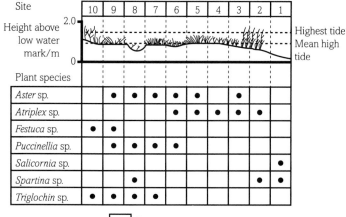

Site	10	9	8	7	6	5	4	3	2	1
Plant species										
Aster sp.		•	•	•	•	•		•		
Atriplex sp.					•	•	•	•	•	
Festuca sp.	•	•								
Puccinellia sp.		•	•	•	•					
Salicornia sp.										•
Spartina sp.			•						•	•
Triglochin sp.	•	•	•	•						

☐ • Species present

(a) Describe a suitable sampling method that could be used to obtain the plant species data recorded in the table above. [2]

(b) Compare and contrast the data for sites 1 and 9. [2]

(c) The differences in plant species found at the different sites are thought to be the result of competition between species leading to succession.

Explain the role of competition in succession. [3]

(d) Site 8 includes a hole which fills with seawater when the area floods. This makes the soil waterlogged for longer periods than the surrounding area. As the water evaporates it leaves very high concentrations of salt in the soil.

The photograph below shows *Spartina* sp. growing in the hole filled with seawater.

(i) Using all the information given, deduce why *Spartina* sp. is found at site 8. [2]

(ii) Explain how the hole at site 8 may change over a long period of time. [3]

[Total: 12]

3 For many centuries, sheep have grazed on the grasslands on many of the islands off the coast of Scotland. These grassland communities are examples of plagioclimax.

As the demand for wool from the sheep reduced, the farming of sheep stopped on many of the islands. Within a few years, the grasslands on some of these islands developed into shrub communities.

(a) Explain what is meant by the term **plagioclimax**. [2]

(b) Explain why the grassland on the islands might develop into shrub communities after sheep farming stopped. [4]

(c) On some of the islands where sheep farming stopped, the grasslands remained for a longer time before any change started to be noticed.

Give **two** reasons for this. [2]

[Total: 8]

4 An area of abandoned grassland was studied over a period of more than 100 years. During that time, there were changes to the plant communities which resulted in changes to the number of species and population density of small birds. The table below shows the changes.

Time since abandoned/years	1–10	10–25	25–100	100+
Plant community	Grass	Shrubs	Pine trees	Mixed woodland

Number of species of small birds	2	8	15	19
Population density of small birds/number of birds per 40 hectares	54	246	226	466

(a) State the name that is given to the process by which communities change over time. [1]

(b) Describe the effect of the change in the plant community on the number of species of small birds. Give reasons for your answer. [5]

(c) The population density of birds drops when shrubs are replaced by pine trees, but then increases with the change to mixed woodland. Give reasons for these changes in population density. [2]

(d) The mixed woodland is the final climax community. Describe **two** ways in which woodland can be damaged by human activities. In each case, explain how the damage is caused. [4]

[Total: 12]

CHAPTER

10.2 **The efficiency of ecosystems**

Introduction

The collection of leaf litter is a widely used way of measuring the productivity of an ecosystem. Leaf fall measurements range from around 0.6 t ha^{-1} year^{-1} in the Arctic tundra to 25.4 t ha^{-1} year^{-1} in lowland tropical rainforests. Rainforests, with their dense leafy canopies, are very productive plant communities, so they have a high level of leaf fall. However, not all of the leaves that fall end up as leaf litter. Much plant material is eaten by animals, including birds, monkeys and insects, before it reaches the ground. In the warm moist climate of a rainforest, decomposers work fast. In spite of the high productivity of the plants, the forest floor does not disappear under a metre-deep layer of old leaves and fruits, because fungi, bacteria, many different types of arthropods and worms break the plant material down and return the nutrients to the soil to sustain the continual growth.

In this chapter you will learn about the efficiency of ecosystems and how light from the Sun provides the energy for the production of plant material, which in turn is passed through different trophic levels. Biologists try to calculate the efficiency of these processes and calculate where and how biomass is used at every level. Biomass contains the energy stored within the bonds of the biological molecules. New biomass is made constantly through photosynthesis, but the inorganic components of biological molecules are finite. You will consider the vital role of microorganisms and other decomposers in the recycling of nutrients within an ecosystem.

All the maths you need

- Recognise and make use of appropriate units in calculations (*e.g. the amount of carbon stored in different carbon sinks*)
- Use ratios, fractions and percentages (*e.g. calculating energy transfers*)
- Estimate results (*e.g. to see if biomass calculations through trophic levels are appropriate*)
- Use appropriate number of significant figures (*e.g. when calculating NPP*)
- Construct and interpret frequency tables and diagrams, bar charts and histograms (*e.g. pyramids of numbers*)
- Translate information between graphical, numerical and algebraic forms (*e.g. biomass transfers*)

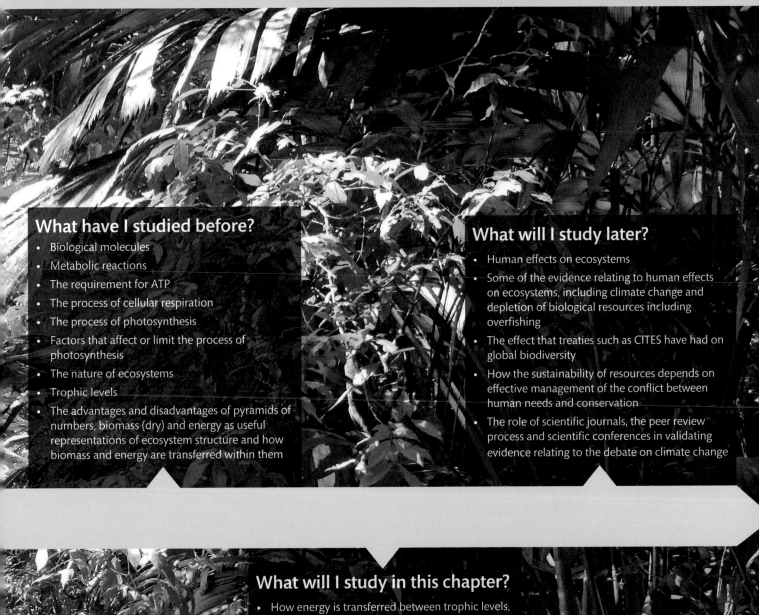

What have I studied before?

- Biological molecules
- Metabolic reactions
- The requirement for ATP
- The process of cellular respiration
- The process of photosynthesis
- Factors that affect or limit the process of photosynthesis
- The nature of ecosystems
- Trophic levels
- The advantages and disadvantages of pyramids of numbers, biomass (dry) and energy as useful representations of ecosystem structure and how biomass and energy are transferred within them

What will I study later?

- Human effects on ecosystems
- Some of the evidence relating to human effects on ecosystems, including climate change and depletion of biological resources including overfishing
- The effect that treaties such as CITES have had on global biodiversity
- How the sustainability of resources depends on effective management of the conflict between human needs and conservation
- The role of scientific journals, the peer review process and scientific conferences in validating evidence relating to the debate on climate change

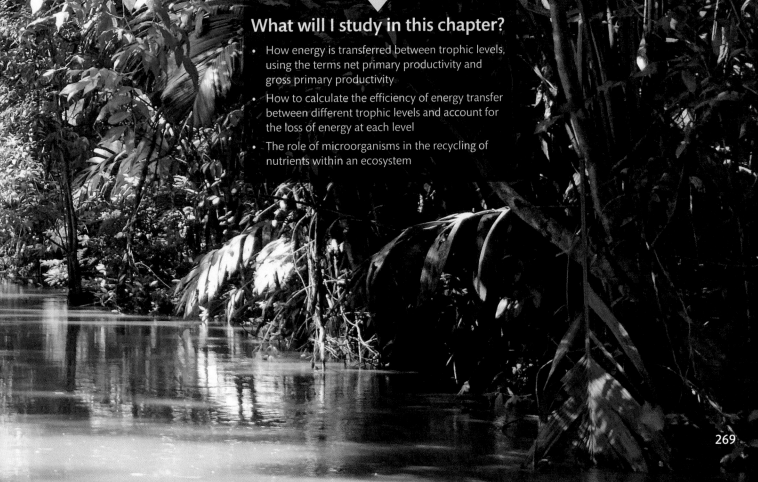

What will I study in this chapter?

- How energy is transferred between trophic levels, using the terms net primary productivity and gross primary productivity
- How to calculate the efficiency of energy transfer between different trophic levels and account for the loss of energy at each level
- The role of microorganisms in the recycling of nutrients within an ecosystem

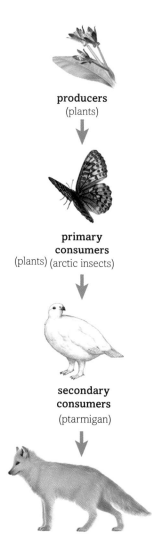

producers
(plants)

↓

primary consumers
(plants) (arctic insects)

↓

secondary consumers
(ptarmigan)

↓

tertiary consumers
(arctic foxes)

fig A A food chain such as this one from Elton's work on Bear Island is the simplest representation of the feeding relationships within an ecosystem. Decomposers are rarely shown in food chains such as this.

In the 1920s, Charles Elton, a young Oxford biologist, began to study the relationships between the animals on Bear Island (off the northern coast of Norway) and their scarce food resources. The island had limited numbers of plant and animal species, and this made his studies easier and more effective than if he had chosen to study a more diverse habitat. The main animals he observed were arctic foxes, and the birds they ate such as sandpipers and ptarmigan. The birds ate the leaves and berries of plants or, in some cases, ate the insects that fed on the plants. Elton called these feeding interactions a food chain and proposed a general model to explain the flow of resources through a community. Each link in the food chain represents a specific **trophic level** (see **fig A**).

A model for a food chain

Elton proposed a general model for the food chain based on the following trophic levels.

- **Producers** make food. In photosynthesis, plants and algae trap light from the sun and this drives the production of ATP, which they then use to make glucose from carbon dioxide and water (see **Chapter 5.2** to remind you of the process of photosynthesis).

- **Primary consumers** are the organisms, mainly animals, that eat producers. They are herbivores. They use the molecules in plants to supply the raw materials needed for their metabolic reactions.

- **Secondary consumers** are the animals that feed on herbivores. They are carnivores. They use the molecules in the herbivores to supply the raw materials needed for their metabolic reactions.

- **Tertiary consumers** are animals that feed on other carnivores. They are usually the top predators in an area, unless there is a quaternary consumer. They use the molecules in the carnivores to supply the raw materials needed for their metabolic reactions.

- **Decomposers** are the final trophic level in any set of feeding relationships. They are the microorganisms, such as bacteria and fungi, that break down the remains of animals and plants and return the mineral nutrients to the soil.

Food webs and beyond

The description of a food chain makes sense and is relatively simple to understand. However, it is now recognised as an oversimplification. Few animals eat a single food – giant pandas and koala bears are two of only a few examples. Most animals have a variety of food sources and exist not in simple food chains but as part of complex interconnected food webs. This is exactly what Elton went on to observe on Bear Island, and ecologists have built up similar models from ecosystems all over the world (see **fig B**).

In situations involving a single food chain, such as the giant panda with its diet made up exclusively of bamboo, the ecosystem is easily disrupted. Any event that reduces the availability of bamboo will also threaten the panda, making pandas very vulnerable to habitat destruction, whether as a result of human actions or natural disasters. When an organism is part of a complex food web, a change in any one component, whilst potentially affecting the balance of the ecosystem, is far less likely to have catastrophic effects and so the system will be more stable (see **fig B**).

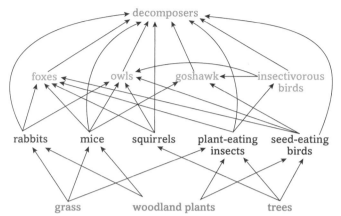

fig B The English deciduous woodland in this simplified food web represents a relatively stable system. When rabbits almost died out from myxomatosis, owls survived by eating more of their other prey species. Although mice and squirrels were more heavily predated, there was reduced competition for the plant food resources, so they could reproduce more successfully. The rest of the community was relatively unaffected by the loss of the rabbits.

Ecological pyramids

Although a food chain is a highly simplified model of the trophic levels within an ecosystem, it has helped us to develop an understanding of the energy economy of ecosystems. We can also use pyramids to illustrate what is happening in an ecosystem.

Pyramids of numbers

In many food chains, the number of organisms decreases at each trophic level. There are more producers than primary consumers, more primary consumers than secondary consumers, and so on. These observations are represented at their simplest by a **pyramid of numbers** (see **fig C**).

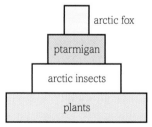

fig C A pyramid of numbers for the simple Bear Island food chain shown in **fig A**. The number of individuals decreases as you move up the chain.

Pyramids of biomass

In many situations a pyramid of numbers does not accurately reflect what is happening in an ecosystem. For example, a single rosebush will support a very large population of aphids, which will be eaten by a smaller population of ladybirds and hoverfly larvae, that are themselves the prey of relatively few birds. This gives a very distorted pyramid of numbers. You get a much more realistic model by using a **pyramid of biomass**. This shows the combined mass of all the organisms in a particular habitat.

Counting the plants and animals for pyramids of numbers can be difficult and measuring biomass is even harder. Either wet or dry biomass may be used; however, wet biomass is very inaccurate. It is affected by water uptake in the soil, transpiration in plants, and drinking, urinating, defaecating and in some cases sweating in animals. Using dry mass eliminates the inaccuracy of variable water content in organisms, but involves destroying the material. To avoid the destruction of the habitat, a small sample of all the organisms involved is taken and the dry mass obtained. The total biomass of the population is then calculated from this sample. It is much more time consuming to produce a pyramid of biomass than a pyramid of numbers, but it gives much more accurate information about what is happening in an ecosystem (see **fig D**).

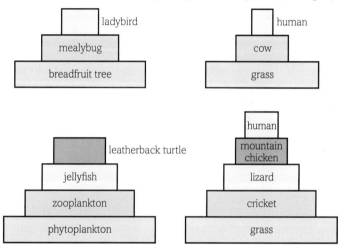

fig D Pyramids of biomass give a more accurate picture of what is happening in a food chain than pyramids of numbers do.

Pyramids of energy

Even pyramids of biomass have their limitations. For example, if the biomass of the organisms in a sample of water from the English Channel is analysed, there appears to be a greater biomass of zooplankton than of the photosynthetic phytoplankton on which it feeds. The sample is only a snapshot of the ecosystem. What it fails to show is that the phytoplankton reproduces much more rapidly than the zooplankton. Although the biomass of the total phytoplankton population at any one time (the standing crop) is smaller than that of the zooplankton, the turnover of the phytoplankton is much higher and so the biomass over a period of time is much greater. A pyramid made up of observations over time gives the most accurate model of what is happening in an ecosystem. This is what we try to do with **pyramids of energy**.

The energy in an ecosystem remains the same at every level, and it is the size of the different type of energy stores that changes. As you move along a food chain, less energy is stored in the organisms and more is stored in the surrounding atmosphere.

Losses along a food chain

Plants use light to catalyse photosynthesise, making a chemical store in their cells that is available to herbivores, but a relatively small proportion of this ends up as new animal material:

- Some is lost to the animal as undigested and therefore unused material in the faeces.

- Much of the material that is digested is used to drive respiration, a series of reactions that result in the production of ATP. This is an exothermic process which heats the tissues of the animal and the surrounding atmosphere.

- Some of the plant material is lost in metabolic waste products such as urea (see **fig E**).

Only a small amount of the chemical store of the plant becomes new animal material and therefore part of the energy store of the animal. The rest is dissipated to the surroundings, increasing the internal energy store of the universe. The process of making new animal biomass is known as **secondary production**.

fig E Relatively little of the energy store of the plants becomes part of the energy store of a primary consumer such as this cow. Much of it becomes part of the internal energy store of the surrounding universe. Similar losses occur at each trophic level in a food chain.

Similar energy transfers to the surroundings occur between animal trophic levels, from herbivores to primary consumers, and so on up the food chain. The energy store in the biomass of one organism compared with the energy store that ends up in an organism in the next trophic level is a measure of the efficiency of energy transfer. This measure is often quoted as being around 10%, but it varies greatly. Studies have shown it as low as 0.1% for some small herbivorous mammals and as great as 80% for some microorganisms. It is most commonly between about 2 and 24%. The causes of such wide variation depend on many factors, including the effort required to find food, the digestibility of the food and the metabolic rate of the organism.

Did you know?

Endotherms use food resources to maintain their body temperatures at a constant level, regardless of the external temperature. Through radiation, convection and conduction, they increase the internal energy store of their surroundings! As a result, endotherms need to eat a lot more food than exotherms to get the same increase in biomass. This is an important consideration in sustainable food production. It takes a lot less plant material to produce fish or insect protein than it does to produce beef or lamb. However, particularly in the more economically developed countries, many people eat more meat than fish, and few people want to eat insects.

The most accurate way to represent a food chain is to use a pyramid of energy (see **fig F**). The amount of energy stored in the organisms decreases at every trophic level along a food chain, while the energy store of the surroundings increases. However, pyramids of energy are extremely difficult to measure practically and often involve a rather outdated model of energy, so pyramids of biomass are widely used.

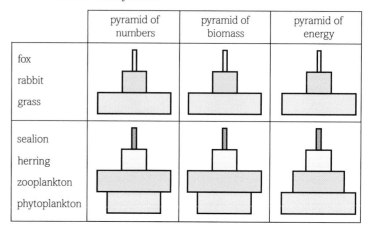

fig F This diagram shows you how pyramids of numbers, biomass and energy compare for two different food chains.

Questions

1. What are trophic levels in an ecosystem and why are they useful?

2. Summarise the advantages and disadvantages of using pyramids of numbers, pyramids of biomass and pyramids of energy as representations of ecosystem structure.

Key definitions

A **trophic level** describes the position of an organism in a food chain or web and describes its feeding relationship with other organisms.

Producers make food by photosynthesis or chemosynthesis.

Primary consumers are organisms that eat producers, either plants or algae.

Secondary consumers are animals that feed on primary consumers.

Tertiary consumers feed on secondary consumers – they eat other carnivores. They are usually the top predators in an area.

Decomposers are the final trophic level in any set of feeding relationships. They are the microorganisms, such as bacteria and fungi, that break down the remains of animals and plants and return the mineral nutrients to the soil.

A **pyramid of numbers** represents the numbers of organisms at each trophic level in a food chain.

A **pyramid of biomass** represents the biomass of the organisms at each trophic level in a food chain.

A **pyramid of energy** represents the total energy store of the organisms at each trophic level in a food chain.

Secondary production is the process of making new animal biomass from plant material that has been eaten.

By the end of this section, you should be able to...

- explain how energy is transferred between trophic levels using the terms 'net primary productivity' and 'gross primary productivity'

- calculate the efficiency of energy transfer between different trophic levels

As you saw in **Section 10.2.1**, the starting point in most ecosystems is the catalytic effect of light from the Sun on the reactions of photosynthesis. This determines the rate at which producers make organic material and this in turn determines the production of biomass within those ecosystems. Only a very small percentage of the light from the Sun actually catalyses the production of plant material (see **fig A**). Figures vary, but between 1 and 3% is generally accepted.

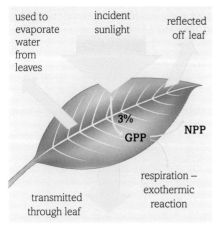

fig A Most of the light that reaches a leaf is not available for photosynthesis to produce new plant tissue.

Gross and net primary productivity

Gross primary productivity (GPP) in plants is the rate at which light from the Sun catalyses the production of new plant material. It may be measured in units of biomass/area/time, such as simple $g\,m^{-2}\,year^{-1}$ or $g\,C$ (grams of carbon assimilated) $m^{-2}\,year^{-1}$. It may also be measured in units of energy, for example, $kJ\,m^{-2}\,year^{-1}$. Measuring productivity is very difficult. It usually involves finding the mass of representative samples of biomass and then multiplying up to represent a whole ecosystem. Converting biomass to energy is done through theoretical calculations rather than empirical measurements, so biomass is the better measure to use.

Plants use at least 25% of the material they produce for their own metabolic needs. Most importantly they respire, breaking down glucose to produce ATP. This is an exothermic process, so it increases the internal energy store of the surroundings (heats the surroundings). The rest of the material is stored as new plant body tissues. This energy store is known as the **net primary productivity (NPP)**.

$$NPP = GPP - R \text{ (respiration)}$$

The NPP of different ecosystems has been estimated. It will depend on all the abiotic factors and biotic factors that affect plant growth within the ecosystem. Latitude is important, because the light within a specific area is lower at latitudes nearer the pole than closer to the equator (see **fig B**).

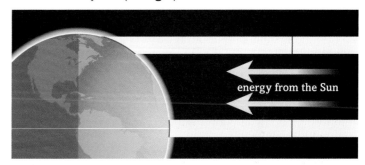

fig B More light falls on a square metre of the Earth's surface at the equator than near the pole, because the curvature of the Earth means the light is spread out over a larger area of the surface at the poles.

In **fig C** you can clearly see the importance of factors such as the availability of water within an ecosystem, as well as ambient temperatures, on the NPP of an area. If you combine the NPP of each type of ecosystem with the area of the Earth's surface it covers, you can see how much each type contributes to the overall NPP of the Earth. For example, tropical rainforests cover only about 5% of the Earth's surface, but yield over 30% of the global NPP. The oceans contribute a similar proportion to the global NPP, but only because they are so vast (see **fig C**).

The human population of the world is growing at a rapid rate, and even at its current level humans are consuming up to 40% of the NPP of the planet.

Energy transfers to higher levels

When we look at the efficiency of energy transfer between trophic levels, it is important to include the decomposers as well as the more obvious herbivores and carnivores. The bodies of dead animals and plants, and the faeces of animals, are broken down and accumulated by decomposers. So when the transfer of biomass or energy to decomposers is taken into account, the calculations look different again. As you have seen elsewhere, the rate of decomposition and therefore the rate of transfer of biomass and energy varies, depending on factors such as temperature, water availability and, especially in temperate zones, the season of the year. These considerations simply add to the difficulty of calculating the energy transfer through an ecosystem.

273

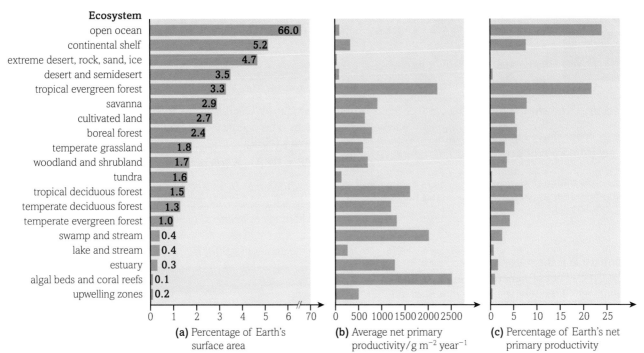

fig C Different ecosystems contribute different amounts to the NPP of the Earth.

Measuring energy transfers

How is the percentage energy or biomass transfer calculated? This involves a simple percentage calculation. So for example:

WORKED EXAMPLE

96,680 kJ m^{-2} year^{-1} of energy are available in a plant source per annum. 6780 kJ m^{-2} year^{-1} per annum are transferred into new herbivore material at the next trophic level – the rest is transferred to the surroundings. What percentage of energy is transferred from the plants to the herbivore per year?

$$\frac{6780}{96\,680} \times 100 = 7\%$$

The limitations of efficiency calculations

It is very difficult to measure energy transfers for whole ecosystems since many assumptions need to be made in calculating the energy in all the organisms, so it is not done very often. Not only do you need to be certain that you have identified all the most abundant species in the ecosystem, you also need to know how many there are, what an average body size is, how much energy that body size represents, how much of the biomass is transferred into decomposers at any stage of the life of the animal or plant, and so on.

In temperate regions there is likely to be more biomass and therefore a bigger energy store in plant and animal bodies during summer, and more in decomposers during the winter. Even in tropical regions, there are substantial differences between the wet and the dry seasons. All of these factors will affect estimates of energy at each trophic level.

Measuring energy transfers from one level to another is difficult and, as you have seen, is often based on biomass measurements. The values that are quoted from studies are the mean values from the calculations and usually have large standard errors, a statistical way of showing how close the mean is to the real value – small is close, large is considered not very reliable. This is not surprising, given the many assumptions that are made during their calculation.

In 1942, the American ecologist R.L. Lindeman published the results of his study of energy transfers in an ecosystem at Lake Mendota, Minnesota, in the journal *Ecology*. He looked at all the organisms at each trophic level. His results showed the following levels of efficiency in the trophic levels of that particular system: producers 0.4%, primary consumers 8.7%, secondary consumers 5.5% and tertiary consumers 13%.

For some years scientists considered that the efficiency of transfer between trophic levels followed specific patterns. Many different studies on energy transfer efficiency between trophic levels have provided very different and wide-ranging values. Scientists now recognise that the efficiency of energy transfer varies greatly and depends on the physiology and behaviour of the organisms involved and the climatic conditions, rather than specific positions in a food chain.

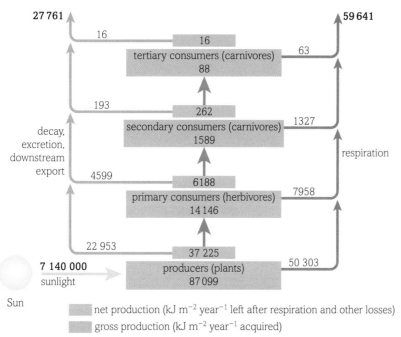

fig D The complex energy interactions between different trophic levels in the river at Silver Springs, Colorado, as measured by E.P. Odum in 1957.

Energy transfer and food chain length

As you have seen, relatively little of the light falling on the Earth is used to produce plant material, and increasingly small amounts of that material are passed from one trophic level to the next in a food chain. One of the main effects of the relatively inefficient transfers through food chains and webs is to limit the number of trophic levels. At higher trophic levels, the organisms usually need to range over larger distances so, by the fourth or fifth trophic level, it could take more energy to get food and a mate than is needed for growth and reproduction, and so survival is impossible. Some scientists suggest this explains why food chains in ecosystems in tropical regions tend to be longer than food chains in polar regions – because tropical regions receive more light. However, the complexity of food webs in tropical systems makes it difficult to isolate individual food chains, so it is hard to confirm or refute the accuracy of this idea.

Questions

1 (a) Looking at the data in **fig C**, which type of ecosystem is most productive by percentage of the area of the Earth that it covers?

   (b) Which type of ecosystem is least productive, based on the same comparison?

   (c) Why is primary production in the open oceans so important?

2 (a) Calculate the efficiency of the net energy transfers between the trophic levels in the ecosystem shown at Silver Springs (to 1 decimal place) see **fig D**.

   (b) Calculate the average energy transfer for this system.

3 (a) Look at **fig D**. Suggest why Odum's diagram shows only trophic levels and not the individual species in the food web. Give as many reasons as you can.

   (b) What assumptions would Odum have made in order to create this diagram?

Key definitions

Gross primary productivity (GPP) in plants is the rate at which light from the Sun catalyses the production of new plant material, measured as $g\,m^{-2}\,year^{-1}$, $g\,C\,m^{-2}\,year^{-1}$ or $kJ\,m^{-2}\,year^{-1}$.

Net primary productivity (NPP) is the material produced by photosynthesis stored as new plant body tissues, that is NPP = GPP – R.

The Sun will provide light for photosynthesis for the foreseeable future, but the supply of the other resources needed by producers to build molecules such as carbohydrates, proteins and lipids is not an inexhaustible one.

The need for nutrient recycling

Plants constantly take minerals such as nitrates from the soil and compounds such as carbon dioxide from the air to build macromolecules in their cells. These are passed on into animals through the food chains and food webs that link all living organisms. If this was a one-way process, the resources of the Earth would have been exhausted long ago. Fortunately, complex cycles exist that ensure the chemical constituents of life are continually recycled within ecosystems. These cycles involve a biotic phase, where the inorganic ions are incorporated in the tissues of living things, and an abiotic phase, where the inorganic ions are returned to the non-living part of the ecosystem.

The nutrients held in the bodies of dead animals and plants, and in animal droppings and urine, are released back into the soil and the air by the action of the bacteria and fungi that make up the decomposers. These decomposers feed on the droppings and dead bodies, digesting them and using the nutrients for respiration, to build their own cells, and for reproduction. The microorganisms also release waste products, and these provide the nutrients in a form that plants can use once more.

fig A The dead body of this tree is slowly being broken down by the action of decomposers. Some of the fungi are clearly visible, but the billions of bacteria are too small to be seen.

Conditions for recycling nutrients

The chemical reactions that take place in microorganisms, like those in many other living things, work faster in warm conditions. However, since the reactions are controlled by enzymes, if the temperature gets too hot they stop altogether, as the enzymes denature. They also stop if conditions are too cold.

Most microorganisms grow better in moist conditions, which make it easier to dissolve their food and also prevent them from drying out. So the decay of dead plants, animals and dung takes place far more rapidly in warm, moist conditions than it does in cold, dry ones. A leaf takes about a year to decompose in the UK, but can take only 6 weeks in a tropical rainforest. The majority of decomposers respire like any other organism, so decay takes place more rapidly when there is plenty of oxygen available.

People use the processes of decomposition in the treatment of sewage and the production of compost. However, it is in the natural world where the role of the decomposers is most important, and where the cycling of resources plays a vital role in maintaining the fertility of our soil and the health of our atmosphere. In a stable community of plants and animals living in an environment, the processes that remove materials from the soil are balanced by processes that return materials. In other words, the materials are constantly cycled through the environment. By the time the microbes and detritus feeders have broken down the waste products and the dead bodies of organisms in ecosystems, most of the energy originally captured by green plants in photosynthesis has been transferred to other organisms or back into the environment. Here are two examples of the importance of microorganisms in nutrient recycling.

Microorganisms in the nitrogen cycle

The recycling of nitrogen between the biotic and abiotic phases of an ecosystem is called the **nitrogen cycle**. Nitrogen is a vital part of the structure of many biologically important molecules, including amino acids and proteins, and it is also part of the molecules of inheritance, DNA and RNA.

Plants can make carbohydrates by photosynthesis, but carbohydrates do not contain nitrogen. Almost 80% of the air is nitrogen, but plants cannot use it because nitrogen is an inert gas and therefore unreactive. Only nitrogen in the form of nitrates is useful to plants and they absorb this from the soil water. They use these nitrates to make proteins, and this protein is passed along the food chain. In this way the nitrogen taken from the soil becomes incorporated into the bodies of all types of living organisms.

The nitrates taken out of the soil by plants are returned in a number of ways. Urine contains urea, a breakdown product of proteins, and proteins are also passed out in the faeces, so the waste passed out of animal bodies contains many nitrogen-rich compounds.

Similarly, when animals and plants die, their bodies contain a large proportion of protein. Some of the decomposers act specifically on the proteins, breaking them down to form **ammonium compounds**. These ammonium compounds are then oxidised by **nitrifying bacteria** that convert them to nitrates. These are returned to the soil to be absorbed by plants through their roots again.

Not all of the nitrates in the soil come from the process of decay. **Nitrogen-fixing bacteria** in the soil can convert nitrogen from the soil air into ammonia, and this is then converted into nitrates by the nitrifying bacteria of the nitrogen cycle.

One group of plants plays a particularly important role in the nitrogen cycle. The **legumes**, plants such as peas, beans and clover, have nodules on their roots that are full of nitrogen-fixing bacteria. This is an example of mutualism, where two organisms live together and both benefit. The bacteria get protection and a supply of organic food from the plant, while the plant gets ammonia that it can use to form amino acids. The bacteria produce far more ammonia than their host plant needs and the excess passes into the soil to be used and turned into nitrates by the nitrifying bacteria.

Not all the bacteria in the soil are helpful in the nitrogen cycle. One group, known as the **denitrifying bacteria**, actually use nitrates as an energy source and break them down again into nitrogen gas. Denitrifying bacteria reduce the amount of nitrates in the soil.

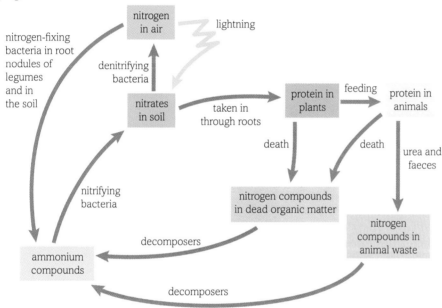

fig B The cycling of nitrogen in an ecosystem.

Microorganisms in the carbon cycle

Microorganisms are also involved in the constant cycling of carbon in the carbon cycle. Carbon is vital for living organisms, because all of the main molecules of life – carbohydrates, proteins, fats and DNA – are based on carbon atoms in combination with other elements. There is a vast pool of carbon in the form of carbon dioxide in the air and dissolved in the water of rivers, lakes and oceans. At the same time, carbon is constantly recycled between living things and the environment. This is known as the **carbon cycle**.

Carbon dioxide is removed from the air by green plants in the process of photosynthesis. It is used to make the carbohydrates, proteins and fats that make up the body of the plant. When the plants are eaten by animals, and those animals are eaten by predators, the carbon is passed on and becomes part of the animals' bodies. This is how carbon is taken out of the environment. How is it returned?

When green plants themselves respire, any carbon dioxide not used for photosynthesis is returned to the atmosphere. Similarly, when animals respire, they release carbon dioxide as a waste product into the air. Finally, when both plants and animals die, their bodies are broken down by the action of decomposers and when these microbes respire, they release carbon into the atmosphere as carbon dioxide, ready to be taken up again by plants for photosynthesis.

Carbon dioxide is also released into the atmosphere by combustion when anything that has been living is burnt – whether wood, straw or fossil fuels made from animals and plants which lived millions of years ago.

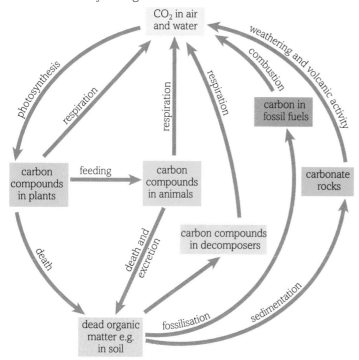

fig C The role of organisms, including microorganisms, in the cycling of carbon in the natural world.

Some of the bacteria involved in the carbon cycle produce cellulase. This enzyme breaks down the cellulose in plant cell walls to produce sugars that can then be used as food by a wide range of other organisms. Decomposers release 60×10^{15} g of carbon per year into the atmosphere compared with the $6–7 \times 10^{15}$ g of carbon released per year by the burning of fossil fuels – although this latter figure is rising.

fig D The actions of decomposing microorganisms break down organic material, including any cellulose in undigested plant material, and release carbon dioxide into the atmosphere.

Carbon sinks

Fig C shows that there are massive abiotic and biotic **carbon sinks** in nature. These are reservoirs where carbon is removed from the atmosphere and 'locked up' in organic or inorganic compounds. In the biotic part of the system, carbon is removed from the atmosphere by photosynthesis and stored in the bodies of living organisms. Soil also contains great amounts of organic, carbon-rich material in the form of humus. In the abiotic part of the system, rocks such as limestone and chalk, and fossil fuels such as coal, oil and natural gas, hold vast stores of carbon.

The oceans also act as massive reservoirs of carbon dioxide and contain around 50 times more dissolved inorganic carbon than is present in the atmosphere. The carbon dioxide is in continual exchange at the air–water interface. Carbon dioxide dissolved in the water is taken up in photosynthesis by the phytoplankton that live in the surface waters of the oceans. Large amounts of carbon are also stored in the calcium carbonate shells produced by many different marine organisms and in coral reefs. By lowering the concentration of dissolved carbon dioxide, they make it possible for more carbon dioxide from the air to be absorbed by the water.

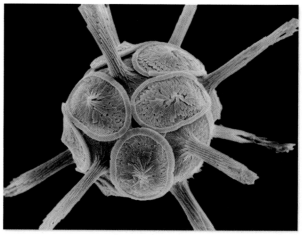

fig E Each year, over half of all the photosynthesis takes place in microscopic phytoplankton in the oceans.

The Atlantic Ocean is a particularly important ocean sink, absorbing up to 23% of human-produced carbon each year. The Southern Ocean covers a much bigger area, but only contains 9% of the total carbon. The differences are due to a variety of factors, including water temperature and ocean currents in the Northern (Atlantic) Ocean that move carbon-rich water downwards and bring more water up from the depths to absorb more carbon.

The quantity of carbon stored in the different carbon sinks is measured in petagrams: 1 petagram is 10^{15} g or 1 billion tonnes. Around 110 petagrams of carbon are removed each year from the atmosphere into the bodies of living organisms by photosynthesis. About 50 petagrams are returned to the atmosphere by the respiration of living organisms, and another 60 petagrams are returned through decomposition. The carbon becomes part of the organic material in the soil – the soil sink – before eventually being released back to the atmosphere as carbon dioxide. Left to itself, the carbon cycle is self-regulating. The amounts of carbon released in respiration and other natural processes and absorbed in photosynthesis remain in balance, so that atmospheric carbon dioxide levels remain relatively steady.

Sink	Amount in petagrams
atmosphere	from 578 (in 1700) to 766 (in 1999)
soil organic matter	1500–1600
ocean	38 000–40 000
marine sediments and sedimentary rocks	66 000 000–100 000 000
terrestrial plants	540–610
fossil fuel deposits	4000

table A Estimated major stores of carbon on the Earth.

The human influence?

Until the last few hundred years, humans were probably fairly carbon-neutral. However, the evidence is building that the enormous increase in the production of carbon dioxide by people, since the Industrial Revolution in the eighteenth and nineteenth centuries, coupled with the development of the internal combustion engine and its use in cars from the nineteenth century onwards, is now threatening the balance of the carbon cycle. Scientists are collecting evidence that the level of carbon dioxide in the atmosphere is increasing and many predict that this could have major effects on climate, geology and the distribution of organisms. You will be looking at human effects on ecosystems in **Chapter 10.3** and an understanding of the carbon cycle will help you consider models of ways in which atmospheric carbon dioxide levels have increased and may be reduced.

Questions

1 Why is the recycling of nutrients in ecosystems so important?

2 Compare the roles of microorganisms in the recycling of nitrogen and of carbon in ecosystems.

3 In **table A**, the amounts of carbon in many of the different carbon sinks are shown as ranges. Suggest reasons why the quantities in the different sinks have changed over time.

Key definitions

The **nitrogen cycle** is the recycling of nitrogen between living things and the environment by the actions of microorganisms.

Ammonium compounds contain the ammonia group $-NH_3$.

Nitrifying bacteria oxidise ammonium compounds to form nitrites and nitrates.

Nitrogen-fixing bacteria in the soil can convert nitrogen from the soil air into ammonia, and this is then converted into nitrates by the nitrifying bacteria of the nitrogen cycle.

The **legumes** are plants, such as peas, beans and clover, which have nodules on their roots that are full of nitrogen-fixing bacteria.

Denitrifying bacteria break down nitrates to power the production of ATP, producing nitrogen gas as a waste product.

The **carbon cycle** is the series of reactions by which carbon is constantly recycled between living things and the environment.

A **carbon sink** is a reservoir where carbon is removed from the atmosphere and 'locked up' in organic or inorganic compounds.

1 Only approximately 10% of energy is transferred between trophic levels. Which of these is **not** a reason for the remaining energy being lost?
   A respiration
   B roots
   C undigested food
   D photosynthesis [1]

   [Total: 1]

2 These types of bacteria are involved in the nitrogen cycle. Choose which of these bacteria carry out the reactions that follow.
   A nitrifying bacteria
   B nitrogen-fixing bacteria
   C denitrifying bacteria
   D decomposing bacteria
   (a) Releasing ammonia from proteins. [1]
   (b) Converting ammonia into nitrates. [1]
   (c) Converting nitrogen into ammonia. [1]
   (d) Converting nitrates into nitrogen. [1]

   [Total: 4]

3 The table below shows the fresh biomass of green plants and consumers on an area of grassland.

Organism	Fresh biomass/g
Green plants	2250.0
Primary consumers	240.0
Secondary consumers	38.0

   (a) (i) Calculate the percentage loss in fresh biomass between the green plants and the primary consumers. Show your working. [2]
       (ii) Give **two** reasons to explain the loss in biomass between the primary and secondary consumers. [2]
   (b) Only a small percentage of the light energy that falls on the green plants is used in photosynthesis. Explain why blue and red light would be more useful to a plant than green light. [3]
   (c) Explain how acid rain could lead to a decline in the biomass of green plants growing on the grassland. [4]

   [Total: 11]

4 A group of students carried out an investigation to determine the energy flow in a rainforest food chain. They collected invertebrate animals from trees, identified them and placed them in the appropriate trophic level.
   For each trophic level the animals were counted, weighed and the energy content determined. Using their data and other sources of information the students produced a diagram to show the energy flow along the food chain. The diagram is shown in **fig A**.

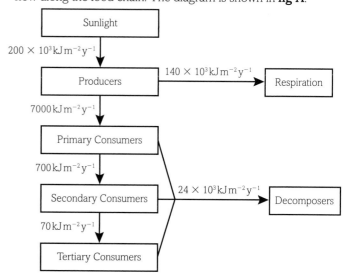

fig A

Some farmers clear plots of rainforest to use for crops. The trees are felled and then burnt. This practice is called slash and burn. The soil is left covered in ash, which is rich in nutrients. However, the nutrients are soon used up by the growing crops. Within two or three years the plot is abandoned and the farmer moves on to a new plot.

The abandoned plot is colonised by tree species and eventually the land is covered by secondary rainforest. The sequence of events is shown in the table below. This table also shows that the total biomass of the rainforest trees is made up of leaves, stems and branches, roots and leaf litter. These components of the biomass change as the rainforest is cleared, farmed and then abandoned.

Biomass component	Primary rainforest	Farming		Secondary rainforest	
		Slash and burn	Plot abandoned	10 years	25 years
Leaves	50	10		50	50
Leaf litter	400	50		200	300
Branches	50	10		50	50
Roots	200	50		50	75

The biomass components are shown in million tonnes per hectare.

(a) State what is meant by the term **net primary production** (**NPP**). [2]

(b) (i) Explain why only a small percentage of the light energy falling onto a leaf is converted into chemical energy. [2]

(ii) Explain why only 10% of the energy locked up in the secondary consumers is transferred to the tertiary consumers. [3]

(c) The energy shown entering the decomposers in **fig A**, is much greater than that entering all of the consumers. Give an explanation for this difference. [1]

(d) (i) Using the table on the previous page, compare and contrast the components of the biomass of the primary rainforest with those of the secondary rainforest after 25 years. [3]

(ii) Using all of the information provided, justify why slash and burn farming is considered to be unsustainable. [3]

(e) Give reasons on how the biodiversity of the surrounding rainforest could be changed by the practice of slash and burn. [4]

[Total: 18]

5 After harvesting, wheat seeds may be stored for several months. During storage, a proportion of the seeds will be lost. Most of this loss is due to two causes. The flour beetle, *Tribolium* sp., feeds on live dormant seeds and bacteria decompose dead seeds. Many of the flour beetles are eaten by carnivorous insects called *Xylocoris* sp.

The diagram below shows the relative energy flow within a wheat store.

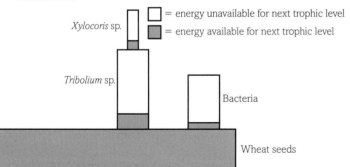

(a) State a suitable unit used to record the energy content of the wheat seeds. [1]

(b) Using the information above, copy and complete the table below by writing appropriate terms in the spaces. [2]

Organism	Trophic level	Mode of nutrition
Wheat seeds		
Tribolium sp.	Primary consumer	Heterotrophic
Xylocoris sp.	Secondary consumer	
Bacteria		Heterotrophic

(c) (i) Explain why some of the energy that passes to *Tribolium* sp. from the wheat seeds is not available to be passed on to *Xylocoris* sp. [3]

(ii) Explain why all of the energy in the wheat seeds is available to be passed on to the next trophic level. [2]

[Total: 8]

6 In summer, algal blooms are a common sight on ponds and lakes. One type of algal bloom is caused by cyanobacteria. The cyanobacteria release poisonous chemicals that can kill animals that drink the water.

(a) Cyanobacteria are prokaryotes. Draw a diagram of a typical prokaryotic cell. On your drawing, label the following **four** structures: cell wall, cell surface (plasma) membrane, flagellum and plasmid. [4]

(b) Under normal conditions, the numbers of cyanobacteria in a lake are controlled by zooplankton (microscopic animals) and fish. A typical food chain involving cyanobacteria is shown below.

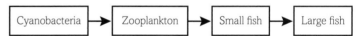

Cyanobacteria → Zooplankton → Small fish → Large fish

Explain how a sudden increase in the level of nitrates in the lake could disrupt this food chain. [3]

(c) The numbers of cyanobacteria can be controlled by using mats of barley straw which float on the water. The barley straw releases growth inhibitors. Explain how these inhibitors could interfere with the growth of the cyanobacteria. [2]

(d) It may be possible to reduce the numbers of cyanobacteria by introducing predators or parasites of the cyanobacteria. Give **one** reason why it would be preferable to use biological control, rather than chemical control, to reduce the cyanobacteria. [1]

[Total: 10]

CHAPTER

10.3 ▶ Human effects on ecosystems

Introduction

The human population of the planet is over 7 billion and growing. Every individual contributes some greenhouse gases to the atmosphere, by actions from breathing and growing the food they need to eat, to using electricity or driving a car. Everyone also produces bodily waste. It is estimated that over 200 million tonnes of human waste is untreated every year and much of that goes into our seas, oceans, rivers and lakes. This causes disease on a massive scale, killing up to 1.4 million children alone each year. The pollution produced also has a massive effect on ecosystems all around the world. This is just one of many ways in which people affect their environment.

In this chapter you will be considering the effect humans have on ecosystems. You will look at some of the evidence that the climate of the world is changing, and the apparent correlations with human activities that increase the emissions of greenhouse gases. You will also look at the human impact on biodiversity and population sizes both on the land and in the oceans. You are going to discuss the idea that the sustainability of the world's resources depends on effective management of the conflict between human needs and the conservation of biodiversity. You will consider evidence for how people are attempting to solve these problems by reducing the production of greenhouse gases that appear to be causing climate change on a global scale, and ways of conserving fish stocks across the world. As part of this process, you will learn about the effects of international treaties such as CITES in helping to conserve global biodiversity. You will also see the role played by scientific journals, the peer review process and scientific conferences in validating evidence relating to the debate on climate change – an area which until recently has been very contentious and where there is still work to be done.

All the maths you need

- Use ratios, fractions and percentages (*e.g. percentage contributions of different greenhouse gases to the greenhouse effect*)
- Construct and interpret frequency tables and diagrams, bar charts and histograms (*e.g. link between number of lactations and methane production in cows*)
- Translate information between graphical, numerical and algebraic forms (*e.g. evidence on carbon dioxide levels from different sources*)

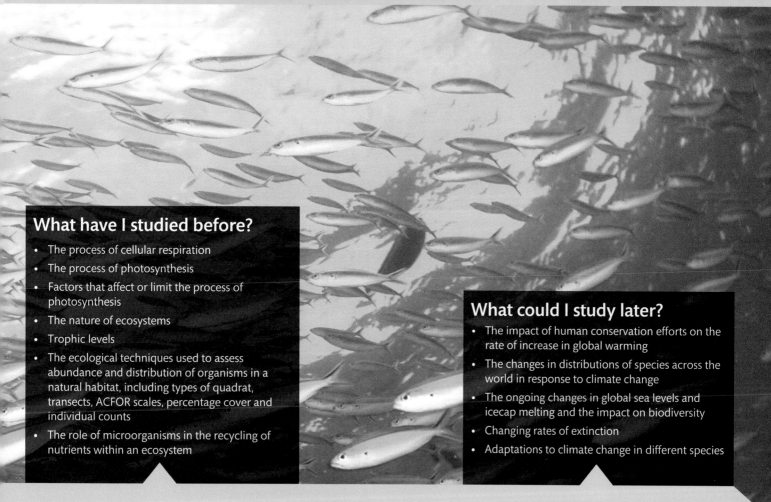

What have I studied before?

- The process of cellular respiration
- The process of photosynthesis
- Factors that affect or limit the process of photosynthesis
- The nature of ecosystems
- Trophic levels
- The ecological techniques used to assess abundance and distribution of organisms in a natural habitat, including types of quadrat, transects, ACFOR scales, percentage cover and individual counts
- The role of microorganisms in the recycling of nutrients within an ecosystem

What could I study later?

- The impact of human conservation efforts on the rate of increase in global warming
- The changes in distributions of species across the world in response to climate change
- The ongoing changes in global sea levels and icecap melting and the impact on biodiversity
- Changing rates of extinction
- Adaptations to climate change in different species

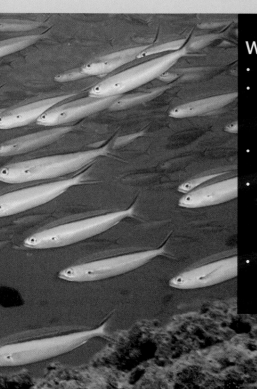

What will I study in this chapter?

- Human effects on ecosystems
- Some of the evidence relating to human effects on ecosystems, including climate change and depletion of biological resources, including overfishing
- The effect that treaties such as CITES have had on global biodiversity
- How sustainability of resources depends on effective management of the conflict between human needs and conservation, using our attempts to conserve fish stocks and to reduce the possible causes of climate change as examples
- The role of scientific journals, the peer review process and scientific conferences in validating evidence relating to the debate on climate change

All over the world people are constantly taking resources from the biosphere. Global ecosystems supply food, water, building materials, clothing, medicines, and more, for a population of over 7 billion people. As the human population gets ever larger and we take more from the environment than we need simply for survival, biological resources are increasingly being depleted. In addition, the amount of human waste produced also has an impact on ecosystems, causing problems for other organisms. The extinction of species and loss of biodiversity in many areas of the world are clear examples of human influences on a wide range of ecosystems.

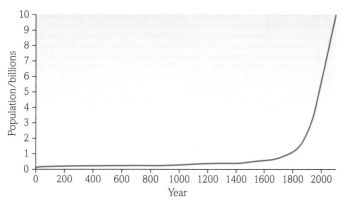

fig A This graph shows the enormous increase in the world population in recent centuries.

The human population explosion

As people learned to farm plants and animals to provide a reliable supply of food, more children survived and the populations grew, but on a relatively small and local scale. Later, tools and then machines enabled us to farm on a much bigger scale. Now we can change the environment with reservoirs, roads, canals, towns and cities. We have developed medicines that keep us and our children alive and allow many of us to survive to great old age. We have also developed engines that burn fossil fuels and release large quantities of exhaust gases into the atmosphere. We clear huge areas of rainforest to grow crops or rear cattle. We have developed floating factories that can clear whole areas of the oceans of fish – a long way from artisanal fishing boats that provide for a family, a village or a local market.

All of these activities have been driven by basic desires to provide ourselves with food, shelter and goods. As an unintended side effect we appear to be having a major impact on many of the ecosystems of the Earth. Scientists have increasing evidence that we are affecting the global climate, biological resources and biodiversity around the world.

fig B A city ecosystem is very different from a natural ecosystem – and Barcelona is not even one of the biggest cities in the world.

Human influences on ecosystems

Human effects on ecosystems are many and far reaching. You are going to consider three in particular:

Climate change

From individual observations, to major scientific studies, there is a growing consensus that the Earth's climate is changing. Global temperatures are rising. This is seen both in overall climate patterns and in the increasing number and frequency of 'extreme weather events' all over the world.

The world's climate has changed regularly over time. The fossil record shows us how often the world has gone through ice ages and periods of tropical heat or desertification. The difference is that this time it is happening fast and there is a growing body of evidence that the changes are the direct result of human activities (see **Section 10.3.2**).

Depletion of biological resources

As the population of the world has grown, so has the demand for resources such as firewood, land for housing, and food. In the more economically developed parts of the world, people have gone way beyond simply fulfilling the basic needs of life. They eat far more food than they need and want a great variety of different foods. As a result, biological resources are being depleted and ecosystems destroyed, both on the land and in the seas and oceans (see **Section 10.3.3**).

Loss of biodiversity

As you have already learned, biodiversity is extremely important, in terms of both variety of species and genetic diversity within each species. Human influences appear to be affecting, and often reducing, biodiversity in a number of ways, through activities such as overfishing, habitat reduction and as a result of climate change.

The burden of proof

As you know scientists cannot simply state that one thing causes another. They look for correlations in data and carry out further investigations to see if there is a causal relationship or whether the correlation is simply coincidence.

When a scientist or a group of scientists has carried out research, they publish their findings. Scientific findings are not widely accepted by the scientific community unless they are published in a reputable scientific journal. This is because a paper submitted for publication in a journal goes through a process of **peer review**. The paper is sent to other scientists, who are experts in the field, for them to read and assess. They may even try out any techniques described. If the peer reviewers all agree that the paper is **valid** and **reliable**, it will be published.

Some journals are well known and very well regarded, while some are smaller and specialised but still respected. Some journals have a reputation for being slightly less rigorous in the content they admit. Wherever science is published the authors have an ethical responsibility to ensure the highest standards of research design, data analysis and interpretation of data are applied. Cases where scientists are less than honest and the peer review process fails to detect this are fortunately very rare and published data is widely regarded as reliable.

Scientists who are working in the same field meet up regularly at scientific conferences. Here they share their data, discuss ideas, and listen to a number of presentations on the same area of work. Conferences give scientists somewhere they can talk, take a critical look at other work in their own field, meet up with others, collaborate and move their investigations and understanding forward.

When you are evaluating a study to decide whether or not to take notice of the conclusions, there are a number of factors to take into account. You need to examine the methodology to see if it is valid, that is, properly designed to answer the questions being asked. You need to know the size of the study and to see if the measurements have been carried out with precision and accuracy. It is also important to find out if other scientists have been able to repeat the methodology and have had similar results – if so, the results are considered more reliable. Knowing who carried out the research, who funded the study and where it was published can also help you decide whether anything might have affected or **biased** the study. You then need to evaluate the data and conclusions from the study in the light of all these factors.

In the next few pages you are going to be presented with some of the evidence scientists have been gathering on the impact of humans on the ecosystems of the world. Much of the data you will study comes from a report by the **Intergovernmental Panel on Climate Change** (**IPCC**), published in 2013. It was gathered and analysed by scientists from countries including Australia, Canada, Chile, China, France, Germany, India, Japan, New Zealand, Norway, Russia, Switzerland, the USA and the UK.

The IPCC is a body arising from the United Nations Environmental Programme and the World Meteorological Organization. This panel analyses research, from scientists all over the world, on climate change and produces regular unbiased reports based on all the available data to be used by politicians and decision makers globally. The role of the IPCC is 'to provide the world with a clear scientific view on the current state of knowledge in climate change and its potential environmental and socio-economic impacts'.

Questions

1 Look at **fig A**. Suggest three developments that might help explain the shape of the graph from around 1800 onwards.

2 Without the benefit of scientific evidence, suggest three reasons why in the twenty-first century the human population might have a major impact on the ecosystems of the Earth.

3 Do some research and find one piece of evidence to show the effect of human influence on each of the following:
   (a) global climate
   (b) the depletion of biological resources
   (c) loss of biodiversity.
   For each of the pieces of evidence you find, assess their reliability.

Key definitions

Peer review is the process by which scientific papers are sent to other scientists who are experts in the field, for them to read and assess before publication in a journal.

A **valid** study has been properly designed to answer the questions being asked.

In a **reliable** study, other scientists can repeat the methodology and obtain similar results.

If a study is **biased**, it has been influenced in some way to deliver a particular conclusion.

The **Intergovernmental Panel on Climate Change (IPCC)** analyses research from scientists on climate change and produces regular unbiased reports based on all the available data to be used by politicians and decision makers globally.

By the end of this section, you should be able to...

● explain some of the data relating to human influences on ecosystems, including climate change

Over the last few years a large body of evidence has been built up to show that:

- global temperatures are increasing, having an inevitable effect on global climate

- the levels of carbon dioxide and other greenhouse gases in the atmosphere are increasing at an unprecedented rate.

Scientists all over the world have been collecting data on the temperature changes and carbon dioxide levels associated with climate change, and looking to see if there is a causal relationship between them.

The greenhouse effect

Greenhouse gases reduce heat loss from the surface of the Earth in a way that is similar to how glass panels reduce heat loss from a greenhouse. These gases have a very important role in the atmosphere, without which life on Earth as we know it would not be possible. They maintain the temperature at the surface of the Earth at a level suitable for life. Without greenhouse gases in the atmosphere, the surface of Earth would probably be more like that of Mars, with an average surface temperature of –63 °C instead of 14 °C.

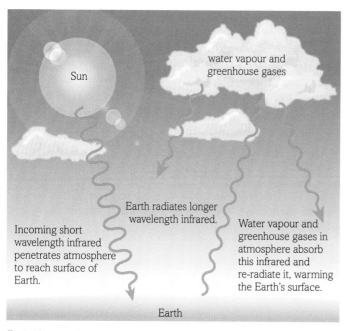

fig A The greenhouse effect.

Carbon dioxide is one of the 'greenhouse gases'; others include methane and water vapour. When radiation from the Sun reaches the Earth, some is reflected back into space by the atmosphere and by the surface of the Earth and some is absorbed by the atmosphere. The key wavelength is infrared, the radiation that makes us feel warm. Infrared radiation that reaches the Earth's surface is of a fairly short wavelength. It is absorbed by the surface of the Earth and then radiated from the surface at a longer wavelength. Some of this radiation is absorbed and re-radiated back to the Earth's surface by greenhouse gas molecules in the atmosphere. This is known as the **greenhouse effect** (see **fig A**) and it is vital to life on Earth. In theory, if the levels of greenhouse gases increase, the 'greenhouse' becomes more effective and the temperature of the atmosphere rises.

Evidence for global temperature increases

The UK Met Office has daily weather records going back to 1869, but written evidence from diaries and ships' logs goes back over 100 years more. Recent weather records also suggest that the Earth's surface temperature is increasing. In 1998 the Intergovernmental Panel on Climate Change (IPCC) gathered together a lot of data to produce a graph of global temperature (see **fig B**).

Observed globally averaged combined land and ocean surface temperature anomaly 1850–2012

fig B The IPCC data published in 2001 uses data from a variety of sources to show how the temperature at the surface of the land and the oceans has changed since 1850.

We have data of measured temperatures only since the mid-1800s. Scientists have used a number of techniques to produce other data that can give an indication of the temperature, but not an exact value – the temperature is inferred. These other sources of data are called temperature proxies, and when they are used on graphs, we use error lines to show how accurate these values are thought to be. To see how temperature has changed over the centuries, scientists have taken readings using temperature proxies including tree rings, corals, ice cores and peat bog data. This has resulted in a famous graph known as the 'hockey stick graph' (see **fig C**) – the fluctuating black line indicates the mean data.

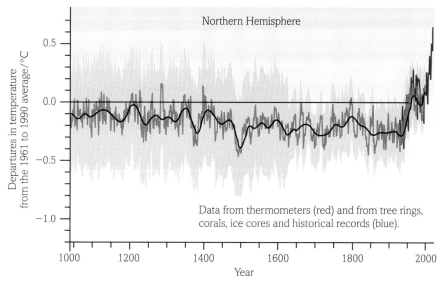

fig C The IPCC 'hockey stick graph' published in 2001 uses data from a variety of sources to show how the temperature has changed over centuries. The term 'hockey stick' comes from the shape of the graph. Note that the grey is the error on each measurement, that is the range in which the real value is thought to lie, as we cannot actually be very accurate!

Frozen isotopes

Antarctic and Greenland ice cores are a widely used source of temperature proxies. Scientists drill deep down into the ice and then analyse the air trapped in the different layers. This provides a record that goes back thousands of years. The oxygen isotopes in melted ice (the proportion of ^{18}O to ^{16}O) reflect the air temperature when the ice layer was laid down. Atmospheric carbon dioxide levels can also be measured.

The results of the analysis of air from ice cores for over 300 000 years is shown in **fig D**. It appears that about 140 000 years ago, the surface of the Earth was about 6 °C cooler than it is today and the Earth was in an ice age. On the other hand, about 120 000 years ago the climate was 1–2 °C warmer than it is now. These warm periods are known as interglacials. Since then we have had another ice age and some more warming.

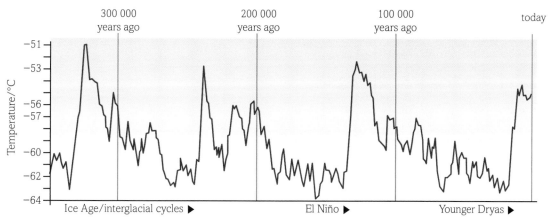

fig D Data from ice cores are used to show how the temperature of the Earth has fluctuated over time.

Increasing data reliability

Both **dendrochronology** (the dating of past events using tree-ring growth) and peat bog dating (using cores taken from peat bogs that show growth patterns over hundreds or even thousands of years) are used to confirm radiocarbon dating. For example, wood or peat bog samples of known age can be dated from radiocarbon measurements, using remains of plants and pollen grains, which give an indication of climatic conditions when those plants are alive, and the results compared to give a form of calibration. This gives scientists clear reference points that they can use to determine the accuracy of their estimations of age.

Data like these were used to produce the IPCC graph you saw in **fig B**. In 2008, scientists recalculated the figures using more than 1200 temperature proxy records going back 1300 years, without using tree-ring data, and used two different statistical methods. The hockey stick graph (**fig C**) was shown to be valid, with all statistical methods used and whether tree-ring data were included or not. All of the evidence points the same way – global temperatures are rising.

Evidence for increasing levels of carbon dioxide

Scientists have found evidence for the increasing levels of carbon dioxide in the atmosphere in many different ways. Some of the most famous evidence comes from what is known as the Mauna Loa curve, a series of readings taken at regular intervals at the Mauna Loa observatory on Hawaii (see **fig E(a)**). The air is sampled continuously at the top of four 7-metre tall towers and an hourly average of carbon dioxide concentration is taken (along with a number of other readings). The air in the area is relatively free from local pollutants and scientists believe it is representative of the air in the Northern Hemisphere. Measurements started in 1958 and the monitoring methods and instruments used have remained very similar throughout that time. The records show that the level of atmospheric carbon dioxide has increased from 315.98 ppmv (parts per million by volume of dry air) in 1959 to 381.74 ppmv in 2006. The annual fluctuations in the levels of carbon dioxide seem to be the result of seasonal differences in the fixation of carbon dioxide by plants, as in temperate regions plants lose their leaves in winter and take up less carbon dioxide.

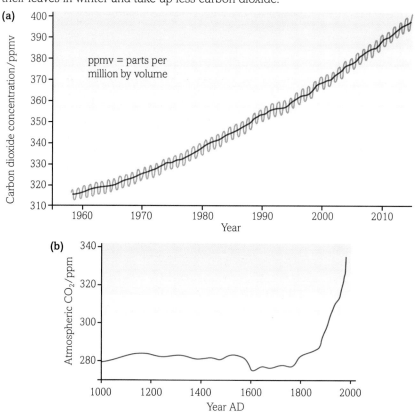

fig E (a) The Mauna Loa curve. Dr Pieter Tans, NOAA/ESRL (www.esrl.noaa.gov/gmd/ccgg/trends/) and Dr Ralph Keeling, Scripps Institution of Oceanography (scrippsco2.ucsd.edu/ (b) Atmospheric carbon dioxide levels from the Law Dome ice cores. Many regard data such as this as compelling evidence that atmospheric carbon dioxide levels are rising steadily. Law Dome Ice Core 2000-Year CO_2, CH_4, and N_2O Data, contributor David Etheridge, CSIRO Marine and Atmospheric Research; ftp://ftp.ncdc.noaa.gov/pub/data/paleo/icecore/antarctica/law/law2006.txt.

Ice core data also show clear changes in carbon dioxide concentration. **fig E(b)** shows data taken from the Law Dome ice cores – particularly pure and undisturbed ice in the Antarctic. The shape of the curve is similar to that of the temperature curves in **figs B** and **C**.

Scientists are also measuring carbon dioxide levels in the oceans and seas, both by the amount of carbon dioxide dissolved in the surface water and the changes in pH. The more carbon dioxide is dissolved in the water, the lower the pH. Look back to **Section 10.2.3** to remind yourself of the role of the oceans as a carbon sink.

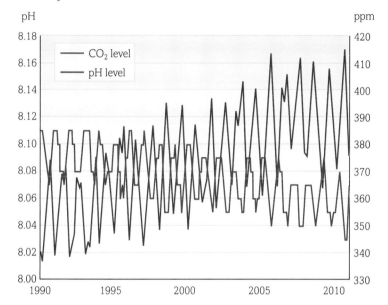

fig F The changing carbon dioxide levels in the oceans and the fall in pH that results reflect the increasing carbon dioxide concentration of the air above the water. Data courtesy of NOAA Coral Reef Watch.

The role of methane

Methane (CH_4) is a potent greenhouse gas, and over a period of 20 years has a 72 times greater effect on warming the atmosphere than carbon dioxide. However, much less of it is produced than of CO_2. Its main sources are from the decay of organic material by some species of bacteria, particularly in wet conditions, and from the digestion of ruminant herbivores, such as deer, sheep and cows. Methane breaks down naturally high in the atmosphere in a series of reactions that eventually form carbon dioxide and water molecules.

Methane levels have risen by about 150% since 1750 for several reasons. Rice paddy fields are waterlogged during much of the time the rice is growing and bacteria in this waterlogged soil release methane as they grow. Levels of rice production have been increasing steadily to feed the ever-increasing world population, and so more methane is produced. In addition, as the human population grows, so do the numbers of animals that we depend on for food, including cattle, who release methane from their digestive systems. Scientists have calculated that up to 60% of the methane in the atmosphere is now produced as a result of human activity in some way.

Greenhouse gas	Percentage estimated contribution
water vapour	36–72
carbon dioxide	9–26
methane	4–9
ozone	3–7

table A The estimated contributions of the main greenhouse gases to the greenhouse effect. The variation depends on the amount of water vapour present and the concentrations of other greenhouse gases.

Did you know?

Cows belch a lot and every time they burp they release methane gas. Estimates of the amount of methane produced per cow per day vary from 100 to 700 dm³. This varies depending on factors such as the breed of cow, the type of food eaten and whether the cow is giving milk. There are an estimated 1.2 billion cows in the world, so a lot of methane is being produced. The IPCC estimate that 16% of the methane produced as a result of human activities comes from livestock, and dairy cows produce the most.

A number of research teams have set out to breed or engineer new strains of grass that can be digested more easily by cows, reducing methane emissions. And an Irish research team looked at whether changing the way cows are farmed could reduce overall methane emissions. Older cows produce more milk, but less methane per pint. By keeping cows alive, healthy and giving milk for longer, the average methane emissions of the entire herd can be reduced.

Adding concentrates to the diet also reduces methane emissions per cow, because the concentrates are easier to digest and help prolong the cow's working life. However, generating the electricity needed in the manufacture of the concentrates produces carbon dioxide, so a balance needs to be struck.

Current evidence suggests that with a combination of good husbandry, careful breeding and possible genetic engineering of food plants, people may still be able to drink milk and eat beef, and reduce the production of methane at the same time.

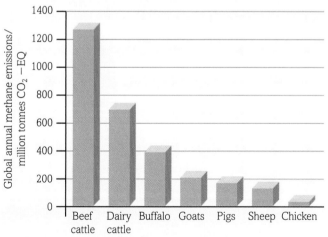

fig G Global annual methane emissions by species. Source: Gerber, P.J., Steinfeld, H., Henderson, B., Mottet, A., Opio, C., Dijkman, J., Falcucci, A. & Tempio, G. *2013 Tackling climate change through livestock – A global assessment of emissions and mitigation opportunities.* Food and Agriculture Organization of the United Nations (FAO), Rome.

Correlation or causation?

According to the IPCC, atmospheric concentrations of carbon dioxide, methane and nitrogen oxides have increased to levels that have not been seen for at least 800 000 years. Carbon dioxide levels have increased by 40% since pre-industrial times, primarily from fossil fuel emissions and secondarily from changes in land use, for example deforestation. The ocean has absorbed about 30% of the emitted **anthropogenic** (produced by people) carbon dioxide and this has caused ocean acidification. A lot of evidence from many studies now suggests a clear correlation between the rise in carbon dioxide and other greenhouse gases in the atmosphere and the increase in global temperatures. However, the correlation is so close (see **fig B**) that it can be difficult to decide whether increases in greenhouse gases are *causing* the rising temperatures or are the *result* of rising temperatures.

Models of global warming

To say that there is a causal relationship between rising carbon dioxide levels and rising temperatures, with their associated climate change, we need a mechanism to explain how one factor changes the other. From our understanding of the greenhouse effect and because of the timing of the changes, a logical step is to consider that humans are responsible. Since the Industrial Revolution we have burnt increasing quantities of fossil fuels for energy and for transport, and more recently to generate electricity, and all this produces carbon dioxide. Alternatively, some scientists have proposed a mechanism where solar activity affects cloud formation and therefore surface temperature. Some data (**fig I**) seems to show a correlation between solar activity and atmospheric temperature, rather than between carbon dioxide concentrations and temperature. However, after looking at all the evidence, the IPCC reached the conclusion that sunspot and solar flare activities over the past 50 years would most likely have produced cooling rather than warming, and that any influence they have on global climate is relatively small.

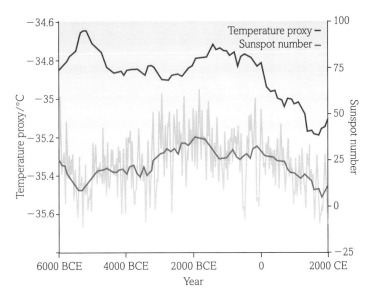

fig I The relationship between solar activity, temperatures high in the atmosphere and carbon dioxide levels is a complex one.

All these theories about global warming and climate change are based on data that require detailed interpretation and the use of computers to model very complex systems. Proving a causal link is almost impossible. However, many studies on different aspects of global warming, such as polar ice melting and climate change in different regions of the world using a wide variety of different computer models, suggest that the increase in atmospheric carbon dioxide and other greenhouse gases is increasing surface temperature, and that human carbon dioxide emissions are responsible for at least some of the current global warming and associated climate changes. The IPCC believe there is sufficient evidence now to state that there is a causal link. However, it will almost certainly turn out that global warming is multifactorial, with many different inputs – it is not only about carbon dioxide levels.

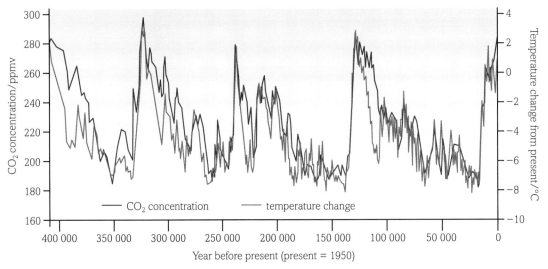

fig H Data from ice cores from Vostok (in the Antarctic) showing carbon dioxide content of the air trapped in the ice cores and the air temperature change relative to modern temperature based on measurements of the isotopic composition of the gases. Vostok Ice Core Data for 420,000 Years, contributor Jean Robert Petit, LGGE-CNRS; ftp://ftp.ncdc.noaa.gov/pub/data/paleo/icecore/antarctica/vostok/readme_petit1999.txt.

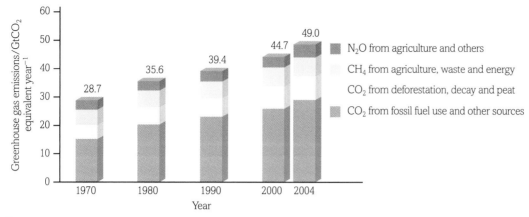

fig J Global anthropogenic greenhouse gas emissions between 1970 and 2004.

In 2007, the IPCC looked at data and models of climate change presented by scientists from all around the world. They saw that anthropogenic carbon dioxide levels increased by 80% between 1970 and 2004, mainly due to the use of fossil fuels. The IPCC decided that the balance of the evidence shows a 95% probability that human activities resulting in the build-up of greenhouse gases are at least partly responsible for the observed increase in global temperatures. In their 2013 report they state that it is *extremely likely* that human influence has been the dominant cause of global warming since the mid-twentieth century.

The IPCC use language carefully designed to indicate how strongly the evidence backs up the hypothesis. They have decided that it is *very likely* that human activities have contributed to the rise in sea level in the second half of the twenty-first century, but only *likely* that they have also influenced the changes in rainfall patterns that have been observed. At the moment climate changes and environmental damage are occurring far faster than anyone imagined. It appears that some of the effects of human influences on global ecosystems through anthropogenic climate change are already irreversible.

Questions

1   The term 'greenhouse effect' is widely used to suggest something negative. Why is this an inaccurate use of the term?

2   Discuss the evidence that global temperatures are steadily rising.

3   (a) What is the overall percentage increase in atmospheric carbon dioxide from 1959 to 2006 based on the Mauna Loa data (**fig E(a)**)?
     (b) Why is the data from Mauna Loa regarded as reliable?

4   What can the data from the Law Dome ice cores (**fig E(b)**) tell us that the Mauna Loa data cannot and how reliable is this data?

5   There is one very simple way of reducing the methane emissions from cattle. What is it and why do you think it is not widely suggested?

6   How does the data shown in **fig H** support the theory that carbon dioxide levels and the temperature at the Earth's surface are linked? How reliable is the data?

7   Using **fig J**, calculate the percentage increase in carbon dioxide from fossil fuel use between 1970 and 2004 and compare it with the overall percentage increase in greenhouse gases from all sources over the same period.

Key definitions

The **greenhouse effect** is the process by which gases in the Earth's atmosphere absorb and re-radiate the radiation from the Sun, which has been reflected from the Earth's surface, maintaining a temperature at the surface of the Earth that is warm enough for life to exist.

Dendrochronology is the dating of past events using tree-ring growth.

Anthropogenic means produced by people.

By the end of this section, you should be able to...

● explain some of the data relating to human influences on ecosystems, including climate change

Some of the biggest concerns about the impact of human activities on ecosystems centre on climate change and how it will affect the distribution of species around the world.

Climate change

Weather describes the state of the atmosphere at a particular time and place, with regard to temperature, rainfall, humidity and windiness. Climate is the average weather pattern in an area over many years. Rising temperatures affect weather and rainfall patterns and can also cause long-term changes in the climate. It is impossible to link any one weather event to global warming, but statistical evidence suggests that there is an increase in extreme weather events linked to the rise in global temperatures.

Rainfall patterns are complex, but they also seem to be changing. For example, there have been a number of years of lower than expected rainfall in Africa. In 2013, around 200 million people (25% of the African population) were experiencing high water stress. If the current trend for low rainfall continues in Africa, it is predicted that by the year 2050 between 350 and 600 million people will be short of water for their crops and to drink. In contrast, in some areas rainfall has been both higher than average and extremely heavy, leading to flooding, which causes devastation and carries away the vital topsoil. Areas of China, Pakistan and India have already seen a clear increase in torrential rainfall leading to severe flooding. An international group of scientists have taken monthly recorded rainfall from around the world from 1925 to 1999 and compared what really happened to various computer models. They found that many of the changes corresponded to those expected if global warming was a factor.

fig A Rising sea levels threaten low-lying communities all over the world. Venice in Italy is just one of many places that could disappear completely if global warming continues.

Risk of flooding

Many scientists believe that the thinning of polar ice is a clear indication of global warming and could result in flooding.

In 2002, 500 billion tonnes of ice broke away from the Antarctic peninsula and eventually melted into the sea. Also, Antarctic temperatures have increased by an average of 2.5 °C in the past 50 years – faster than anywhere else on the Earth. In the Arctic, the sea ice has been retreating by about 2.7% each decade since 1978 and many glaciers are also retreating at a rate of about 50 metres a year.

As the ice melts, the volume of water in the seas and oceans of the world will increase, causing sea levels to rise. Also, as the water gets warmer, its volume increases, resulting in an even bigger impact on sea levels. The implications for human life as sea levels rise are immense. Around 100 million people live less than 1 metre above current sea levels. For example, in the UK, large areas of the east coast could be lost for good, and the Netherlands might disappear completely!

The effect on organisms

Temperature has an effect on enzyme activity, which in turn affects the whole organism. There is an optimum temperature for many enzyme-controlled reactions and if the temperature increases beyond that point the enzyme starts to denature and the reaction rate falls, eventually stopping completely.

Increased temperature could have different effects on processes, including the rate of growth and reproduction. If plants grow faster they will take up more carbon dioxide and may therefore reduce atmospheric carbon dioxide levels. In other places, temperature may exceed the optimum for some enzymes, and organisms there will die. The majority of plant and animal species are found in the tropics and many have very little tolerance for change, because conditions in the tropics tend to vary very little throughout the year. Experimental data suggest that a change of just 1 °C could threaten the survival of up to 10% of all species. The insects, which are vital as pollinators of the many flowering plants, are particularly vulnerable and if they go, so do the plants, followed by the animals that feed on them, in a mass extinction.

At higher latitudes, seasonal cycles affect life cycles. Global warming appears to be affecting the onset of the seasons, affecting both life cycles and the distribution of species. Warmer temperatures mean that plants grow and flower earlier and insects such as moths and butterflies become active earlier as the plant food they need for their caterpillars is available. Some birds can adapt to these changes. For example, the breeding cycle of the great tits in Wytham Woods near Oxford in the UK has moved forward, triggered by the same temperature changes that have resulted in winter moth larvae, the main food supply for their chicks, being available. The UK great tits lay eggs about

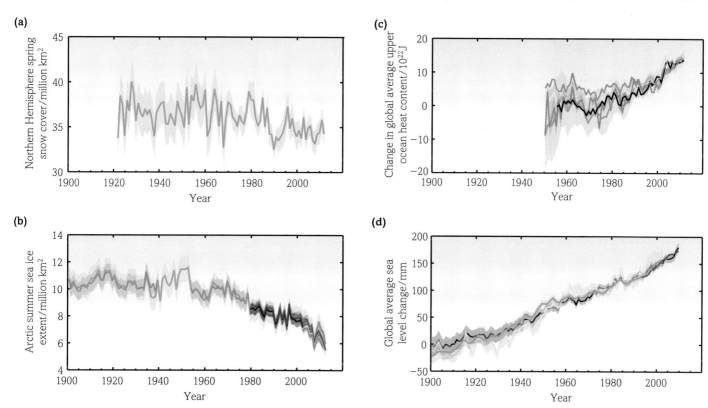

fig B The melting of the sea ice, the reduction in snow cover and the rise in sea levels all seem linked to anthropogenic climate change.

2 weeks earlier now than they did 47 years ago. However, great tit populations in the Netherlands are not doing so well. The breeding time is getting earlier every year but the caterpillars are emerging even earlier so the birds are missing the peak population, and raising fewer chicks. For some animal species, breeding earlier in the year may mean they can fit more than one breeding cycle in, so those populations will increase.

Changes in temperature could have an even more drastic effect on some organisms. For example, the embryos of some reptiles are sensitive to temperature as they develop. Male crocodiles develop only if the eggs are incubated at 32–33 °C. If the eggs are cooler or warmer, females develop. If global warming results in only female crocodiles developing, it could be the end of a species that has survived virtually unchanged for millions of years.

Changes in species distribution

A change in climate could affect the range of many different organisms. For example, alpine plants in mountainous parts of the UK are becoming rarer. Most animals can move more easily than plants, so they can often survive change more easily. As areas become warmer, some animals may be able to extend their ranges northwards, but may become extinct at the southern end of their current range. Others may be able to colonise a bigger area. In a study by Parmesan *et al.* in 1999 of 35 species of non-migratory European butterflies, the ranges of 63% of the species had shifted northwards by 35–240 km in the past 100 years and only 3% (one species) had shifted south. The shift in butterfly populations paralleled a 0.8 °C warming over Europe during this time.

If organisms involved in the spread of disease are affected, patterns of world health could change as well. The World Health Organization (WHO) has warned that global warming could be responsible for a major increase in insect-borne diseases in Britain and Europe. The prediction is that by 2100, conditions could be ideal for disease-carrying organisms such as mosquitoes, ticks and rats. The WHO is urging countries to make plans so that preventative measures can be put in place as the climate changes.

Questions

1. The models for global carbon dioxide stabilisation show a great deal of uncertainty and variety. Explain why.

2. (a) Using data from **fig B**, calculate the increase in average global heat content of the sea per year between 1950 and 2010, and the percentage reduction in Arctic summer sea ice extent between 1950 and 2010.

   (b) What was the average increase in average sea levels over the same period of time?

   (c) Suggest how these data might be linked to rising carbon dioxide levels, to the falling snow cover in the Northern Hemisphere and to each other.

3. Discuss how a change in atmospheric carbon dioxide levels can have an impact on disease in animals, plants and humans.

4. Draw up a table to summarise the main effects of global warming. Use this information to develop a flowchart of events that illustrate the best current model of the events linked to global warming.

By the end of this section, you should be able to...

● explain data relating to human influences on ecosystems, including the depletion of biological resources, such as overfishing

● explain how the sustainability of resources depends on effective management of the conflict between human needs and conservation

As the human population grows, so does our influence on ecosystems. All around the world, biological resources are being depleted at a very rapid rate.

Depletion of resources – farming

When we farm we remove the crop before the plants die and decompose and therefore break the natural cycles that return minerals to the soil. As a result, soil mineral concentrations can decrease rapidly, especially when a monoculture (where one crop is grown over a large area) absorbs large quantities of particular minerals. In some regions, monocultures are the mainstay of farming, from huge wheat fields to massive oil palm and banana plantations. In many other parts of the world, small-scale and family farms are common, but these too can deplete biological resources in the soil and in the surrounding ecosystems.

Artificial fertilisers can replace the minerals used by plants, but they do not support the structure of the soil. Once soil biodiversity is lost, the soil structure breaks down and it becomes infertile even when fertilisers are used.

Science and practical experience provide evidence of the impact of different farming methods and suggest ways in which land can be sustainably managed, but each way has cost and ethical implications. Science cannot dictate which is the best way for a particular community to farm; that is for each society to choose and the choice will depend on the needs and priorities of the people involved. Maximum yield from the land may be the priority for financial reasons or because without it a family will starve.

fig A (a) Virgin tropical rainforest, and (b) oil palm plantation. The forest is a source of immense biodiversity that is lost when it is destroyed. A farmed monoculture has no biodiversity and also continually depletes the soil of the same minerals.

Depletion of resources – fishing

Fishing, the harvesting of fish and other aquatic organisms such as crustaceans from coastal areas, seas and inland waters, provides food and employment for around 820 million people worldwide. Sadly, the fish stocks of the ocean are fast becoming one of the most depleted biological resources. If we take too many fish, or fish at the wrong times of year, the fish cannot breed and replenish the populations, and fishing becomes unsustainable. The problem is particularly acute in coastal areas, but is also seen in deeper oceans and in inland waters. The Food and Agriculture Organization of the United Nations (UN FAO) has published data showing that up to 25% of the major marine fish stocks are being depleted or over-exploited, putting the fish populations at risk of extinction. Another 44% are being fished right up to their safe biological limit.

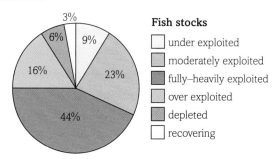

Fish stocks
☐ under exploited
☐ moderately exploited
☐ fully–heavily exploited
☐ over exploited
☐ depleted
☐ recovering

fig B Data from the UN FAO document *Fisheries at the limit?* showing the depletion of fish stocks throughout the world.

More Peruvian anchovies are caught than any other fish in the world, but in 2012 catches were down by 46%. Often fishing is local, carried out on a relatively small scale by local people, but it is also carried out on an industrial scale by large fishing fleets. Atlantic cod was traditionally caught by the UK fishing fleet and widely eaten in the UK, but the development of factory fishing fleets by other countries, and overfishing by the UK fleet, has seen such a dramatic decline in cod numbers that the population may never fully recover.

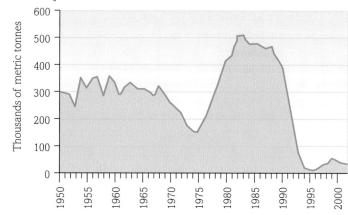

fig C There has been a 90% decline in the number of Atlantic cod caught by Canada's fishing fleet.

A number of factors appear to be causing the large-scale depletion of fish stocks around the world. These include:

- the size of the global fishing fleet – the fleet is currently almost twice the size that would be needed to take a sustainable supply of fish
- open-ocean factory ships that take huge catches of fish, often including many species that are not wanted as food
- techniques such as bottom trawling, where nets are dragged along the seabed, damaging the vulnerable seabed habitat and catching a wide variety of species, many of which are not wanted
- enormous drift nets that are almost invisible and catch and kill many species accidentally
- nets with small mesh sizes that catch immature fish as well as adult fish
- fishing through the breeding seasons.

Anthropogenic changes in the ocean

Changes in fish populations are not only related to the amount and method of fishing. There was a marked fall in the cod populations around 1975 and a peak around 1980, followed by steady decline. A number of studies suggested that these changes were the result of changes in environmental conditions as much as in the human fishing quotas. Global water temperatures, levels of pollution and numbers of natural predators such as seals can all vary considerably. Sea temperatures are affected by global warming through the cooling effect of melting sea ice. Rises and falls in sea temperature can affect the amount of phytoplankton, which are the producers in most marine food chains, and this can affect the food available for fish higher up the food chain. Another factor is that many fish spawn in relatively shallow coastal waters, which are more affected by temperature changes than the deep ocean.

The conservation conundrum

The scientific evidence is growing steadily to show that the sustainability of biological resources is dependent on human beings changing their behaviour. It is easy for a country with plenty of food, readily available education and healthcare and a strong infrastructure to condemn the felling of a rainforest. However, when people have very little, the drive to make money to buy the food, healthcare and education they so badly need for themselves and their children is understandably more important. Farming on an industrial scale, or fishing the oceans, provide food and a way of earning a living for many millions of people.

We need to find ways to halt the depletion of biological resources before it is too late. As you saw in **Book 1 Section 3.3.4**, a great deal of work is being done to conserve the biodiversity of individual species and of ecosystems and there is more on this in **Section 10.3.5**.

People are becoming increasingly aware that the **sustainability** of our resources depends on the effective management of the conflict between human needs and conservation. Sustainability demands a decent standard of living for everyone now, without

compromising the needs of future generations or of the ecosystems around us. This is not easy to achieve. There are too many human needs, vested interests and conflicts of interest to make this a simple problem to solve.

The evidence is building all the time for the effect of human activities on climate change. You have looked at some of the evidence showing how carbon dioxide levels in the atmosphere and the oceans are increasing steadily and the planet is getting warmer. When you look at **fig D**, you can see that the biggest producers of greenhouse gas emissions are not the cars on the road or the farmer, but the companies producing electricity, the industries that support the economic stability of the world, and forestry – especially deforestation and burning. The problem is that we all want electricity, industries are needed for global economic success and the countries that are cutting down rainforests need the resources to move above the basic survival line. It will take a lot of international cooperation as well as the effort of millions of individuals to make the changes needed to slow the production of greenhouse gases and replace as much of the lost vegetation as possible.

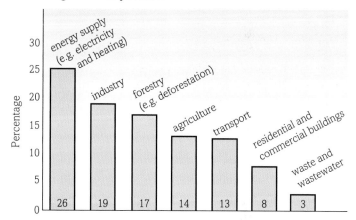

fig D More evidence of the way people are affecting the climate and how we produce our greenhouse gases.

Conserving fish stocks

Various methods of protecting fish populations have been introduced and are already having a small but measurable impact on fish numbers:

- controlling the size of the mesh in the fishing nets, so only the largest fish are caught
- banning fishing during the breeding seasons of different fish
- imposing very strict quotas on fishing fleets and individual fishing vessels
- encouraging the use of fishing methods that are less damaging to the ecosystems
- banning the catching of severely endangered species of fish altogether.

However, these controls need to be policed and introducing them deprives people of their livelihoods. In the UK alone, many fishing communities have been badly hit by unemployment, as the fishing boats have been forced to stay at home to protect our stocks of cod, haddock, plaice, sole and other fish. Encouragingly, there is evidence that this sustainable approach to fishing is enabling stocks to recover, and that it is now becoming possible to fish a sustainable harvest from the sea each year.

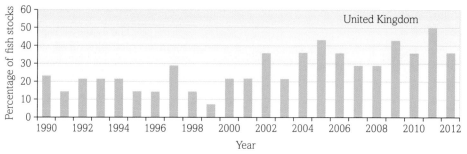

fig E The percentage of fish stocks harvested sustainably has gradually increased over the last 25 years or so, although there are still fluctuations.

Fish farming and aquaculture

fig F A fish farm – they produce tonnes of fish a year that does not have to come from the river or coastal ecosystems of the UK.

Another way of protecting the wild fish stocks in our coastal, ocean and lake ecosystems is to farm fish and other seafood such as mussels, prawns and other crustaceans. Fish farming, or **aquaculture**, is becoming a very successful way of providing people with the fish they want and therefore the protein that they need. More fish are eaten in China than anywhere else in the world, and around two-thirds of all commercial fish farming in the world takes place there – it is a successful business. In Africa people are increasingly farming fish like *Tilapia* in a new but growing industry. In countries such as the UK and USA, fish such as salmon, trout and shellfish, including mussels, are farmed.

In 2012, around 47% of all the fish eaten by people worldwide was produced in aquaculture. This will have an impact on protecting the wild stocks of fish in global ecosystems. Fish farms are not carbon neutral, they use electricity and produce greenhouse gases. Fish farms also feed their fish. Unfortunately their food is often made from other fish. So fish farming does nor necessarily prevent overfishing. Increasingly, alternative foodstuffs based on ingredients such as marine algae are being used in fish farms to reduce their environmental impact. Fish farms are not an ideal answer, but they are certainly an important part of the solution, producing sustainable fish stocks for the future and preserving the biological resources in our coastal waters and oceans.

Questions

1. (a) Give two examples of how the need to provide food for the human population is depleting biological resources in land ecosystems.
   (b) Give two examples of how the need to provide food for the human population is depleting biological resources in aquatic ecosystems.

2. Suggest two reasons why problems with the depletion of fish stocks are particularly noticeable in coastal areas.

3. Describe and discuss the data in **fig E** on the impact of sustainable fishing measures on UK catches.

4. Investigate fish farming and write a short report on the benefits and problems of this solution to the problem of overfishing.

Key definitions

Sustainability is the production of a decent standard of living for everyone now, without compromising the need of future generations or of the ecosystems around us.

Aquaculture is the farming of fish or shellfish in tanks or in containers in rivers, lakes or the sea.

By the end of this section, you should be able to...

● explain the effect that treaties such as CITES have had on global biodiversity

One of the main effects of human influences on ecosystems is a loss of biodiversity around the world. Loss of biodiversity may be a direct effect of human activities such as deforestation, farming, overfishing, building and the introduction of alien species. Equally, much biodiversity is being lost as an indirect result of human influences, such as soil, air and water pollution, and global warming and climate change.

The nature of extinction

Extinction means the permanent loss of all members of a species. It may be localised, when a species becomes extinct in an area or country, or total, when it is completely lost from the Earth. Well-known examples of extinct species include the dinosaurs, sabre-toothed tigers and the dodo. Extinction is a common event. Of the billions of species that have evolved since life appeared on Earth, only an estimated 2 million are still in existence today. Individual species become extinct because something changes in their habitat, for example changes in temperature may affect food supplies, a rise in sea level can cause flooding, or a new predator or disease may arrive.

Scientists have calculated the 'natural' or 'background' extinction rate, based on the fossil record. They estimate that this is between 1 and 100 species each year. The current extinction rate is often quoted as being much higher than this, with quotes varying from 100 times higher to 10 000 times higher than the background rate. However, these figures are largely theoretical and nobody knows exactly what is happening. The number of actual extinctions recorded and confirmed remains relatively small, although many species are endangered.

The rise in the extinction rate is thought to be mainly the result of human activity. Before species become extinct, both population sizes and genetic diversity fall. Scientists try to evaluate what is happening to species, to try and prevent them from becoming extinct. Threatened species are classified as follows:

• Critically Endangered – facing an extremely high risk of extinction in the wild

• Endangered – facing a very high risk of extinction in the wild

• Vulnerable – facing a high risk of extinction in the wild.

The **International Union for Conservation of Nature (IUCN)** is the largest global environmental organisation. It has 1200 member organisations, including over 200 governments and over 900 non-governmental organisations (NGOs). It produces a regularly updated Red List of Threatened Species™. This takes account of data from over 11 000 voluntary scientists and experts from all over the world to list species that are endangered and close to extinction. Species come onto the list as their endangered status becomes apparent. Species may move off the list for a number of reasons. Conservation measures such as habitat protection may remove the threat of extinction, the species may be reclassified or an error in the data collection may be identified. The Red List is widely used by conservation organisations, governments and NGOs. The IUCN are also developing Red List Indices to show the trends in the levels of threat to different taxonomic groups over time. It also works to develop practical solutions to international conservation challenges.

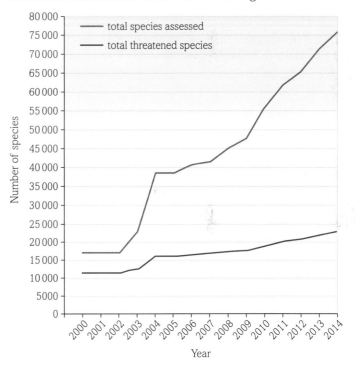

fig A The increase in the numbers of species assessed for the IUCN Red List of Threatened Species™ 2000–2014.

The natural rate of extinction varies between different groups of organisms. Mammals on average become extinct relatively rarely (one every 200 years). Worryingly, 89 species of mammals have become extinct in the last 400 years, many more than would be predicted. What is more, another 169 mammal species are on the Critically Endangered list. Based on current trends, some scientists have predicted that up to 30% of the species on Earth could become extinct in the next 100 years. If that is true, the consequences for biodiversity would be severe. It is important to remember that although the species that we are aware of tend to be plants, mammals, birds, fish, reptiles and amphibians, most extinctions will affect fungi, invertebrates, protists and bacteria.

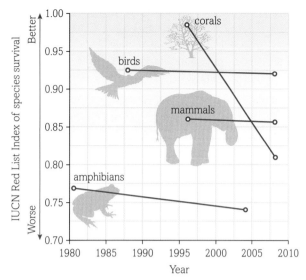

fig B IUCN Red List Indices showing the level of threat to reef-forming corals, birds, mammals and amphibians over time.

Human causes of extinction

Human activities cause many of the threatened extinctions being recorded at present. Habitat destruction is a major issue. Human activity also pollutes seas, rivers and lakes with sewage and with chemicals from industries and farms.

People eating plants and animals can sometimes drive them to extinction, as in the case of the dodo which lived on Mauritius. Species also become extinct as a result of other animals, plants or diseases introduced by people. For example, the prickly pear cactus, *Opuntia*, was introduced into Australia in the 1830s. By the early twentieth century, the cactus had taken over 62 000 km^2 of land, considerably reducing the natural biodiversity of endemic species. In 1924, caterpillars of *Cactoblastis cactorum* (the cactus moth) were deliberately introduced to Queensland from South America, as they devour the cactus and eat nothing else. Moth numbers soared rapidly, reducing *Opuntia* to a few small isolated patches and allowing the natural Australian flora to return. The moth was later introduced to some of the Caribbean islands to control invading *Opuntia*. Unfortunately, some moths have been transferred accidentally to the US mainland, where they are beginning to cause serious problems in the natural cactus populations of Florida. Here the cacti are part of the natural biodiversity, so the moths are destructive rather than helpful.

fig C In the desert states of the US, *Opuntia* are part of the natural biodiversity. They are threatened by the moths that are used to protect biodiversity in other regions of the world where the cacti are alien invaders.

There are fears that climate change as a result of human activity will also cause extinctions. So far no one really knows what the impact will be. It may just move the ranges of species rather than causing extinctions. However, extinction is inevitable when a species has nowhere else to move to, for example alpine plants in mountainous areas of the UK. Climate change may also interfere with food webs within an environment. For example, some breeding populations of sea birds on the North Sea coast depend on sand eels to feed their chicks, but the temperature of the North Sea has risen by about 1 °C, so sand eel populations are peaking earlier and are no longer synchronised with the arrival of the chicks. This may cause extinction, or the birds may breed earlier too as temperatures increase. There is still much to find out.

Conserving biodiversity

Increasing numbers of international organisations, including the United Nations, are working to stem the loss of biodiversity in ways that are sustainable and realistic.

Over the last 60 years there have been a number of international agreements designed to support conservation, encourage the sustainable use of resources and protect endangered species from extinction. One of the best known of these is **CITES**, the **Convention on International Trade in Endangered Species of Wild Fauna and Flora**.

The role of CITES

CITES is an international agreement designed to protect endangered animals and plants by regulating trade in both the living organisms and products made from them. It was drafted after a meeting of the IUCN in 1963, and it is one of the oldest and largest international treaties on conservation and sustainable use of biological resources, with 180 nations signed up to CITES. It is a voluntary agreement so it is not legally binding, but trade sanctions are used against parties who have signed up to the convention and then break the agreement.

How does CITES work?

CITES states strict rules about the import or export of threatened animals or plants and of materials made from them, such as powdered rhino horn or ivory from elephant tusks. Some entire groups of organisms are protected, including primates and cetaceans (dolphins, porpoises and whales). In some cases it is simply a sub-species that is protected, or the population of an animal or plant in a particular country. For example, the population of *Panax ginseng* found in the Russian Federation is protected by CITES, but no other population of the same plant is protected. Almost 40 000 groups of organisms are now protected by CITES, including over 30 000 plants and around 6000 animals.

The protected organisms are split between three appendices:

* Appendix I: Organisms in Appendix I are threatened with extinction and no trade in wild-collected specimens or their products is allowed. They include the gorilla species, tigers, cheetahs, Asian elephants, and dugongs.

- Appendix II: Organisms in Appendix II are not yet threatened by extinction, but they are endangered and could become threatened if they are not protected. CITES applies strict trade controls to these species. Any organisms from Appendix I that are bred in captivity are treated as members of Appendix II by CITES, making it possible for severely endangered animals and plants to be moved internationally for breeding programmes. Examples include the bigleaf mahogany tree, the African grey parrot and the queen conch.

- Appendix III: Organisms in Appendix III are there because one country has asked other CITES members to help to protect a local population. The organisms may be threatened with extinction in that country, but are not threatened globally. For example, the two-toed sloth of Costa Rica, walruses and the African civet are listed in Appendix III.

The organisms included in the appendices are reviewed regularly and the IUCN Red Lists of endangered species are used to help inform the decisions.

CITES successes

How successful is CITES in conserving biodiversity and preventing extinctions? There are a variety of opinions on this, but most organisations seem to think that CITES has some successes, and world efforts to conserve biodiversity are stronger with CITES than without it. Here are some of the CITES success stories:

- CITES banned the international ivory trade in 1989 and since then poaching levels have fallen and elephant numbers have increased in some areas.

- In 2013, five species of sharks and two species of rays were added to the CITES lists, seen as vital to prevent the extinction of sharks due to the production of shark's fin soup.

- Several turtle species have been brought back from the brink of extinction as a result in part of CITES bans on trade of any turtle products, either meat or shell.

Limitations of CITES

There are many limitations to CITES and other international agreements that attempt to protect organisms threatened by extinction. These include:

- CITES deals exclusively with trade agreements. It does not protect ecosystems or attempt to legislate for protection of breeding areas and threatened habitats.

- Many countries have not signed up to CITES.

- Strong commercial interests can override scientific evidence for the need for protection. For example, Canada prevented the addition of polar bears to Appendix I, as polar bears are hunted there for their fur and other body parts, which are sold on the international market. This is in spite of the fact that numbers are becoming dangerously low as a result of climate change.

- There is no legal obligation to abide by the CITES appendices, so some countries that have signed up to CITES, including China, still have a thriving trade in tiger and rhino body parts, even though these animals are given the highest level of protection on Appendix I.

Much of the depletion of biological resources that has been seen in recent years appears to be the result of human influences on the ecosystems of the world. The sustainability of biological resources depends on us and the effective management of conflicts between human needs and conservation. The publication of scientific research, resources such as the IUCN Red Lists and the CITES agreements, the work of international organisations and governments and the efforts of individuals will all be needed to enable the world population to survive and to thrive in a sustainable way.

fig D Human influence holds the key to the continuing biodiversity of the ecosystems around us.

Questions

1 Consider the estimates of natural and human-induced extinction rates and how they may have been calculated. How reliable do you think the estimates are, and how much reliance should we put on these values when comparing them?

2 Look at the information carried in the data in **fig B**.
   (a) Suggest reasons for the trend seen in reef-forming corals.
   (b) Which vertebrate group gives most cause for concern and how might human influences be affecting this group?

3 Is CITES a success or a failure? Investigate and write a short report, backing up your opinions with evidence.

Key definitions

The **International Union for Conservation of Nature (IUCN)** is the largest global environmental organisation and produces the regularly updated Red List of Threatened Species™.

CITES, the **Convention on International Trade in Endangered Species of Wild Fauna and Flora**, is an international agreement designed to protect endangered animals and plants by regulating trade in both the living organisms and products made from them.

THE TURTLES OF TORTUGUERO

In the beautiful but tiny Costa Rican area of Tortuguero, a major effort by scientists, conservationists, the local population and the government has produced a 500% increase in the nesting population of rare green turtles. As a result of their work, the Tortuguero green turtle colony is by far the largest in the Western Hemisphere.

TORTUGUERO SEA TURTLE PROGRAM

For 50 years, the Sea Turtle Conservancy has conducted annual sea turtle nest monitoring studies on the 21 mile black sand beach of Tortuguero, Costa Rica, the nesting site of more endangered green turtles than anywhere else in the Western hemisphere. Since being initiated by Dr Archie Carr in the 1950s, this monitoring program has provided much information on the reproductive ecology and migratory habits of sea turtles. A recent peer-reviewed analysis showed an encouraging trend in green turtle nesting activity. Through this five-decade-long conservation initiative, STC has reversed the decline of green turtles in the Caribbean Sea.

The overall goal of STC's sea turtle research and conservation work in Tortuguero is to conserve the area's nesting green and leatherback turtle populations, so that these species fulfill their ecological roles. The strategies used to achieve this goal include the following: (1) monitoring and studying Tortuguero's nesting turtles; (2) working with the Costa Rican government, the community of Tortuguero and others to protect nesting turtles from poachers; (3) training young scientists, conservationists, and others to help ensure the continuation of sea turtle protection efforts in Tortuguero and elsewhere; and (4) educating the public about sea turtles and the threats to their survival.

Research methods include turtle tagging, turtle track surveys, collection of biometric data, fibropapilloma examination, determination of nest survivorship and hatching success, collection of physical data, and collection of data on human impacts to the nesting beach and the turtles. Protection methods include a cooperative effort with Tortuguero National Park officials and law enforcement to reduce poaching of eggs and turtles. Training methods include training research assistants, recruited heavily from Latin American countries, and training Tortuguero National Park guards as well as local eco-tour guides in sea turtle biology and conservation. Public outreach methods include teaching Tortuguero school children, local adults and tourists about sea

turtles and working with the international media to raise awareness about sea turtles and threats to their survival.

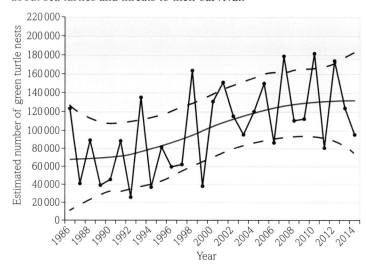

fig A Trend in green turtle nests in Tortuguero, Costa Rica between 1986 and 2013.

fig B Tourists who visit the turtle beaches at Tortuguero learn about the turtles and how they are protected, and the money they pay is used to help fund the conservation efforts.

Where else will I encounter these themes?

Book 1   5.1   5.2   6.1   6.2   6.3   7.1   7.2

This is an extract from the website of the Sea Turtle Conservancy, www.conserveturtles.org. The website provides information on sea turtles, the global threats they face and the work of the Sea Turtle Conservancy, as well as updates on the latest efforts to save these amazing reptiles. Parts of the site are designed to inform the general public of what is going on, for example: www.conserveturtles.org/seaturtleinformation.php

Other parts are reports aimed at the scientific community, the Costa Rican and other governments, and potential donors, for example: www.conserveturtles.org/pdf/reports/Tortuguero%20Green%20Turtle%20Report%202011.pdf

1. Read the extract from the Sea Turtle Conservancy website you have been given and make a judgement on whether it is aimed at the general public or the scientific community. Give reasons for your decision.

2. Visit the two webpages listed above. Compare the writing styles and explain why they are so different.

3. The extract says 'A recent peer-reviewed analysis showed an encouraging trend in green turtle nesting activity.' What is peer review and why is it important for an organisation such as the Sea Turtle Conservancy to be able to quote a peer-reviewed analysis of its work?

Now let us examine the biology. You already know about biodiversity, ecology, the effects of different factors on population size, ways of measuring population sizes, human influences on ecosystems, how sustainability of resources depends on effective management of the conflict between human needs and conservation, and the importance of the peer review process in validating scientific evidence. Use this knowledge as well as the Sea Turtle Conservancy website to help you answer the questions below.

The skills that you have developed throughout your course will be valuable, whether or not you continue your study of biology beyond A level. Whenever you read a scientific article, remember to consider the source, think critically and, above all, think like a scientist!

4. Many different aspects of human behaviour have had an effect on sea turtle population numbers.

   a. Discuss three ways human activities can have a negative effect on sea turtle numbers.

   b. Discuss two ways human activities can have a positive effect on sea turtle numbers.

5. Look at the data in **fig A**.

   a. What is the lowest number of turtle nests recorded in the period 1986–2013, and when was it recorded?

   b. What is the highest number of turtle nests recorded and when was it?

   c. What is the percentage difference between the highest and the lowest figures?

   d. Although the overall trend has been a steady increase in the number of turtle nests over the recorded period, there have been fluctuations from year to year. Suggest reasons for these variations in nest numbers.

Activity

Develop a news report, a presentation or a public information poster on one of the following topics:

- The natural history of sea turtles
- Threats to the future of sea turtles
- Saving sea turtles from extinction – a conservation success story?

● The website of the Sea Turtle Conservancy, www.conserveturtles.org

1 Which of these contributors to the greenhouse effect is anthropogenic?
   A methane from cows
   B methane from rice fields
   C carbon dioxide from volcanic activity
   D carbon dioxide from burning fossil fuels [1]
   [Total: 1]

2 Which of the following statements provide evidence that global warming is happening?
   (1) Earlier hatch times for winter moths.
   (2) An increase in the carbon dioxide concentration in ice core samples.
   (3) Wider tree rings.
   A Statement 1 only
   B Statements 1 and 2
   C Statements 1, 2 and 3
   D Statements 1 and 3 [1]
   [Total: 1]

3 Increasing the efficiency of fishing methods has reduced the stocks of fish in the oceans. The belief for a long time was that the reduced stocks would lead to reduced catches so that a sustainable level of fishing would be reached.

   However, the response to the reduction in fish numbers has been to fish in deeper oceans, with bigger nets that can reach the bottom of very deep seas in international waters outside government control.

   Some governments have responded by imposing limits on the size of fish catches, limiting the number of days that fishing is allowed in the areas that they control and providing compensation for fishermen who stop fishing.

   (a) Explain what is meant by a 'sustainable level of fishing'. [1]

   (b) Discuss how effective the actions of governments described above are likely to be in protecting fish stocks. [3]

   (c) Populations of fish may become isolated in different protected regions. Explain how this could eventually lead to the evolution of new species of fish. [3]
   [Total: 7]

4 A study of mammals in West Africa has found that populations have decreased by up to 76% in 41 different species. Some species have become extinct in the area, reducing the biodiversity.

   (a) Explain why extinction of species reduces biodiversity. [2]

   (b) It is thought that this loss of biodiversity is an indirect result of European legislation which limits fishing in European waters. Rather than lose their way of life, European fishermen now fish off the West African coast, and this has reduced fish stocks dramatically. The highest density of human populations in West Africa is along the coast.
      (i) Give **one** reason why fish stocks are important for the human population in West Africa. [1]
      (ii) Give reasons why changes in European legislation have had an effect on the wild mammal populations of West Africa. [2]

   (c) With reference to the information given, describe the importance of considering cultural issues when using legislation for conservation of organisms. [2]

   (d) Explain why local legislation is likely to be less effective than international agreements for major conservation projects. [3]
   [Total: 10]

5 The orang-utan is an endangered species found in Indonesia. Some estimates suggest that its population is falling by 5000 animals each year.

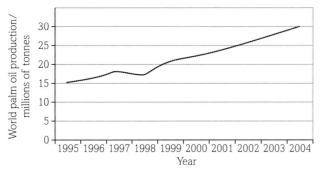

Much of the forest area inhabited by orang-utans has a suitable climate for the production of palm oil, and the demand for palm oil is increasing for use in food production and biofuels. Orang-utans avoid palm oil plantations. Demand is predicted to be double the 2004 value by 2020 as biofuels become more widely used, replacing fossil fuels.

The graph below shows changes in world palm oil production between 1995 and 2004. 85% of this palm oil is produced in Indonesia and Malaysia.

(a) (i) Describe the changes in palm oil production between 1995 and 2004. [1]
     (ii) If Indonesia and Malaysia continue to produce 85% of the world supply of palm oil, calculate how many tonnes of palm oil they will produce in 2020 if predictions are correct. [2]

(b) The map below shows some areas of Borneo where orang-utans were found in 1999 and in 2004. It also shows areas where palms have been planted to produce palm oil.

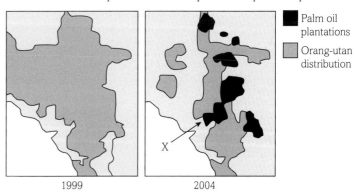

■ Palm oil plantations
▨ Orang-utan distribution

1999   2004

(i)   Describe the changes in the distribution of orang-utans between 1999 and 2004 shown by the map. [2]
(ii) Explain how the palm oil plantations labelled X might affect the genetic variety of the overall orang-utan population. [2]
(iii) Discuss the benefits and problems caused by the planting of palms to produce oil. Your answer should refer to the effects on humans, orang-utans and the wider environment. [5]

[Total: 12]

Maths skills

In order to be able to develop your skills, knowledge and understanding in biology, you will need to have developed your mathematical skills in a number of key areas. This section gives more explanation and examples of some key mathematical concepts you need to understand. Further examples relevant to your A level Biology studies are given throughout the book and in Book 1.

Using logarithms

Calculating logarithms

Many formulae in science and mathematics involve powers. Consider the equation:

$$10^x = 62$$

We know that the value of x lies between 1 and 2, but how can we find a precise answer? The term logarithm means index or power, and logarithms allow us to solve such equations. We can take the 'logarithm base 10' of each side using the **log** button of a calculator.

WORKED EXAMPLE

$10^x = 62$

$\log_{10}(10^x) = \log_{10}(62)$

$x = 1.792392...$

We can calculate the logarithm using any number as the base by using the **log** button.

WORKED EXAMPLE

$2^x = 7$

$\log_2(2^x) = \log_2(7)$

$x = 2.807355...$

Many equations relating to the natural world involve powers of e. We call these exponentials. The logarithm base e is referred to as the natural logarithm and denoted as **ln**.

Using logarithmic plots

An earthquake measuring 8.0 on the Richter scale is much more than twice as powerful as an earthquake measuring 4.0 on the Richter scale. This is because the units involved in measuring earthquakes use the concept of logarithm scales in charts and graphs. This helps us to accommodate enormous increases (or decreases) in one variable as another variable changes.

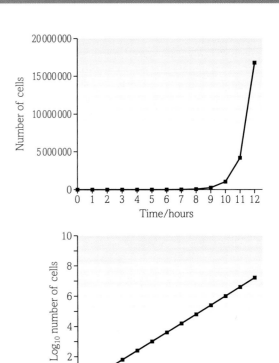

fig A Logarithmic scales are useful when representing a very large range of values, such as in the case of bacterial growth.

Selecting and using a statistical test

In Book 1, you learned how to use measures of average (mean, median and mode) and measures of dispersion (range and standard deviation). We will now look at some statistical tests that can be used to analyse data.

Deciding on a null hypothesis

When you use a statistical test, you need to be clear what you are testing for. We do this by setting a null hypothesis, which is the thing you are trying to prove or disprove. You should clearly state your null hypothesis before you begin a test.

After running your chosen statistical test, you will be left with a number known as the observed value. To know whether or not you should accept your null hypothesis, you need to compare the observed value with a table of critical values. To find the correct value to compare it with, you will usually need to calculate the 'degree of freedom' for your results.

In essence, the critical values tell you whether or not you can accept your null hypothesis. If your observed result fits with the 5% significance level given in the table, you can be 95% certain that your null hypothesis is true – and you should therefore accept it. If not, reject it!

Choosing a test

There are many different statistical tests that can be used. The tests that you may meet in your A level Biology studies are shown in **table A**.

Using the Spearman's rank correlation coefficient to test for correlation

Step 1: State the null hypothesis
The null hypothesis for this test is: 'r_s is equal to zero, meaning that there is no correlation between the two variables.'

Step 2: Calculate the observed value
The formula for the Spearman's rank correlation coefficient is:

$$r_s = 1 - \frac{6\sum d^2}{n(n^2 - 1)}$$

where d is the difference in rank between each pair of variables, n is the number of pairs and the symbol $\sum$ means 'the sum of'. To use the Spearman's rank correlation coefficient, find the difference in rank between each pair, square these differences and add them all together. Now substitute this value into the formula in place of $\sum d^2$.

Step 3: Decide whether to accept the null hypothesis
To decide whether you can accept the null hypothesis, you need to compare you observed value of r_s with a critical values table.

Find the critical value that corresponds to the 5% significance level for your number of pairs. If your value of r_s (ignoring whether it is positive or negative) is less than the critical value, accept the null hypothesis. If not, reject it.

Using the chi squared test

Step 1: State the null hypothesis
The null hypothesis for this test is: 'The observed results are consistent with the expected distribution, meaning that differences between observed and expected results are due to chance.'

Step 2: Calculate the observed value
The formula for the chi squared test is:

$$\chi^2 = \sum \frac{(O - E)^2}{E}$$

where O is your observed result, E is the expected result and the symbol $\sum$ means 'the sum of'. For each result, you need to find the difference between the observed and expected values, square this difference then divide by the expected value. You then add together the results of all these calculations to get the observed value for χ^2.

Step 3: Decide whether to accept the null hypothesis
To decide whether you can accept the null hypothesis, you need to compare you observed value of χ^2 with a critical values table. To do this you need to know the degree of freedom, which for the chi squared test is the number of categories minus 1:
$df = n - 1$

Find the critical value that corresponds to the 5% significance level for your calculated degree of freedom. If your value of χ^2 is less than the critical value, accept the null hypothesis. If not, reject it.

Using the Student's t-test

It might seem obvious that you could easily compare the means of the two categories by simply subtracting one from the other. However, can you be sure that the means calculated from the two sample data sets are representative of the whole population? The Student's t-test takes into account the degree of overlap between the two sets of data and allows you to judge whether any difference between the means is statistically significant or just due to chance.

Step 1: State the null hypothesis
The null hypothesis for this test is: 'The means of the interval variable for the two categories are equal.'

Step 2: Calculate your observed value
In order to calculate t, you first need to find the mean $\bar{x}$ and the variance s^2 for both of the categories.

$$\bar{x} = \frac{\sum x}{n}$$

$$s^2 = \frac{\sum x^2 - \frac{(\sum x)^2}{n}}{n - 1}$$

You can now use the following formula to calculate the value of t:

$$t = \frac{\bar{x}_1 - \bar{x}_2}{\sqrt{\frac{s_1^2}{n_1} + \frac{s_2^2}{n_2}}}$$

where:

$\bar{x}_1$ = mean of the first set of data

$\bar{x}_2$ = mean of the second set of data

s_1^2 = variance of the first set of data

s_2^2 = variance of the second set of data

n_1 = number of items in first set of data

n_2 = number of items in second set of data.

Name	Purpose	Example	Null hypothesis
Spearman's rank correlation coefficient	To test whether two variables display correlation.	Test whether there is a correlation between finishing positions in a race and age of runners.	There is no correlation between the two variables. (Spearman's rank correlation coefficient equals zero.)
Chi squared test	To test how likely it is that any differences between observed and expected results are due to chance.	Test whether a ratio of phenotypes from mating supports a particular inheritance model.	The observed results are consistent with the expected distribution. (The differences between observed and expected results are due to chance.)
Student's T-test	To test whether the difference in the mean value of a variable for the two categories is significant.	Test whether a drug treatment has been effective compared with a placebo.	The means of the interval variable for the two categories are equal.

table A

Step 3: Decide whether to accept the null hypothesis
To decide whether you can accept the null hypothesis, you need to compare you observed value of t with a critical values table. To do this you need to know the degree of freedom, which for the Student's t-test is the total number of data values minus 2:
$df = n_1 + n_2 - 2$

Find the critical value that corresponds to the 5% significance level for your calculated degree of freedom. If your value of t is less than the critical value, accept the null hypothesis. If not, reject it.

Applying your skills
You will often find that you need to use more than one maths technique to answer a question. In this section, we will look at three example questions and consider which maths skills are required and how to apply them.

WORKED EXAMPLE

Microorganisms can be grown quickly using a fermenter. If the conditions of the fermenter are set correctly (for example, there are enough nutrients and sufficient space to grow) the division of bacteria can become exponential. One such microorganism, a species of yeast called S. cerevisiae, can multiply every 100 minutes.

(a) *Assuming there is no cell death, and two yeast cells are introduced into the fermenter, how many cells will there be after 300 minutes?*

(b) *Write an equation to show how the population of yeast over a period of time can be calculated, using d to represent the number of divisions.*

(c) *Using your equation from (b), calculate the population of yeast generated after 32 cell divisions in a fermenter.*

(d) *How many cell divisions would have taken place if the population of yeast was only allowed to reach 65 536 cells?*

(a) In 300 minutes, the cells will divide three times. After the first division, the population will have doubled to four from the two original yeast cells. The second division will double the population again, to eight cells. The third division will double the population yet again, resulting in 16 cells.

(b) If we started with just one cell, the number of cells after time d would be given by 2^d. However, because we are starting with two cells, we need to double this, so $N_d = 2 \times 2^d$ where N_d is the number of cells after d divisions. Using the rules of indices, we could also write this as $N_d = 2^{(d + 1)}$.

(c) To find the number of cells after 32 divisions, we can just substitute $d = 32$ into our equation:

$N_{32} = 2^{(32 + 1)} = 2^{33} = 8\,589\,934\,592$ cells

(d) To find the number of divisions necessary to produce 65 536 cells, we can substitute $N_d = 65\,536$ into our equation:

$2^{(d + 1)} = 65\,536$

Now we take logarithm base 2 of both sides to simplify the equation:

$\log_2(2^{(d + 1)}) = \log_2(65\,536)$

$d + 1 = 16$

$d = 15$

So there will be 65 536 cells after 15 divisions.

WORKED EXAMPLE

A student investigates the effect of two newly developed fertilisers (A and B) on the growth of potato crops. Multiple 10 m^2 areas of a field were sectioned off and treated with either fertiliser A or B. The table below shows the yields of potatoes from the test areas following harvest.

(a) *Explain why a Student's t-test would be the most appropriate statistical test to see whether there is a difference in efficacy between fertilisers A and B.*

(b) *Is there a significant difference between potato yields for plots treated with fertiliser A or B?*

Fertiliser	Test plot yield/kg							Mean
	Plot 1	Plot 2	Plot 3	Plot 4	Plot 5	Plot 6	Plot 7	
A	25	27	34	18	21	26	28	25.6
B	17	35	42	19	35	22	44	30.6

(a) The Student's t-test can be used to determine whether there is a statistically significant difference between the means of two categories (fertiliser A and B).

(b) To answer this question, we need to perform a Student's t-test.

Step 1: State the null hypothesis
Our null hypothesis is: 'There is no difference in potato yield between plots treated with fertiliser A or B.'

Step 2: Calculate the observed value
We have already been given the means of the two data sets:

$\bar{x}_A = 25.6$

$\bar{x}_B = 30.6$

In order to calculate the variance of the data sets, we need to calculate the sum of squared values, $\sum x^2$, and the sum of the values squared, $(\sum x)^2$.

$\sum x^2$ for A $= (25)^2 + (27)^2 + (34)^2 + (18)^2 + (21)^2 + (26)^2 + (28)^2 = 4735$

$\sum x^2$ for B $= (17)^2 + (35)^2 + (42)^2 + (19)^2 + (35)^2 + (22)^2 + (44)^2 = 7284$

$(\sum x)^2$ for A $= (25 + 27 + 34 + 18 + 21 + 26 + 28)^2 = 32\,041$

$(\sum x)^2$ for B $= (17 + 35 + 42 + 19 + 35 + 22 + 44)^2 = 45\,796$

We can now calculate the variance (s^2) of each data set:

$$s_A^2 = \frac{\left[4735 - \left(\frac{32\,041}{7}\right)\right]}{6} = 26.29$$

$$s_B^2 = \frac{\left[7284 - \left(\frac{45\,796}{7}\right)\right]}{6} = 123.61$$

And so we can now calculate t:

$$t = \frac{\bar{x}_A - \bar{x}_B}{\sqrt{\frac{s_A^2}{n_A} + \frac{s_B^2}{n_B}}} =$$

$$t = \frac{(25.6 - 30.6)}{\sqrt{\left[\left(\frac{26.29}{7}\right) + \left(\frac{123.61}{7}\right)\right]}}$$

And so we can now calculate t:

$$t = \frac{\bar{x}_A - \bar{x}_B}{\sqrt{\frac{s_A^2}{n_A} + \frac{s_B^2}{n_B}}}$$

$$t = \frac{(25.6 - 30.6)}{\sqrt{\left[\left(\frac{26.29}{7}\right) + \left(\frac{123.61}{7}\right)\right]}}$$

$$t = \frac{5}{\sqrt{[(3.76) + (17.66)]}}$$

$$t = \frac{5}{4.63}$$

$$t = 1.08$$

Step 3: Decide whether to accept the null hypothesis
The number of degrees of freedom is $n_1 + n_2 - 2 = 12$

The critical value for a 5% significance level with 12 degrees of freedom is 2.18.

The observed value of 1.08 is less than the critical value. Therefore we accept the null hypothesis that there is no significant difference in crop yield between plots treated with fertiliser A or B.

WORKED EXAMPLE

The inheritance of a recessive genetic disease, A, is being investigated.

(a) *If the faulty gene that causes disease A is not sex-linked (i.e. the gene is not found on the X or Y chromosome), out of 300 people affected by the disease, how many would you expect to be female?*

(b) *In a sample of 300 patients affected by the disease, 295 were male and 5 were female. Use an appropriate statistical test to show that the difference between the observed and expected results is not due to chance.*

(c) *What can you conclude about the inheritance of disease A?*

(a) If the disease is not found on either the X or Y chromosome, it would be reasonable to assume that the proportion of males and females inheriting the disorder is equal. Therefore, we would expect 150 females in a sample of 300 patients.

(b) Since the question is asking to test whether an observed frequency is different to an expected frequency, we should use a chi squared test.

Step 1: State the null hypothesis
The null hypothesis is: 'The observed number of male and female patients with disease A is consistent with the inheritance model.'

Step 2: Calculate the observed value
The formula for the chi squared test is:

$$\chi^2 = \sum \frac{(O - E^2)}{E}$$

where O is your observed result, E is the expected result and the symbol $\sum$ means 'the sum of'.

$$\chi^2 = \frac{(295 - 150)^2}{150} + \frac{(5 - 150)^2}{150} = 140.2 + 140.2 = 280.3$$

Step 3: Decide whether to accept the null hypothesis
The number of degrees of freedom is $n - 1 = 2 - 1 = 1$

The critical value for a 5% significance level with 1 degree of freedom is 3.841.

The observed value is greater than the critical value. Therefore we must reject the null hypothesis. The difference between the observed and expected numbers of male and female patients is not down to chance.

(c) A patient with disease A is much more likely to be male than female. This suggests that the disease is found on the X chromosome. Since males have only one X chromosome, they only need to inherit a single copy of the gene to have symptoms. Females would need to inherit two copies, which is far less likely.

Preparing for your exams

Introduction

It is important to be familiar with the format of the Edexcel A level Biology B examinations, so that you can ensure you are well prepared. Here are some key features:

- You will sit three exam papers, each covering content from both years of your course. The third paper will include synoptic questions that may draw on two or more different topics and questions that assess your conceptual and theoretical understanding of experimental methods.
- You teacher will assess your competency in key practical skills, in order for you to gain the Science Practical Endorsement. The endorsement will not contribute to your overall grade but the result (pass or fail) will be recorded on your certificate.

The table below gives details of the three exam papers.

A level exam papers

Paper	Paper 1: Advanced Biochemistry, Microbiology and Genetics	Paper 2: Advanced Physiology, Evolution and Ecology	Paper 3: General and Practical Principles in Biology
Topics covered	Topics 1–7	Topics 1–4 Topics 8–10	Topics 1–10
% of the A level qualification	30%	30%	40%
Length of exam	1 hour 45 minutes	1 hour 45 minutes	2 hours 30 minutes
Marks available	90 marks	90 marks	120 marks
Question types	multiple-choice short open open-response calculation extended writing	multiple-choice short open open-response calculation extended writing	short open open-response calculation extended writing synoptic
Experimental methods?	No	No	Yes
Mathematics	A minimum of 10% of the marks across all three papers will be awarded for mathematics at Level 2 or above.		
Science Practical Endorsement	Assessed by teacher throughout course. Does not count towards A level grade but result (pass or fail) will be reported on A level certificate.		

Exam strategy

Arrive equipped

Make sure you have all of the correct equipment needed for your exam. As a minimum you should take:

- pen (black ink or ball-point pen)
- pencil (HB)
- ruler (ideally 30 cm)
- rubber (make sure it's clean and doesn't smudge the pencil marks or rip the paper)
- calculator (scientific).

Ensure your answers can be read

Your handwriting does not have to be perfect but the examiner must be able to read it! When you're in a hurry it's easy to write key words that are difficult to decipher.

Plan your time

Note how many marks are available on the paper and how many minutes you have to complete it. This will give you an idea of how long to spend on each question. Be sure to leave some time at the end of the exam for checking answers. A rough guide of a minute a mark is a good start, but short answers and multiple choice questions may be quicker. Longer answers might require more time.

Understand the question

Always read the question carefully and spend a few moments working out what you are being asked to do. The command word used will give you an indication of what is required in your answer.

Be scientific and accurate, even when writing longer answers. Use the technical terms you've been taught.

Always show your working for any calculations. Marks may be available for individual steps, not just for the final answer. Also, even if you make a calculation error, you may be awarded marks for applying the correct technique.

Plan your answer

In questions marked with an *, marks will be awarded for your ability to structure your answer logically showing how the points that you make are related or follow on from each other where appropriate. Read the question fully and carefully (at least twice!) before beginning your answer.

Make the most of graphs and diagrams

Diagrams and sketch graphs can earn marks – often more easily and quickly than written explanations – but they will only earn marks if they are carefully drawn.

- If you are asked to read a graph, pay attention to the labels and numbers on the x and y axes. Remember that each axis is a number line.
- If asked to draw or sketch a graph, always ensure you use a sensible scale and label both axes with quantities and units. If plotting a graph, use a pencil and draw small crosses or dots for the points.
- Diagrams must always be neat, clear and fully labelled.

Check your answers

For open-response and extended writing questions, check the number of marks that are available. If three marks are available, have you made three distinct points?

For calculations, read through each stage of your working. Substituting your final answer into the original question can be a simple way of checking that the final answer is correct. Another simple strategy is to consider whether the answer seems sensible. Pay particular attention to using the correct units.

Sample answers with comments

Question type: multiple choice

The cerebellum is an area of the human brain. What is the main function of the cerebellum?

☐ **A** *Maintaining homeostasis*

☐ **B** *Control of breathing and heart rate*

☐ **C** *Decision making and memory*

☐ **D** *Control of balance and coordination* [1]

This question relies on recall of knowledge. While this question requires a choice from a list of statements, other questions may be based around a choice of rows from a table or a list of letters linked to labels on a diagram.

Question analysis

- Multiple choice questions may require simple recall, as in this case, but sometimes a calculation or some other form of analysis will be required.

- In multiple choice questions you are given the correct answer along with three incorrect answers (called distractors). You need to select the correct answer and put a cross in the box of the letter next to it.

- The three distractors supplied will feature the answers that you are likely to arrive at if they make typical or common errors. For this reason, multiple choice questions aren't as easy as you might at first think. If possible, try to answer the question before you look at any of the answers.

- If you change your mind, put a line through the box with the incorrect answer (☒) and then mark the box for your new answer with a cross (☒).

- If you have any time left at the end of the paper, multiple choice questions should be put high on your list of priority for checking answers.

As with many multiple choice questions, the answer choices contain alternatives that can be easily confused. This may be because the words themselves are similar or because the alternatives refer to key terms linked to one topic. To avoid being misdirected to an incorrect answer, one approach is to try to think of the answer initially without reading the choices given and then look for your answer in the list. Some students skim read notes and learn to recognise technical terms only by the initial letters, which can lead to errors. If you sometimes have difficulty with spelling or recognising key words, try reading them out loud one syllable at a time when preparing for exams.

Average student answer

☒ C Decision making and memory

Verdict

This is an incorrect answer because:

- The student has confused the two brain areas of the cerebellum and the cerebrum. These are similar-sounding words. While areas of the cerebrum are responsible for decision making and memory, the main function of the cerebellum is control of balance and coordination (answer D).

Question type: short open

Stem rust fungus (Puccinia graminis) is a pathogen that affects cereal crops. Describe how this fungus may be transmitted to newly planted cereal plants. *[2]*

The command word 'describe' requires you to give an account and provide the reasoning behind it. As there are up to 2 marks available, the answer should have two clear points relating directly to the question. The key word here is 'transmitted' – how the disease is passed from one host to another. Comments on how the pathogen then invades tissues would not gain credit.

Question analysis

- Short open questions usually require simple short answers, often one word. Generally they will require simple recall of the biology you have been taught.

- Short open questions require succinct and clear answers. They may be worth 1 or 2 marks. For 1-mark questions it is not always necessary to write in full sentences. For a 2-mark question there may be credit for two distinct points or alternatively for one main idea with further detail or elaboration.

Average student answer

Stem rust fungus can infect cereals such as wheat and barley and is a pathogen because it causes disease. Spores transmit it from plant to plant.

The student has wasted time and space repeating information from the question and giving irrelevant detail. The first sentence does not answer the question and would not gain credit. On short open questions it is easy to fill the available space with comment that is superfluous and then fail to add sufficient relevant detail.

Verdict

This is an average answer because:

- Spores are correctly identified as the mode of transmission and would be credited with 1 mark.

- Further detail, such as spore transfer by wind, would be needed for 2 marks.

Question type: open response

A laboratory experiment was carried out to investigate the effect of different wavelengths of light on the rate of photosynthesis in the pondweed Elodea. Pieces of Elodea of equal mass were placed in experimental chambers and exposed to light at wavelengths in the red, blue, yellow and green regions of the spectrum. The oxygen produced in the first hour was collected at each of the colours of light, and this was used as an indication of the rate of photosynthesis. The experiment was repeated several times and the mean and standard deviation for each light treatment was calculated.

Using the results from the graph in fig A, describe the results of this investigation. *[3]*

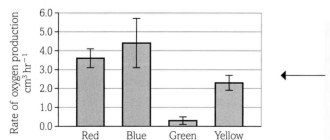

The command word here is 'describe', so you will need to give an account of the results shown. You do not need to explain the scientific reasons behind any patterns. It is important to note the command word carefully in graph-based questions, as you may be asked to analyse, explain or compare and contrast results, all of which require a very different approach.

fig A The mean rate of oxygen production by the pondweed *Elodea* under light of different wavelengths. The error bars show the standard deviation of the means.

Question analysis

- Questions based around graphs are frequently encountered. Make sure that you check the axes scales carefully before describing results or reading from graphs, as scales are often more complex at A level.

- In questions that ask for a description, start by stating the general pattern or overall trends. Then develop your statements, giving more detail on the extent of any differences and more detail of the patterns shown. In some questions credit is also given for manipulating the data; this is not simply reading figures from the graph but actually using figures to make a comparison, usually by way of a simple calculation.

Average student answer

The blue light produced the highest rate of oxygen production. However, the error bars for blue light show a large degree of overlap with those of red light, so there may be no real difference between the results from these two treatments. The standard deviation for blue light was greater than for the other treatments, which indicates that the results from this treatment may be less reliable. The lowest rate of oxygen production was shown with green light, which had very little photosynthesis. Yellow light was associated with an intermediate rate of oxygen production, having approximately half the rate shown by blue light.

The student has given a detailed answer and has considered both the mean values and the standard deviation in describing the results of the experiment and has correctly avoided explaining any differences seen.

Verdict

This is a good answer because:

- There are at least three clear points made about the patterns shown in the results.

- The answer is logically structured.

- The general pattern and extent of any difference between treatments is clear, and there has been some attempt to manipulate the data.

Question type: extended writing

* Explain how nerve impulses are propagated along an axon once an action potential has been initiated. [6]

The command word here is 'explain', so there must be some reasoning and detail of how an impulse is conducted along the axon. A simple description of an action potential would not suffice.

Question analysis

- In questions marked with an asterisk (*), marks will be awarded for your ability to structure your answers logically showing how the points that you make are related or follow on from each other where appropriate.

- Four marks are available for making valid points in response to the question. In mark schemes, these are referred to as 'indicative marking points'. To gain all four marks, your answer needs to include six indicative working points. The remaining two marks are awarded for structuring your answer well, with clear lines of reasoning and linkages between points.

- In extended writing questions, as in any question, look at the number of marks and try to think of at least that number of distinct points to make. In longer questions it may help to plan out your answer in brief bullet points on a blank space on the paper. This will help you to order your points more logically.

- It is important that you use correct scientific terminology to demonstrate your biological knowledge. Vague answers will only achieve the lowest mark band.

Average student answer

When an action potential is initiated, this causes localised currents along the axon membrane. These currents cause depolarisation of the adjacent section of membrane. This causes gated sodium channels to open. The ions move in and there is positive feedback on the sodium channels. A wave of depolarisation and repolarisation repeats itself along the axon and so travels along it. If the axon is myelinated then conduction will be faster.

The answer demonstrates adequate knowledge and some relevant biological facts have been selected and used. There is some linkage of facts and a reasonably logical structure, and all material is relevant to the question. As such, this answer would achieve a moderate mark band.

Verdict

This is an average answer because:

- While there are at least six points made in this answer, some are not adequately explained and it is not clear if the student really understands all of the underlying biological processes. The student does not explain the movement of ions that results in depolarisation and repolarisation. Sodium ions should always be referred to (not simply sodium). The nature and mechanism of saltatory conduction in particular is not explained.

- The points made are all relevant, but the student does not consider the concept of the refractory period and how this results in propagation in one direction only.

Question type: calculation

Gregor Mendel crossed round yellow-seeded pea plants with wrinkled green-seeded plants. The wrinkled green traits were lost in the F_1 generation but reappeared in the F_2 generation. The offspring in the F_2 generation had four phenotypes: most had round yellow seeds, some had round green seeds, some had wrinkled yellow seeds while only a few had wrinkled green seeds. The ratios indicated by these early experiments were $9:3:3:1$.

A geneticist decided to repeat these genetic cross experiments. She used a chi squared test to determine whether the results demonstrated the patterns predicted by Mendel's work. The results obtained are shown in the table below.

Seed type	Expected ratio	Observed	Expected	$o - e$	$(o - e)^2$	$\dfrac{(o - e)^2}{e}$
round yellow	9	587	567	+20	400	0.71
wrinkled yellow	3	197	189	+8	64	0.34
round green	3	168	189	-21	441	2.33
wrinkled green	1	56				

The formula for the chi squared statistic is:

$$\chi^2 = \sum \frac{(o - e)^2}{e}$$

Complete the table then calculate the chi squared statistic for these results. Show your working. **[3]**

Question analysis

- The command word here is 'calculate'. This means that you need to obtain a numerical answer to the question, showing relevant working. If the answer has a unit, this must be included.
- Always have a go at calculation questions. You may get some small part correct that will gain credit.
- In multipart questions such as this, you may be awarded marks for applying the correct method, even if earlier stages (for example, completing the table) are incorrect.
- Always do a 'common sense' check on answers by looking to see if your values seem in line with the data as a whole.
- Make sure that you understand any symbols used, such as the 'sum of' symbol ($\sum$) here.
- Take an approved calculator into every exam and make sure that you know how to use it!

> The first part of this question, completing the table, should be straightforward, as a lot of help is given by the examples in the rows that are already completed. Check your approach on a completed row if you are not sure it is correct, before calculating the missing values. Calculation of the 'expected' value is usually carried out using the ratios and the total number of individuals in all categories (so $\frac{1}{16}$th of 1008 in this case). However, in this question it could also be worked out from the values already entered in the expected column.

Average student answer

Seed type	Expected ratio	Observed	Expected	$o - e$	$(o - e)^2$	$\dfrac{(o - e)^2}{e}$
wrinkled green	1	56	63	-7	49	0.78

$\chi^2 = 4.16$

> The student has answered this uncomplicated question correctly. Make sure that you also understand how the chi squared statistic is used, including the use of degrees of freedom, probability and the interpretation of significance.

Verdict

This is an average answer because:

- The student has filled in the table and calculated the chi squared statistic correctly, so this answer would gain all 3 marks. However, the student would have been wise to show the main steps of their working out. This is because, had they made an error in their calculation, they still might have gained some credit for demonstrating the correct mathematical approach.

Question type: synoptic

A student wanted to carry out an investigation on the effect of aluminium ions on the growth of lawn grass. He carried out a pilot study in which he grew seeds from a commercial grass species mix in the laboratory. Liquid growth medium was used, to which aluminium sulfate was added at a variety of concentrations that were representative of clean to moderately polluted ground water. Three germinating seeds were grown in each concentration. The student allowed the seedlings to grow for a few weeks then measured the length of the longest shoot. His pilot study results are shown in the table below.

fig B Diagram of a grass seedling showing roots and shoots.

Aluminium ion concentration /micromoles dm^{-3}	Shoot length/mm
50	98
50	81
50	90
200	80
200	89
200	89
400	77
400	86
400	83

When measuring the seedlings, the student noticed that the amount of root growth seemed to be stunted at higher aluminium ion levels, while shoots seemed relatively unaffected.

He decided that his final project would investigate the relative amount of roots and shoot growth at different aluminium ion concentrations.

His hypothesis was: 'The higher the aluminium concentration, the smaller the roots will be relative to the shoots.'

Devise a plan that would allow this hypothesis to be tested. Your plan should extend and modify methods from the pilot study in order to allow valid and reliable results to be gathered. [6]

The command word here is 'devise'. This means that you must plan or invent a procedure from existing principles or ideas. There are lots of clues to suitable methods given in the text, but no marks will be awarded for simply repeating these – you will need to do some fine-tuning or amending to gain credit. A data table is given but you are not asked to do anything with the figures; you can therefore assume that the table must provide further prompts and ideas for your plan. Again, you should critically assess what has been done already and modify where necessary. The diagram (fig B) also gives some hints: it shows that there are several roots and shoots to each seedling and that roots have a complex shape.

Question analysis

- A synoptic question is one that addresses ideas from different areas of biology in one context. In your answer, you need to use the different ideas and show how they combine to explain the biology in the context of the question.

- Questions in Paper 3 may draw on any of the topics in this specification. This question is assessing your understanding of experimental methods.

- This is a synoptic question because, even though the main content is based around Core Practical 16, it covers skills and knowledge that have been acquired over a wider area of the A level course. For example, to answer this question you will need to apply the investigation planning skills that you have developed throughout your core practical work. The question also has links to Topics 1 and 5 (knowledge of factors affecting plant growth) and requires mathematical understanding of the use of ratios or percentages.

- When a question asks you to plan and experiment, you should start by thinking about each type of variable. For instance: How could the independent variable (in this case aluminium ion concentration) be changed – how many different levels and what range? Will you need a control (a null or comparison level of the independent variable)? How will the dependent variable be measured and will any calculations or conversion of data be required? What variables should be controlled and how? Once you have considered these points, make sure you list any equipment that you will need (not beakers or spatulas, only special items such as microscopes, water baths or reagents). Finally, you will need to think about sample size and repeats.

Average student answer

At least five different concentrations of aluminium ions should be used. For example: 100, 200, 300, 400 and 500 micromoles dm^{-3}. At least five grass seedlings should be grown in each concentration. Factors that should be controlled include time, light levels and the concentration of mineral ions in the growth medium. The amount of growth solution used should also be the same in each case. The length of the longest shoot and ◄——— longest root should be measured with a ruler to the nearest mm after 3 weeks of growth. A mean should be calculated for each concentration, and graphs of shoot length and root length against aluminium ion concentration drawn and compared.

The answer would gain credit for a suitable number/range of concentrations, for considering repeats and for stating several variables that should be controlled. A stronger answer would include more detail of how variables might be controlled, especially light (for example, by using a light box with a 12/12 hour dark cycle, or by stating the details of randomised design layout). The 'amount' of growth solution is also not specific enough – volume and concentration are more suitable terms. Other suitable controlled variables might include temperature, grass species or pH. The diagram shows that root length might be difficult to measure, even for just the longest root. In addition, simple length does not take into account the overall shoot and root size on the seedling; total shoot mass and total root mass for each seedling would be a more valid measurement. The student has not given a method for determining the relative size of roots to shoots; root mass could be expressed as a percentage (or as a decimal ratio) of shoot mass, for example.

Verdict

This is an average answer because:

- The student has thought about many of the key variables but has not given sufficient detail for some.

- A key requirement of the question has been missed: a measure of the relative size of roots and shoots was required, rather than simply the separate size of each.

- There is no mention of a control: seeds should also be grown without added aluminium ions, for comparison.

Glossary

Abiotic factors are the non-living elements of the habitat of an organism.

Abscission is the shedding of leaves, flower parts or fruits from a plant after the formation of an abscission zone across the stem attaching the organ to the plant.

The **absolute refractory period** is the first millisecond or so after the action potential when it is impossible to re-stimulate the fibre – the sodium ion channels are completely blocked and the resting potential has not been restored.

An **absorption spectrum** is a graph of the amount of light absorbed by a pigment against the wavelength of light.

The **abundance** of an organism refers to the relative representation of a species in a particular ecosystem.

Acetyl coenzyme A (acetyl CoA) is the 2C compound produced in the link reaction which feeds directly into the Krebs cycle, combining with a 4C organic acid to form a 6-carbon compound.

Acetylcholine (ACh) is a neurotransmitter found in the synapses of motor neurones, the parasympathetic nervous system and cholinergic synapses in the brain.

Acetylcholinesterase is an enzyme found embedded in the post-synaptic membrane of cholinergic nerves that breaks down acetylcholine in the synapses after it has triggered a post-synaptic potential.

The **ACFOR scale** is a simple scale used to describe the abundance of a species in a given area.

The **action potential** is when the potential difference across the membrane is briefly reversed to about +40 mV on the inside with respect to the outside for about 1 millisecond.

An **action spectrum** is a graph demonstrating the rate of photosynthesis against the wavelength of light.

Adrenergic nerves use noradrenaline as the neurotransmitter in their synapses.

Aerobic respiration is the form of cellular respiration that takes place in the mitochondria in the presence of oxygen.

Aestivation is an extreme physiological adaptation for avoiding hot, dry conditions, similar to hibernation. The metabolic rate slows right down and the animals become completely inactive and torpid for months, buried in dried mud or in rock crevices.

Agglutination is the clumping of cells caused when antibodies bind to the antigens on pathogens.

Ammonium compounds contain the ammonia group $-NH_3$.

When DNA is **amplified**, it is replicated repeatedly using the polymerase chain reaction to produce a much bigger sample.

Anaerobic respiration is the form of cellular respiration that takes place only in the cytoplasm when there is no oxygen present.

Anthropogenic means produced by people.

An **antibiotic** is a drug that either destroys microorganisms or prevents them from growing and reproducing.

An **antibiotic-resistant** microorganism is not affected by an antibiotic, even one that may have been effective in the past.

An **antibody** is a glycoprotein that is produced in response to a specific antigen.

Antidiuretic hormone (ADH) is a hormone produced in the hypothalamus and stored in the posterior pituitary that increases the permeability of the distal convoluted tubule and the collecting duct to water.

Antigen-presenting cell (APC) is a cell displaying an antigen/MHC protein complex.

Antigens are glycoproteins, proteins or carbohydrates on the surface of cells, toxins produced by bacterial and fungal pathogens, and some whole viruses and bacteria that are recognised by white blood cells during the specific immune responses to infection. They stimulate the production of an antibody.

Aquaculture is the farming of fish or shellfish in tanks or in containers in rivers, lakes or the sea.

An **arteriovenous shunt** is a system which closes to allow blood to flow through the major capillary networks near the surface of the skin, or opens to allow blood along a 'shortcut' between the arterioles and venules, so it does not flow through the capillaries near the surface of the skin.

Artificial active immunity is when the body produces its own antibodies to an antigen acquired through vaccination.

Artificial passive immunity is when antibodies are extracted from one individual and injected into another e.g. the tetanus vaccine.

Attenuated pathogens are viable pathogens that have been modified so that they do not cause disease but still cause an immune response that results in the production of antibodies and immunity.

The **autonomic nervous system** is the involuntary nervous system. Autonomic motor neurones control bodily functions that are not normally dealt with by the conscious area of the brain.

Autotrophic organisms make complex organic compounds from simple compounds in their environment.

Auxins are plant hormones that act as powerful growth stimulants (e.g. indoleacetic acid) and are involved in apical dominance, stem and root growth, and tropic responses to unilateral light.

Axons are the long nerve fibre of a motor neurone, which carries the nerve impulse.

B cells are lymphocytes made in the bone marrow which are found both in the lymph glands and free in the body once they are mature.

B effector cells divide to form the plasma cell clones.

B memory cells provide the immunological memory to a specific antigen, allowing the body to respond very rapidly if you encounter a pathogen carrying the same antigen again.

Bactericidal antibiotics kill bacteria.

Bacteriostatic antibiotics inhibit the growth of bacteria.

Baroreceptors are mechanoreceptors in the aorta and carotid arteries that are sensitive to pressure changes.

A **belt transect** is when two tapes are laid out and the ground between them surveyed.

If a study is **biased**, it has been influenced in some way to deliver a particular conclusion.

Biomes are the major ecosystems of the world.

The **biosphere** is all of the areas of the surface of the Earth where living organisms survive.

Biotic factors are the living elements of a habitat that affect the ability of a group of organisms to survive there.

The **brain** is the area of the CNS where information can be processed and from where instructions can be issued as required to give fully coordinated responses to a whole range of situations.

The **Calvin cycle** is a series of enzyme-controlled reactions that take place in the stroma of chloroplasts and result in the reduction of carbon dioxide from the air to bring about the synthesis of carbohydrate.

The **carbon cycle** is the series of reactions by which carbon is constantly recycled between living things and the environment.

A **carbon sink** is a reservoir where carbon is removed from the atmosphere and 'locked up' in organic or inorganic compounds.

The **cardiac control centre** in the medulla oblongata receives input from a number of different receptors and controls changes to the heart rate and the cardiac volume through parasympathetic and sympathetic nerves.

The **cardiac output** is a measure of the volume of blood pumped by the heart per minute, worked out by the combination of cardiac volume and heart rate.

The **cardiac volume** is the volume of blood pumped at each heartbeat.

Carotenoids are photosynthetic pigments made up of orange carotene and yellow xanthophyll.

Cell differentiation is the process by which a cell becomes specialised for a particular function.

Cellular respiration is the process by which food is broken down to yield ATP, which is used as a source of energy for metabolic reactions.

The **central nervous system (CNS)** is a specialised concentration of nerve cells where incoming information is processed and from where impulses are sent out through motor neurones, which carry impulses to the effector organs.

The **cerebellum** is the area of the brain that coordinates smooth movements. It uses information from the muscles and the ears to control balance and maintain posture.

The **cerebral hemispheres** are the two parts of the cerebrum, joined by the corpus callosum.

The **cerebrum** is the area of the brain responsible for conscious thought, personality, control of movement and much more.

Chemiosmosis is the process that links the electrons that are passed down the electron transport chain and the production of ATP, by the movement of hydrogen ions through the membrane along electrochemical, concentration and pH gradients.

The **chemiosmotic theory** was developed by Peter Mitchell to explain the production of ATP in mitochondria, chloroplasts and elsewhere in living cells.

Chemoreceptors are sensory nerve cells or organs that respond to cemical stimuli.

Chi squared (χ^2) test is a statistical test used to compare differences between sets of data to evaluate whether they differ from each other significantly.

Chlorophyll *a* is a blue-green photosynthetic pigment, found in all green plants.

Chlorophyll *b* is a yellow-green photosynthetic pigment.

Cholinergic nerves use acetylocholine as the neurotransmitter in their synapses.

Chromatin is the DNA/protein complex that makes up the chromosomes.

CITES, the **Convention on International Trade in Endangered Species of Wild Fauna and Flora**, is an international agreement designed to protect endangered animals and plants by regulating trade in both the living organisms and products made from them.

A **climatic climax community** is the only climax community possible in a given climate.

A **climax community** is a self-sustaining community with relatively constant biodiversity and species range. It is the most productive group of organisms that a given environment can support long term.

Clonal selection is the selection of the cells that carry the right antibody for a specific antigen.

A **clone** is a group of identical cells all produced from one cell.

Cobra venom is a substance made by several species of cobra that binds reversibly to acetylcholine receptors in post-synaptic membranes in motor neurones, preventing the production of a post-synaptic action potential.

When alleles are **codominant**, both alleles are expressed and the proteins they code for act together without mixing to produce a given phenotype.

The **collecting duct** takes urine from the distal convoluted tubule to be collected in the pelvis of the kidney. It is the region of the kidney where most of the water balancing needed for osmoregulation takes place.

A **community** is all the populations of all the different species of organisms living in a habitat at any one time.

Complementary DNA (cDNA) is DNA which can act as an artificial gene. It is made by reversing the transcription process from mRNA using reverse transcriptase.

Cones are photoreceptors found in the fovea of the retina which contain the visual pigment iodopsin. They respond to bright light, give great clarity of vision and colour vision.

The **corpus callosum** is the band of axons (white matter) that join the left and right cerebral hemispheres of the brain.

A **countercurrent multiplier** is a system that produces a concentration gradient in a living organism, using energy from cellular respiration.

In a **culture** microorganisms are provided with the nutrients, level of oxygen, pH and temperature that they require to grow large numbers so they can be observed and measured.

Cyclic AMP (cAMP) is a compound formed from ATP that is produced when protein hormones such as adrenaline bind to membrane receptors and acts as a second messenger in cells.

Cyclic photophosphorylation is a process that drives the production of ATP. Light-excited electrons from PSI are taken up by an electron acceptor and passed directly along an electron transport chain to produce ATP, with the electron returning to PSI.

Cytochrome oxidase is an enzyme in the electron transport chain which receives the electrons from the cytochromes and is reduced as the cytochromes are oxidised, with the production of a molecule of ATP.

Cytochromes are members of the electron transport chain. They are protein pigments with an iron group rather like haemoglobin which are reduced by electrons from reduced FAD, which is reoxidised, with the production of a molecule of ATP.

Cytokines are cell signalling molecules with several roles in the immune system, including stimulating other phagocytes to move to the infection site.

Cytokinins are plant hormones that promote cell division in the apical meristems and the cambium through interactions with auxins. They promote lateral bud development, which can overcome apical dominance, and work synergistically with ethene in the abscission of leaves, flowers and fruits.

Day-neutral plants (DNPs) are plants where flowering is not affected by the length of time they are exposed to light or dark.

Deamination is the removal of the amino group from excess amino acids in the ornithine cycle in the liver. The amino group is converted into ammonia and then to urea, which can be excreted by the kidneys.

The **death phase** is when reproduction has almost ceased and the death rate of cells is increasing so that the population number falls.

Decarboxylases are enzymes that remove carbon dioxide.

Decomposers are the final trophic level in any set of feeding relationships. They are the microorganisms, such as bacteria and fungi, that break down the remains of animals and plants and return the mineral nutrients to the soil.

Dehydrogenases are enzymes that remove hydrogen (carry out oxidation reactions).

Dendrites are the slender, finger-like processes that extend from the cell body of a neurone and connect with neighbouring neurones.

Dendrochronology is the dating of past events using tree-ring growth.

Dendrons are the long nerve fibre of a sensory neurone, which carries the nerve impulse.

Denitrifying bacteria break down nitrates to power the production of ATP, producing nitrogen gas as a waste product.

Depolarisation is the condition of the neurone when the potential difference across the membrane is briefly reversed during an action potential, with the cell becoming positive on the inside with respect to the outside for about 1 millisecond.

Diabetes insipidus is a relatively rare condition caused by lack of ADH in the body or an inability of the kidney to respond to ADH. The main symptoms are constant production of large volumes of dilute urine and constant thirst.

Dilution plating is a method used to obtain a culture plate with a countable number of bacterial colonies.

Directional selection is the 'classic' natural selection that occurs whenever environmental pressure is applied to a population, showing a change from one phenotypic property to a new one more advantageous in the circumstances.

Disruptive selection/diversifying selection gives an increase in the diversity of the population rather than a trend in one particular direction. It is common when conditions are diverse and small subpopulations evolve different phenotypes suited to their very particular surroundings, and often results in the evolution of new species.

The **distal convoluted tubule** is the section of the nephron after the loop of Henle that leads into the collecting duct, where balancing the water needs of the body takes place.

The **distribution** of an organism describes where a species of organism is found in the environment and how it is arranged.

DNA or **gene sequencing** is the analysis of the individual base sequence along a DNA strand or an individual gene.

DNA demethylation is the removal of the methyl group from methylated DNA enabling genes to become active so they can be transcribed.

DNA methylation is the methylation of DNA (addition of a methyl $-CH_3$ group) to a cytosine in the DNA molecule next to a guanine in the DNA chain and prevents the transcription of a gene.

DNA profiling is the identification of repeating patterns in the non-coding regions of DNA.

When a phenotype shows **dominance** it is expressed whether the individual is homozygous for the characteristic or not.

An **ecosystem** is an environment including all the living organisms interacting within it, the cycling of nutrients and the physical and chemical environment in which the organisms are living.

Ectotherms are animals that are largely dependent on the external environment for their body temperature.

Edaphic factors relate to the structure of the soil.

Effector cells are specialised cells that bring about a response when stimulated by a neurone.

Effectors are systems (usually muscles or glands) that either work to reverse, increase or decrease changes in a biological system systems.

The **electron transport chain** is a series of electron-carrying compounds along which electrons are transferred in a series of oxidation/reduction reactions, driving the production of ATP.

Embryonic stem cells are the undifferentiated cells of the early human embryo with the potential to develop into many different types of specialised cells.

An **endemic** disease is one in which the active disease, or the pathogen that causes the disease, is constantly present in a particular country or area.

Endocrine glands produce hormones. They do not have ducts. They release the hormones directly into the bloodstream.

Endotherms are animals that warm their bodies through metabolic processes at least in part and usually have a body temperature higher than the ambient temperature.

Endotoxins are lipopolysaccharides that are an integral part of the outer layer of the cell wall of Gram-negative bacteria.

Enhancer sequences are specific regions of the DNA to which transcription factors bind and regulate the activity of the DNA by changing the structure of the chromatin, making it more or less available to RNA polymerase, and so either stimulating or preventing the transcription of the gene.

The **envelope** of a chloroplast is the outer and inner membranes along with the intermembrane space.

Ethanol is an organic chemical with the formula C_2H_5OH produced as a result of anaerobic respiration (fermentation) in fungi and some plant cells.

Etiolated describes the form of plants grown in the dark, with long internodes, thin stems, small or unformed leaves and white or pale yellow in colour.

The **excitatory post-synaptic potential (EPSP)** is the potential difference across the post-synaptic membrane caused by an influx of sodium ions into the nerve fibre, as the result of the arrival of a molecule of neurotransmitter on the receptors of the post-synaptic membrane that makes the inside more positive than the normal resting potential, increasing the chance of a new action potential.

Exocrine glands produce chemicals (e.g. enzymes) and release them along small tubes or ducts.

Exons are the coding regions of DNA (the genes).

Exotoxins are soluble proteins that are produced and released into the body by bacteria as they metabolise and reproduce in the cells of their host.

FAD is a hydrogen carrier and coenzyme. In cellular respiration it accepts hydrogen to form reduced FAD ($FADH_2$), driving the production of ATP.

Far red light has a wavelength of 700–730 nm, which is detected by plants using phytochromes.

Florigen is a hypothetical plant hormone which is involved in the photoperiodic response. It may be FTmRNA.

The **founder effect** is the loss of genetic variation that occurs when a small number of individuals become isolated, forming a new population with allele frequencies not representative of the original population.

Gene flow describes the migration of either whole organisms or genetic material into or out of a population and into another population, tending to make different populations more alike, but changing the allele frequencies within each individual population all the time.

Gene guns are used to produce recombinant DNA by shooting the desired DNA into the cell at high speed on minute gold or tungsten pellets.

Gene linkage is when genes for two characteristics are found on the same chromosome and are close together so they are linked and inherited as a single unit.

The **gene pool** is the sum total of all the genes in a population at a given time.

Gene probes are short DNA sequences labelled with a fluorescent molecule that are complementary to specific DNA sequences which are being sought.

Gene variants are different versions of genes, an alternative term for allele.

Generation time is the time span between bacterial divisions.

Genetic drift describes random changes in the gene pool of a population that occur by chance, not because they confer any advantage or disadvantage on the offspring.

Genetic engineering/genetic modification is the insertion of genes from one organism into the genetic material of another organism or changing the genetic material of an organism.

Gibberellins are plant hormones that act as growth regulators, particularly in the internodes of stems by stimulating elongation of the growing cells; they also promote the growth of fruit and are involved in breaking dormancy in seeds and in germination.

Gluconeogenesis is the synthesis of glucose from non-carbohydrates.

Glyceraldehyde 3-phosphate (GALP) is a 3-carbon sugar produced in the Calvin cycle using reduced NADP and ATP from in the light-dependent stage. GALP is the key product of photosynthesis. It is used to replace the RuBP needed in the first step of the cycle, in glycolysis and the Krebs cycle, and in the synthesis of amino acids, lipids, etc. for the plant cells.

Glycerate 3-phosphate (GP) is a phosphorylated 3-carbon intermediate in the process of glycolysis.

Glycolysis is the first stage of cellular respiration, which takes place in the cytoplasm and is common to both aerobic and anaerobic respiration.

Grana are stacks of thylakoid membranes within a chloroplast.

Green fluorescent protein (GFP) is the product of a gene often used as a marker in the production of recombinant DNA.

The **greenhouse effect** is the process by which gases in the Earth's atmosphere absorb and re-radiate the radiation from the Sun, which has been reflected from the Earth's surface, maintaining a temperature at the surface of the Earth that is warm enough for life to exist.

Grey matter consists of the cell bodies of neurones in the CNS.

Gross primary productivity (GPP) in plants is the rate at which light from the Sun catalyses the production of new plant material, measured as $g\,m^{-2}\,year^{-1}$, $g\,C\,m^{-2}\,year^{-1}$ or $kJ\,m^{-2}\,year^{-1}$.

A **habitat** is the place where an organism lives.

A **haemocytometer** is a thick microscope slide with a rectangular indentation and etched grid of lines that is used to count cells.

Haemophilia is a sex-linked genetic disease in which one of the factors needed for the clotting of the blood is not made in the body.

The **Hardy-Weinberg equilibrium** is the mathematical relationship between the frequencies of alleles and genotypes in a population. The equation used to describe this relationship can be used to work out the stable allele frequencies within a population.

Healthcare-associated infections (HCAIs) are infections that are acquired by patients while they are in hospitals or care facilities. They may be the result of poor hygiene between patients or the result of antibiotic treatment, and may be antibiotic resistant.

Herd immunity is produced when a high proportion of a population is immune to a pathogen, usually by vaccination, lowering the risk of infection to all, including those not vaccinated, as they are less likely to encounter the pathogen.

Heterochromatin is the densely supercoiled and condensed chromatin where the genes are not available to be copied to make proteins.

A **heterogametic** individual produces gametes that contain two different types of sex chromosome.

Heterotrophic organisms obtain complex organic molecules by feeding on other living organisms or their dead remains.

A **heterozygote** is an individual where the two alleles coding for a particular characteristic are different.

Hibernation is the state when an animal goes into a very deep sleep to avoid very cold conditions. The metabolic rate slows and the core body temperature is greatly lowered, making substantial energy savings.

Histamines are chemicals released by the tissues in response to an allergic reaction.

Histone acetylation is the addition of an acetyl group ($-COCH_3$) to one of the lysines in the histone structure, which opens up the structure and activates the chromatin, allowing genes in that area to be transcribed.

Histone methylation is the addition of a methyl group ($-CH_3$) to a lysine in the histone. Depending on the position of the lysine, methylation may cause inactivation or activation of the region of DNA.

Homeostasis is the maintenance of a state of dynamic equilibrium in the body, despite changes in the external or internal conditions.

A **homogametic** individual produces gametes that only contain one type of sex chromosome.

A **homozygote** is an individual where both of the alleles coding for a particular characteristic are identical.

Hormones are organic chemicals produced in endocrine glands and released into the blood and carried through the transport system to parts of the body where they bring about changes, which may be widespread or very targeted. They are usually either proteins, parts of proteins such as polypeptides, or steroids.

Hybridisation is the binding of gene probes to the complementary DNA strands.

A **hydrogen acceptor** is a molecule which receives hydrogen and becomes reduced in cell biochemistry.

A **hypha** (plural hyphae) is a thread-like fungal structure that is a single unit of the mycelium.

The **hypothalamus** is a small area of brain directly above the pituitary gland that controls the activities of the pituitary gland and coordinates the autonomic (unconscious) nervous system.

Immune response is the specific response of the body to invasion by pathogens.

Immunisation is the process of protecting people from infection by giving them passive or active artificial immunity.

Immunoglobulins are antibodies.

Individual counts measure the number of individual organisms in an area.

Induced pluripotent stem cells (iPS cells) are adult cells that have been reprogrammed by the introduction of new genes to become pluripotent again.

Inflammation is a common, non-specific response to infection involving the release of histamines from mast cells and basophils, causing the blood vessels to dilate, giving local heat, redness and swelling.

The **inhibitory post-synaptic potential (IPSP)** is the potential difference across the post-synaptic membrane caused by an influx of negative ions as the result of the arrival of a molecule of neurotransmitter on the receptors of the post-synaptic membrane, which makes the inside more negative than the normal resting potential, decreasing the chance of a new action potential.

Inoculation is the process by which microorganisms are transferred into a culture medium under sterile conditions.

The **Intergovernmental Panel on Climate Change (IPCC)** analyses research from scientists on climate change and produces regular unbiased reports based on all the available data to be used by politicians and decision makers globally.

The **International Union for Conservation of Nature (IUCN)** is the largest global environmental organisation and produces the regularly updated Red List of Threatened Species™.

Interspecific competition occurs when different species within a community compete for the same resources.

Intraspecific competition is competition between members of the same species for a limited resource.

Introns are the large, non-coding regions of DNA that are removed before messenger RNA is translated into proteins.

The **Krebs cycle** is a series of biochemical steps that lead to the complete oxidation of glucose, resulting in the production of carbon dioxide, water and relatively large amounts of ATP.

Lactate is a 3-carbon compound which is the end-product of anaerobic respiration in mammals.

The **lag phase** is when bacteria are adapting to their new environment and are not yet reproducing at their maximum rate.

Lamellae are extensions of the thylakoid membranes which connect two or more grana, acting as a supporting skeleton in the chloroplast, maintaining a working distance between the grana so that they get the maximum light and function as efficiently as possible.

The **law of independent assortment** describes Mendel's second law, which states that different traits are inherited independently of each other. This means that the inheritance of alleles for one phenotype has nothing to do with the inheritance of alleles for another characteristic.

The **law of segregation** describes Mendel's first law, which states that one unit or allele for each trait is inherited from each parent to give a total of two alleles for each trait. The segregation (separation) of alleles in each pair takes place when the gametes are formed and some alleles code for phenotypes that are dominant over others.

Leaching describes the loss of minerals from soil as water passes through rapidly.

The **legumes** are plants, such as peas, beans and clover, which have nodules on their roots that are full of nitrogen-fixing bacteria.

Lidocaine is a drug used as a local anaesthetic that works by blocking the voltage-gated sodium channels in post-synaptic membranes in sensory neurones, preventing the production of an action potential.

Light-dependent reactions are reactions that take place in the light on the thylakoid membranes of the chloroplasts. They produce ATP and split water molecules in a photochemical reaction, providing hydrogen ions to reduce carbon dioxide and produce carbohydrates.

Light-independent reactions are reactions that use the reduced NADP and ATP produced by the light-dependent stage of photosynthesis in a pathway known as the Calvin cycle, which take place in the stroma of the chloroplast and result in the reduction of carbon dioxide from the air to bring about the synthesis of carbohydrates.

A **limiting factor** is the factor needed for a reaction such as photosynthesis to progress that is closest to its minimum value.

Line transects are a way of gathering data more systematically. A tape is stretched between two points and every individual plant (or animal) that touches the tape is recorded.

The **link reaction** is the reaction needed to move the products of glycolysis into the Krebs cycle.

Lipopolysaccharides are large molecules containing a lipid element and a polysaccharide element.

Liposome wrapping is a technique for producing recombinant DNA that involves wrapping the gene to be inserted in liposomes. which fuse with the cell membrane and can pass through it to deliver the DNA into the cytoplasm.

The **log phase** is when the rate of bacterial reproduction is close to or at its theoretical maximum, repeatedly doubling in a given time period.

Long-day plants (LDPs) are plants flowering when days are long and nights are short.

Lymphocytes are granulocytes, made in the white bone marrow of the long bones, that make up the main cellular components of the immune system.

Major histocompatibility complex (MHC) proteins are proteins that display antigens on the cell surface membrane.

Massively parallel sequencing is a very rapid method of sequencing millions of DNA fragments at the same time.

The **medulla oblongata (medulla)** is the most primitive part of the brain that controls reflex centres controlling functions such as the breathing rate, heart rate, blood pressure, coughing, sneezing, swallowing, saliva production and peristalsis.

Methicillin-resistant *Staphylococcus aureus* (MRSA) is a strain of *S. aureus* that is resistant to several antibiotics, including methicillin.

Clostridium difficile is a type of bacteria that often exists in the intestines and causes no problems unless it becomes dominant as a result of the normal gut flora being removed/damaged by antibiotic treatment

A **microclimate** is a small area with a distinct climate that is different to the surrounding areas.

A **microhabitat** is a small area of a habitat.

Microinjection (DNA injection) is a technique for producing recombinant DNA that involves injecting DNA into a cell through a very fine micropipette.

A **micro-satellite** is a section of DNA with a 2–6 base sequence repeated between 5 and 100 times.

A **mini-satellite** is a section of DNA with a 10–100 base sequence repeated 50 to several hundred times.

The **mode of infection** is the way a pathogen causes infection.

Modes of transmission are the different ways a pathogen is spread from one host to another.

A **monogenic cross** is a genetic cross where only one gene is considered.

Anopheles **mosquitoes** are the type of mosquitoes that carry the malaria parasite from one host to another.

If a gene has **multiple alleles** it means there are more than two possible variants at a particular locus.

Multipotent describes a cell that can form a very limited range of differentiated cells within a mature organism.

A **mycelium** is a fungal body made up of a mass of thread-like hyphae.

The **myelin sheath** is a fatty insulating layer around some neurones produced by the Schwann cell.

NAD is a coenzyme that acts as a hydrogen acceptor.

Natural active immunity is when the body produces its own antibodies to an antigen encountered naturally.

Natural passive immunity is when antibodies made by the mother are passed to the baby via the placenta or breast milk.

Negative feedback systems provide a way of maintaining a condition, such as the concentration of a substance, within a narrow range. A change in conditions is registered by receptors and as a result effectors are stimulated to restore the equilibrium.

Nephrons are microscopic tubules that make up most of the structure of the kidney.

Nerve impulses are the electrical signals transmitted through the neurones of the nervous system.

Net primary productivity (NPP) is the material produced by photosynthesis stored as new plant body tissues, that is NPP = GPP – R.

Neurones are nerve cells specialised for the rapid transmission of impulses throughout an organism.

Neurosecretory cells are nerve cells that produce secretions from the ends of their axons. These secretions either stimulate or inhibit the release of hormones from the anterior pituitary, or are stored in the posterior pituitary and then later released as hormones.

A **neurotransmitter** is a chemical which transmits an impulse across a synapse.

Neutralisation is the action of antibodies in neutralising the effects of bacterial toxins on cells by binding to them.

Nicotine is a drug found in cigarettes that mimics the effect of acetylcholine and binds to specific acetylcholine receptors in post-synaptic membranes known as nicotinic receptors.

Nitrifying bacteria oxidise ammonium compounds to form nitrites and nitrates.

The **nitrogen cycle** is the recycling of nitrogen between living things and the environment by the actions of microorganisms.

Nitrogen-fixing bacteria in the soil can convert nitrogen from the soil air into ammonia, and this is then converted into nitrates by the nitrifying bacteria of the nitrogen cycle.

The **nodes of Ranvier** are gaps between the Schwann cells that enable saltatory conduction.

Non-cyclic photophosphorylation is a process involving both PSI and PSII in which water molecules are split using light energy to provide reducing power to make carbohydrates and at the same time produce more ATP.

Noradrenaline is a neurotransmitter found in the synapses of the sympathetic nervous system and adrenergic synapses of the brain.

The **null hypothesis** is the hypothesis that any differences between data sets are the result of chance.

Nutrient agar is a jelly extracted from seaweed and used as a solid nutrient for culturing microorganisms, commonly used in Petri dishes.

Nutrient broth is a liquid nutrient for culturing microorganisms, commonly used in flasks, test tubes or bottles.

Nutrient medium is a substance used for the culture of microorganisms, which can be in liquid form (nutrient broth) or in solid form (usually nutrient agar).

Opsonins are chemicals which bind to pathogens and label them so they are more easily recognised by phagocytes.

Opsonisation is a process that makes a pathogen more easily recognised, engulfed and digested by phagocytes.

The **ornithine cycle** is the series of enzyme-controlled reactions that convert ammonia from excess amino acids into urea in the liver.

Osmoreceptors are sensory receptors in the hypothalamus that detect a change in the concentration of inorganic ions, and therefore changes in the osmotic potential of the blood.

Osmoregulation is the maintenance of a constant osmotic potential in the tissues of a living organism by controlling water and salt concentrations.

Oxidation is the removal of electrons from a substance, e.g. by the addition of oxygen or removal of hydrogen.

Oxidative phosphorylation is the oxygen-dependent process in the electron transport chain where ADP is phosphorylated.

The **parasympathetic nervous system** involves autonomic motor neurones with very long myelinated preganglionic fibres that leave the CNS and synapse in a ganglion very close to the effector organ. They have very short unmyelinated postganglionic fibres. They produce acetylcholine at the synapses and often have a relatively slow, inhibitory effect on an organ system.

Parental phenotypes describe offspring that have the same phenotypes as the parental organisms.

The **pathogenic effect** of a microorganism describes the symptoms of disease it causes.

Pathogens are microorganisms that cause disease.

Peer review is the process by which scientific papers are sent to other scientists who are experts in the field, for them to read and assess before publication in a journal.

Penicillin was the first antibiotic discovered. It affects the formation of bacterial cell walls, and it is bactericidal.

Percentage cover describes the area covered by the above ground parts of a particular species.

The **peripheral nervous system** includes the parts of the nervous system that spread through the body and are not involved in the central nervous system.

Phaeophytin is a grey pigment which is a breakdown product of the other photosynthetic pigments.

A **phagosome** is the vesicle in which a pathogen is enclosed in a phagocyte.

A **photochemical reaction** is a reaction initiated by light.

Photolysis is the splitting of a molecule using light.

Photomorphogenesis is the process by which the form and development of a plant is controlled by the levels of and type of light.

Photorespiration is the alternative reaction catalysed by RUBISCO in a low carbon dioxide environment which uses oxygen and releases carbon dioxide, making photosynthesis less efficient.

Photosynthesis is the process by which living organisms, particularly plants and algae, capture the energy of the Sun using chlorophyll and use it to convert carbon dioxide and water into simple sugars.

Photosystem I (PSI) is a combination of chlorophyll pigments which absorbs light of wavelength 700 nm and is involved in cyclic and non-cyclic photophosphorylation.

Photosystem II (PSII) is a combination of chlorophyll pigments which absorbs light of wavelength 680 nm and is involved only in non-cyclic photophosphorylation.

Phytochrome is a plant pigment that reacts with different types of light, and in turn affects the responses of the plant.

The **pituitary gland** is a small gland in the brain that has an anterior lobe and a posterior lobe and produces and releases secretions that affect the activity of most of the other endocrine glands in the body.

A **plagioclimax** is a climax community that is at least in part the result of human intervention.

Plasma cell clones are clones of identical cells that all produce the same antibody.

Plasma cells produce antibodies to particular antigens at a rate of around 2000 antibodies per second.

Plasmodium **spp.** are the parasitic protozoa that cause malaria. They have a life cycle split between two different hosts, female *Anopheles* mosquitoes and people.

Pluripotent describes an undifferentiated cell that can form most of the cell types needed for an entire new organism.

Polarised describes the condition of a neurone when the movement of positively charged potassium ions out of the cell down the concentration gradient is opposed by the actively produced electrochemical gradient, leaving the inside of the cell slightly negative relative to the outside.

Polygenic phenotypic traits are determined by several interacting genes.

Polymerase chain reaction (PCR) is the reaction used to amplify a sample of DNA, to make more copies of it very rapidly.

A **population** is a breeding group of individuals of the same species occupying a particular habitat and a particular niche.

A **population bottleneck** is the effect of an event or series of events that dramatically reduces the size of a population and causes a severe decrease in the gene pool of the population, resulting in large changes in allele frequencies and a reduction in genetic diversity.

Positive feedback systems are where effectors work to increase an effect that has triggered a response.

Pre-mRNA is the mRNA that is transcribed directly from the DNA before it has been modified.

The **presynaptic membrane** is the membrane on the side of the synapse where the first impulse arrives and from which neurotransmitters are released.

Primary consumers are organisms that eat producers, either plants or algae.

Primary infection is the initial stage of tuberculosis when *M. tuberculosis* has been inhaled into the lungs, invaded the cells of the lungs and multiplied slowly, often causing no obvious symptoms.

Producers make food.

Promoter sequences are specific regions on the DNA to which transcription factors bind to stimulate transcription.

The **proximal convoluted tubule** is the first region of the nephron after the Bowman's capsule, where over 80% of the glomerular filtrate is absorbed back into the blood.

Puccinia graminis/**stem rust fungus** is the fungus that causes wheat stem rust.

A **pyramid of biomass** represents the biomass of the organisms at each trophic level in a food chain.

A **pyramid of energy** represents the total energy store of the organisms at each trophic level in a food chain.

A **pyramid of numbers** represents the numbers of organisms at each trophic level in a food chain.

Pyruvate is the end-product of glycolysis.

A **quadrat** is a square frame divided into sections that you lay on the ground to identify the sample area.

Receptor cells are specialised neurones that respond to changes in the environment.

When a phenotype is **recessive** it is only expressed when both alleles code for the recessive feature, in other words the individual is homozygous recessive for that trait.

Recognition sites are specific base sequences where restriction endonucleases cleave the DNA molecule.

Recombinant DNA is new DNA produced by genetic engineering technology that combines genes from the DNA of one organism with the DNA of another organism.

Recombinant phenotypes describe offspring that have different phenotypes to their parents as a result of recombination of the chromosomes during sexual reproduction.

Red light has a wavelength of 580–660′ nm, which is detected by plants using phytochromes.

Reduced NAD is NAD which has accepted a hydrogen atom in a metabolic pathway.

Reduction is the addition of electrons to a substance, e.g. by the addition of hydrogen or removal of oxygen.

Reflex responses are rapid responses that take place with no conscious thought involved.

The **refractory period** is the time it takes for ionic movements to repolarise an area of the membrane and restore the resting potential after an action potential.

The **relative refractory period** is a period of several milliseconds after an action potential and the absolute refractory period when an axon may be re-stimulated, but only by a much stronger stimulus than before.

In a **reliable** study, other scientists can repeat the methodology and obtain similar results.

Replica plating is the process used to identify recombinant cells that involves growing identical patterns of bacterial colonies on plates with different media.

The **respiratory substrate** is the substance oxidised during cellular respiration.

A **respirometer** is a piece of apparatus used for measuring the rate of respiration in whole organisms or cultures of cells.

The **resting potential** is the potential difference across the membrane of around −70 mV when the neurone is not transmitting an impulse.

Restriction endonucleases are special enzymes used to chop up strands of DNA at particular points in the intron sequences.

Reverse transcriptase is the enzyme used to make artificial copies of a desired gene by taking an mRNA molecule transcribed from the gene and using it to produce the correct DNA sequence.

Rhodopsin (visual purple) is the visual pigment in the rods.

Ribulose bisphosphate (RuBP) is a 5-carbon compound that joins with carbon dioxide from the air in the Calvin cycle to fix the carbon dioxide and form a 6-carbon compound.

Ribulose bisphosphate carboxylase/oxygenase (RUBISCO) is a key, rate-controlling enzyme that catalyses the reaction between carbon dioxide/oxygen and ribulose biphosphate.

Rods are photoreceptors found in the retina which contain the visual pigment rhodopsin. They respond to low light intensities, give black and white vision and are very sensitive to movement.

Saltatory conduction is the process by which action potentials are transmitted from one node of Ranvier to the next in a myelinated nerve.

A **Schwann cell** is a specialised type of cell associated with myelinated neurones. It forms the myelin sheath.

Secondary consumers are animals that feed on primary consumers.

Secondary production is the process of making new animal biomass from plant material that has been eaten.

Selective medium is a growth medium for microorganisms containing a very specific mixture of nutrients, so only a particular type of microorganism will grow on it.

Selective reabsorption is the process by which substances needed by the body are reabsorbed from the kidney tubules into the blood.

Selective toxicity means that a substance is toxic against some types of cells or organisms but not others.

Sense organs are groups of receptors working together to detect changes in the environment.

Sensors/receptors are specialised cells that are sensitive to specific changes in the environment.

Sex linkage refers to genes that are carried on the sex chromosomes.

A **sex-linked disease** is a genetic disease that results from a mutated gene carried on the sex chromosomes – in human beings, largely on the X chromosome.

Short tandem repeats are micro-satellite regions that are now widely used in DNA identification.

Short-day plants (SDPs) are plants flowering when days are short and nights are long.

Sodium gates are specific sodium ion channels in the nerve fibre membrane that open up, allowing sodium ions to diffuse rapidly down their concentration and electrochemical gradients.

Somatic stem cells/adult stem cells are undifferentiated cells found among the normal differentiated cells in a tissue or organ that can differentiate when needed to produce any one of the major cell types found in that particular tissue or organ.

Southern blotting is the name of a process in which DNA fragments are drawn from an electrophoresis gel to a filter, leaving the DNA fragments as 'blots' on the filter. The process also denatures the DNA fragments so the strands separate and the base sequences are exposed.

Spearman's rank correlation coefficient is a statistical tool used to test whether two variables are significantly correlated.

The **spinal cord** is the area of the CNS that carries the nerve fibres into and out of the brain and also coordinates many unconscious reflex actions.

Spliceosomes are enzyme complexes that act on pre-mRNA, joining exons together after the removal of the introns.

Stabilising selection is the natural selection acting to conserve what is already present in a population, reducing variation in a population, so that the frequency of some alleles is very high but other alleles are greatly reduced.

Stalked particles are structures where ATP production takes place on the inner mitochondrial membrane.

The **stationary phase** is when the total growth rate is zero as the number of new cells formed by binary fission is equalled by the numbers of cells dying due to factors including competition for nutrients, lack of essential nutrients, an accumulation of toxic waste products and possibly lack of, or competition for, oxygen.

Sterile is a term used to describe something that is free from living microorganisms and their spores.

A **sticky end** is the name given to the area of base pairs left longer on one strand of DNA than the other by certain restriction endonucleases, making it easier to attach new pieces of DNA.

The **stroma** is the matrix which surrounds the grana and contains all the enzymes needed to complete the process of photosynthesis and produce glucose.

The **Student's t-test** is a statistical test that allows you to judge whether any difference between the means of the two sets of data is statistically significant.

Succession is the process by which the communities of organisms colonising an area change over time.

Sustainability is the production of a decent standard of living for everyone now, without compromising the need of future generations or of the ecosystems around us.

The **sympathetic nervous system** involves autonomic motor neurones with very short myelinated preganglionic fibres that leave the CNS and synapse in a ganglion very close to the CNS. They have long unmyelinated postganglionic fibres. They produce noradrenaline at the synapses and often stimulate a rapid response, activating an organ system.

A **synapse** is the junction between two neurones that nerve impulses cross via neurotransmitters.

The **synaptic cleft** is the gap between the pre and post synaptic membranes in a synapse.

Synaptic knobs are the bulges at the end of the presynaptic neurones where neurotransmitters are made.

Synaptic vesicles are membrane-bound sacs in the presynaptic knob which contain about 3000 molecules of neurotransmitter and move to fuse with the presynaptic membrane when an impulse arrives in the presynaptic knob.

T cells are lymphocytes made in the bone marrow that mature and become active in the thymus gland.

T helper cells are lymphocytes involved in the process that produces antibodies against the antigens on a particular pathogen.

T killer cells are lymphocytes that produce chemicals that destroy pathogens.

T memory cells are very long-lived cells which make up part of the immunological memory.

Terminator bases are modified versions of the four nucleotide bases that halt the production of a DNA molecule as soon as they are incorporated as no more bases can be added.

Tertiary consumers feed on secondary consumers – they eat other carnivores. They are usually the top predators in an area.

Tetracycline is a bacteriostatic antibiotic that inhibits protein synthesis.

Therapeutic cloning is an experimental technique used to produce embryonic stem cells from an adult cell donor.

Thermoregulation is a homeostatic mechanism that enables organisms to control their internal body temperature within set limits.

The **thermoregulatory centre** comprises temperature receptors sited in the hypothalamus in the brain that act as the thermostat of the body and fire when the temperature of the blood flowing through the hypothalamus increases or decreases.

The **threshold** is the point when sufficient sodium ion channels open for the rush of sodium ions into the axon to be greater than the outflow of potassium ions, resulting in an action potential.

A **thylakoid** is a membrane disc found in the grana of a chloroplast.

A **total viable cell count** is a measure of the number of cells that are alive in a given volume of a culture.

Totipotent describes an undifferentiated cell that can form any one of the different cell types needed for an entire new organism.

Transcription factors are proteins that bind to the DNA in the nucleus and affect the process of transcribing the genetic material.

Transgenic animals are animals that have had their DNA modified using gene technology, so that at least some of their cells contain recombinant DNA.

A **trophic level** describes the position of an organism in a food chain or web and describes its feeding relationship with other organisms.

Tropisms are plant growth responses to environmental cues.

A **tubercle** is the result of a healthy immune response to an infection by *M. tuberculosis*. A localised inflammatory response forms a mass of tissue containing dead bacteria and macrophages.

Tuberculosis is a lung disease caused by *Mycobacterium tuberculosis* and *M. bovis*.

Tubular secretion is the process by which inorganic ions are secreted into or out of the kidney tubules as needed to maintain the osmotic balance of the blood.

Turbid is a term used to describe something that is opaque, or thick with suspended matter.

Turbidimetry is a method of measuring the concentration of a substance by measuring the amount of light passing through it.

Ultrafiltration is the process by which fluid is forced out of the capillaries in the glomerulus of the kidney into the kidney tubule through the epithelial walls of the capillary and the capsule.

Vaccination is the introduction of harmless forms of organisms or antigens by injection or mouth to produce artificial immunity.

A **valid** study has been properly designed to answer the questions being asked.

Vasoconstriction is the narrowing of the blood vessels by contraction of their muscle walls, reducing blood flow.

Vasodilation is the widening of the blood vessels by relaxation of their muscle walls, increasing blood flow

Vectors are living organisms or environmental factors that transmit infection from one host to another.

The **voluntary nervous system** involves motor neurones that are under voluntary or conscious control involving the cerebrum.

White matter consists of the nerve fibres of neurones in the CNS.

A **zoonotic infection** is an infection in a person caused by a pathogen that can cross the species barrier from other animals.

Index